Design for Manufacturing: A Structured Approach

Design for Manufacturing:

A Structured Approach

Corrado Poli
Mechanical and Industrial Engineering Department
University of Massachusetts Amherst

Boston Oxford Auckland Johannesburg Melbourne New Delhi

Recognizing the importance of preserving what has been written, Butterworth–Heinemann prints its books on acid-free paper whenever possible.

Library of Congress Cataloging-in-Publication Data
Poli, C., 1935–
Design for manufacturing : a structured approach / Corrado Poli.
p. cm.
Includes bibliographical references and index.
ISBN 0-7506-7341-9 (alk. paper)
1. Engineering design. 2. Design, Industrial. I. Title.
TA174 .P644 2001
620'.0042—dc21

2001025915

British Library Cataloguing-in-Publication Data
A catalogue record for this book is available from the British Library.

The publisher offers special discounts on bulk orders of this book.
For information, please contact:
Manager of Special Sales
Butterworth–Heinemann
225 Wildwood Avenue
Woburn, MA 01801-2041
Tel: 781-904-2500
Fax: 781-904-2620

For information on all Butterworth–Heinemann publications available, contact our World Wide Web home page at: http://www.bh.com

10 9 8 7 6 5 4 3 2 1

Printed in the United States of America

To Ann

Who, along with Peter, Chuck, and Kim, has helped make me the luckiest guy in the world.

Contents

List of Figures

List of Tables

Preface

Design for Manufacturing: A Structured Approach is intended as a text for mechanical, industrial, and manufacturing engineering students at the junior level or higher, and as a continuing education text for manufacturing engineers and engineering design practitioners in industry. Unlike many books on design for manufacturing that provide only broad qualitative guidelines to designers, this book uses a combination of both qualitative and quantitative information to assist readers in making informed decisions concerning alternative competing designs.

The quantitative methods presented here inherently require considerably more attention to geometric detail, more than required by conceptual or even configuration design. If students are to do quantitative DFM (design for manufacturing) evaluations, they simply must take the time and trouble to learn to do them. Although engineering design can be considerably more abstract than DFM, the transition from the abstraction of engineering design to the concrete detail of DFM simply goes with the territory if one wants to perform quantitative DFM evaluations in a concurrent fashion. The part-coding systems that are the basis for some of the DFM methodologies may initially appear cumbersome (there are a lot of new terms to learn), but they are in fact easy to learn. The reward is not only the ability to estimate relative tooling and processing costs, but also a good sense of the design issues that drive part costs.

This book is based on material that first appeared in *Engineering Design and Design for Manufacturing: A Structured Approach*, a book I co-authored with John R. Dixon and published in 1995. *Design for Manufacturing: A Structured Approach* is an updated, revised, reorganized, and stand-alone version of the chapters in the earlier book that dealt primarily with design for manufacturing and materials and process selection. The greatest changes occur in the chapters dealing with stamping. The coding system originally used to estimate the number of active stations required to produce a stamped part has been dropped. Instead, a more accurate and user-friendly algorithm based on the concept of process planning and strip-layout development is used to determine the number of active stations required to produce the tooling for a stamped part.

Another change that has been made in this book occurs when estimating the total relative part cost, that is, the cost of the part relative to a reference part. The original approach allowed the cost of the reference part to vary with production volume. This, in turn, often resulted in increases in the total relative cost of the designed part with increases in production volume. This counterintuitive result appeared to cause confusion among students. The approach used here, as suggested by Professor Larry Murch of the University of Massachusetts Amherst, is to fix the cost of the reference part to $1 and the production volume of the reference part to the production volume that results in a part cost of $1.00. This approach, which results in an increase in relative part cost with increases in production volume, appears to be favored by students, and allows them to use the DFM systems to roughly estimate the production cost of a proposed design.

To assist students in better visualizing and understanding the many orthographic views that appear in the book, isometric views have also been included where it was deemed advantageous. Many new homework problems have also been added.

To help students better envisage and comprehend the rationale behind the DFM methodologies described here, a series of Power Point presentations has been developed, which are on a CD-ROM that is available for use by faculty in courses that require the use of this book. These presentations contain both video clips and quick-times movies of processes such as injection molding, stamping, die casting, forging, rolling, rod drawing, and aluminum extrusion.

Readers of this book may also be interested in visiting the design for manufacturing Web site hosted by the Mechanical and Industrial Engineering Department at the University of Massachusetts Amherst.

Design for Manufacturing: A Structured Approach can be used to support a one-semester course in Design for Manufacturing (or Manufacturing for Design) at the level of the junior or senior year in college. Such a course could be taught independently of a design course, or the two courses could be taught simultaneously. It is strongly recommended that this course include a design for manufacturing project. Although it is true that DFM can be taught without the use of a supplementary DFM project, my experience is that student interest in manufacturing and how parts are made increases significantly when a project is included.

Acknowledgments

The methodologies described in this book rely on the prior work, assistance, support, encouragement, and dedication of many colleagues and students. The development of these methodologies in turn depends on the knowledge base and data made available by various collaborating companies and the financial support provided by both private industry and state and federal funding agencies. Although there are too many individuals, companies, and funding agencies to mention here I do want to mention a few.

Among faculty colleagues at the University of Massachusetts Amherst who have helped are J. Edward Sunderland and Robert Graves (now at Rensselaer), who participated in much of the DFM work described here, and Larry Murch, who suggested the total relative part cost model used in Chapters 5, 7, and 10 and who made many valuable suggestions while reading the manuscript. I am especially appreciative of the many useful discussions and suggestions provided by my good friend and colleague John R. Dixon to the original version of this manuscript, which first appeared in our book *Engineering Design and Design for Manufacturing: A Structured Approach.*

Although many graduate students participated and contributed to the ongoing research in design for manufacturing (DFM), among the students who have contributed the most were Jiten Divgi, Sheng-Ming Kuo, Ricardo Fernandez, Lee Fredette, Ferruccio Fenoglio, Juan Escodero, Shyam Shanmugasundaram, Pratip Dastidar, Prashant Mahajan, and Shrinivasan Chandrasrkran.

The contributions of these students would not have been possible without the financial support of the former Digital Equipment Corporation, Xerox, the National Science Foundation, and the Massachusetts Center of Excellence Corporation.

Most of the knowledge base and data base upon which the DFM methodologies in die casting and stamping are based was acquired with the cooperation of the following New England-based companies, namely, Cambridge Tool and Manufacturing Company, North Billerica, MA; Kennedy Die Castings, Worcester, MA; Larson Tool and Stamping Company, Attleboro, MA; K. F. Bassler Company, Attleboro, MA; Leicester Die and Tool Inc., Leicester, MA; Metropolitan Machine and Stamping Company, Medfield, MA; Thomas Smith Company Inc., Worcester, MA; Hobson and Motzer, Inc., Wallingford, CT; Newton New Haven Co., North Haven, CT. The knowledge base and data base for injection molding was obtained with the cooperation of the following companies: Jada Precision Plastics, Rochester, NY; Sajar Plastics, Midfield, OH; Colonial Machine Company, Kent, OH; Tremont Tool and Gage, Cleveland, OH; Woldring Plastic Mold Technology, Grand Rapids, MI; Sinicon Plastics, Pittsfield, MA; and Tog Mold, Pittsfield, MA. I am deeply indebted to all of these companies, as well as to those companies whom I have forgotten to mention, for their invaluable help and for their willingness to work closely with the students mentioned above.

To help in better visualizing and understanding the logic behind the methodologies described in this book, a series of Power Point presentations have been developed to accompany the book. These presentations make heavy use of animations developed by students from the Center for Knowledge Communication in the Computer Science Department at the University of Massachusetts Amherst. Although many students contributed to the development of these animations I want to especially thank Ryan Moore and Nick Steglich for their work on the stamping and die-casting animations.

Finally, I also want to express my indebtedness to both Geoffrey Boothroyd and Beverly Woolf. It was Geoffrey Boothroyd who introduced me to the entire design for manufacturing mindset, and it is his work in design for assembly that opened the frontier to all the "design-fors." Chapter 12, "Assembly," borrows a great deal from his prior work on manual assembly. Beverly Woolf introduced me to the fascinating world of multimedia and pointed out its usefulness in education, especially in the area of design for manufacturing.

Lastly, to those who contributed but whom I have failed to acknowledge here, I apologize. There just isn't space for you all, but I do sincerely appreciate your help.

To the extent that the book is correct and useful, I owe much to all of you; to the extent that it is not, I am solely responsible.

Responsibilities of Users

What do I say to people who use the information and methods described here and something goes wrong? THEY are responsible, not me.

Chapter 1

Introduction

1.1 MANUFACTURING, DESIGN, AND DESIGN FOR MANUFACTURING

Different uses of the word *manufacturing* create an unfortunate confusion. Sometimes the word is used to refer to the entire product realization process, that is, to the entire spectrum of product-related activities in a firm that makes products for sale, including marketing (e.g., customer desires), design, production, sales, and so on. This complete process is sometimes referred to as "big-M Manufacturing."

But the word manufacturing is also used as a synonym for production, that is, to refer only to the portion of the product realization process that involves the actual physical processing of materials and the assembly of parts. This is sometimes referred to as "little-m manufacturing."

We will use the little-m meaning for manufacturing in this book. In other words, manufacturing here consists of physical processes that modify materials in form, state, or properties. Thus in this book, manufacturing and production have the same meaning. When we wish to refer to big-M Manufacturing, we will call it the product realization process.

Design (as in a design process) is the series of activities by which the information known and recorded about a designed object is added to, refined (i.e., made more detailed), modified, or made more or less certain. In other words, the process of design changes the state of information that exists about a designed object. During successful design, the amount of information available about the designed object increases, and it becomes less abstract. Thus, as design proceeds the information becomes more complete and more detailed until finally there is sufficient information to perform manufacturing. Design, therefore, is a process that modifies the information we have about an artifact or designed object, whereas manufacturing (i.e., production) modifies its physical state.

A *design problem* is created when there is a desire for a change in the state of information about a designed object. Consequently, a design problem exists when there is a desire to generate more (or better) information about the designed object, when we want to develop a new (but presently unknown) state of information. For a simple example, we may know from the present state of information that a designed object is to be a beam, and we desire to know whether it is to be an I-beam, a box beam, an angle beam, or some other shape. Our desire to know more about the designed object defines a new design problem—in this example, determining the beam's shape. Later, once we know the shape (say it is to be an I-beam), another design problem is defined when we want also to know the dimensions. There are many kinds of design problems defined by the present and desired future states of information.

Design for manufacturing (DFM) is a philosophy and mind-set in which manufacturing input is used at the earliest stages of design in order to design parts and products that can be produced more easily and more economically. Design for manufacturing is any aspect of the design process in which the issues involved in manufacturing the designed object are considered explicitly with a view to influencing the design. Examples are considerations of tooling costs or time required, processing costs or controllability, assembly time or costs, human concerns during manufacturing (e.g., worker safety or quality of work required), availability of materials or equipment, and so on. Design for manufacturing occurs—or should occur—throughout the design process.

1.2 FUNCTIONAL DESIGNED OBJECTS

We distinguish among the following types of functional designed objects, though not all of them are mutually exclusive: parts, assemblies, subassemblies, components, products, and machines.

Parts

A *part* is a designed object that has no assembly operations in its manufacture. (Welding, gluing, and the like are considered assembly operations for the purposes of this definition.) Parts may be made by a sequence of manufacturing processes (e.g., casting followed by milling), but parts are not assembled.

Parts are either *standard* or *special purpose*. A standard part is a member of a class of parts that has a generic function and is manufactured routinely without reference to its use in any particular product. Examples of standard parts are screws, bolts, rivets, jar tops, buttons, most beams, gears, springs, and washers. Tooling for standard parts is usually on hand and ready for use by manufacturers. Manufacturers, distributors, or vendors often carry standard parts themselves in stock. Standard parts are most frequently selected by designers from catalogs, often with help from vendors.

Special purpose parts are designed and manufactured for a specific purpose in a specific product or product line rather than for a generic purpose in several different products. Special purpose parts that are incorporated into the subassemblies and assemblies of products and machines are often referred to as piece parts. Special purpose parts that stand alone as products (e.g., paper clips, Styrofoam cups) are referred to as single-part products.

Even though screws, springs, gears, and the like, are generally manufactured as standard parts, a special or unique screw, spring, gear, and any other part that is specially designed and manufactured for a special rather than a general purpose is considered a special purpose part. This is not often done, however, because it is usually less expensive to use an available standard part if one will serve the purpose.

Assemblies and Subassemblies

An assembly is a collection of two or more parts. A subassembly is an assembly that is included within an assembly or other subassembly.

A standard module or standard assembly is an assembly or subassembly that—like a standard part—has a generic function and is manufactured routinely for general use or for inclusion in other subassemblies or assemblies. Examples of standard modules are electric motors, electronic power supplies or amplifiers, heat exchangers, pumps, gear boxes, v-belt drive systems, batteries, light bulbs, switches, and thermostats. Standard modules, like standard parts, are generally selected from catalogs.

Products and Machines

A product is a functional designed object that is made to be sold and/or used as a unit. Products that are marketed through retailing to the general public are called consumer products. Many manufactured products are designed for and sold to other businesses for use in the business; this is sometimes called the trade (or commercial, or industrial) market. For example, a manufacturer may buy a pump to circulate cooling water to a machine tool already purchased. In addition, there are products, including especially standard parts and standard modules, that are sold to other manufacturers for use in products being manufactured; this is called the original equipment manufacturer (or OEM) market. An example is the purchase of a small motor for use in an electric fan. Trade marketing is usually done through a system of regional manufacturer's representatives and distributors.

A machine is a product whose function is to contribute to the manufacture of products and other machines.

1.3 THE PRODUCT REALIZATION PROCESS

Product Realization is the set of cognitive and physical processes by which new and modified products are conceived, designed, produced, brought to market, serviced, and disposed of. That is to say, product realization is the entire "cradle to grave" cycle of all aspects of a product.

Product realization includes determining customers' needs, relating those needs to company strategies and products, developing the product's marketing concept, developing engineering specifications, designing both the product and the production tools and processes, operating those processes to make the product, and distributing, selling, repairing, and finally disposing of or recycling the product and the production facilities.

Product realization also includes those management, communication, and decision-making processes that organize and integrate all of the above, including marketing, finance, strategic planning, design (industrial, engineering, detail, and production), manufacturing, accounting, research and development, distribution and sales, service, and legal operations.

Product realization consists of several overlapping stages including product development, industrial design, engineering design, and production design. These are defined in the paragraphs below. Figure 1.1 also provides a supporting illustration for the definitions of these terms.

Product development is the portion of the product realization process from inception to the point of manufacturing or production. Though product development does not, by this definition, include activities beyond the beginning of production, it does require the use of feedback from all the various downstream product realization activities for use in designing, evaluating, redesigning parts and assemblies, and planning production. This feedback especially includes information about manufacturing issues—that is, information about design for manufacturing. Thus product development (including product improvement and redevelopment) is an ongoing activity even after production has begun.

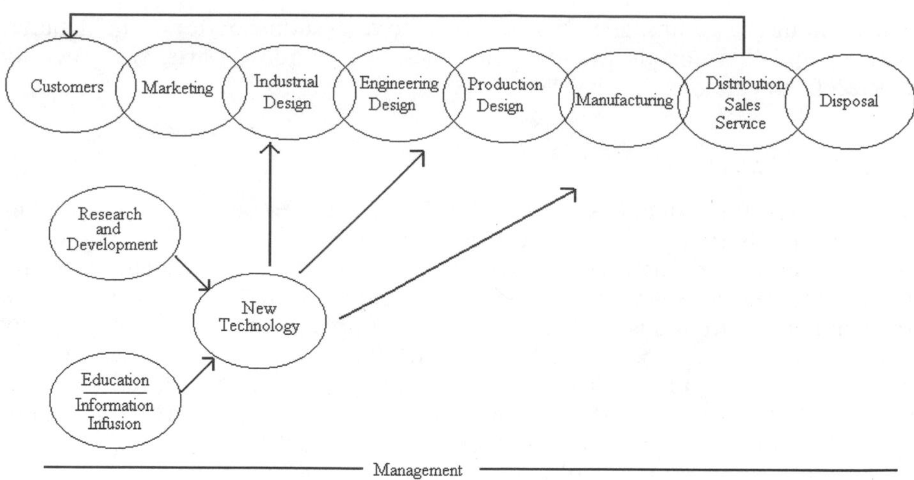

FIGURE 1.1 *View of the product realization process.*

1.4 INDUSTRIAL (OR PRODUCT) DESIGN

The process of *industrial design* (sometimes the phrase *product design* is used) creates the first broadly functional description of a product together with its essential visual conception. Artistic renderings of proposed new products are made, and almost always physical models of one kind or another are developed. The models are often merely very rough, nonfunctional ones showing external form, color, and texture only, but some models at this stage may also have a few moving parts.

There is great variability in the way industrial design is organized and utilized within different firms. In one firm, for example, a very small number of industrial designers are employed who work constantly to generate new or revised product concepts. These industrial designers are essentially a separate department but keep in close communication with colleagues in both marketing and engineering design. When a product concept has been approved by management for further development, then outside consultants in industrial design are brought in to work with the in-house industrial designers and with the firm's engineering designers and manufacturing people to refine and complete the industrial design phase. Marketing remains involved during this phase.

In another firm, no industrial designers per se are employed. In this firm, creative engineering designers working with marketing develop initial product concepts. Once a concept has approval, then outside industrial design consultants are brought in to perform the work described above.

An example of the kinds of issues that must be resolved by cooperation among industrial and engineering designers at this early stage is determination of the basic size and shape of the product. Industrial designers will have aesthetics, company image, and style primarily in mind in creating a proposed size and shape for a product. Engineering designers, on the other hand, will be concerned with how to get all the required functional parts into the (usually) small space proposed. Another issue requiring cooperation may be choices of materials for those parts that can be seen or handled by consumers. And, of course, both design engineers and manufacturing engineers are concerned about how the product is to be made within the required cost and time constraints.

It is helpful to use the phrase *product-marketing concept* to describe the results of industrial design. Sometimes the product-marketing concept is written into a formal (or at least informal) product-marketing specification. Whether written or not, the information available about the product after industrial design includes any and all information about the product that is essential to its marketing. Thus it will include qualitative or fuzzy statements of the marketing rationale and the in-use function of an artifact as perceived by potential customers, including any special in-use features the product is to have. It will certainly include any visual or other physical characteristics that are considered essential to marketing the product.

The product marketing concept or specification should, however, contain as little information as possible about the engineering design and manufacture of the product. This is to allow as much freedom as possible to the engineering design and production phases that follow. Such a policy is called *least commitment*, and it is a good policy at all stages of product realization. The idea is to allow as much freedom as possible for downstream decisions so that designers are free to develop the best possible solutions unconstrained by unnecessary commitments made at previous stages.

Some engineers who like things to be quite precise will fret over the fact that the definition of the product marketing concept, and any associated product marketing specification, is fuzzy. However, the activity of industrial design that produces the product marketing concept is creative, involves aesthetics and psychology, and naturally produces somewhat imprecise results. Industrial design is, nevertheless, extremely important to the product realization process. Engineers should respect it, and learn to live and work cooperatively with it—frustrating though it may be on occasion.

The use of the terms *industrial design* or *product design* to describe this phase of product development is clearly misleading. Both terms are nevertheless in customary use—and there is no hope of changing the custom. Logically, the whole product development process, including engineering design, could be accurately called product design. More accurate terms for industrial or product design would be preliminary product concept design or marketing product concept design, but we will generally adhere to the more commonly used industrial design in this book.

1.5 ENGINEERING DESIGN

Engineering design generally follows but overlaps industrial design, as illustrated in Figure 1.1. Engineering design consists of four roughly sequential but also overlapping stages or subprocesses, each corresponding to a design problem type:

engineering conceptual design;
configuration design of parts;
parametric design; and
detail design.

The first three of these basic engineering design problem types are introduced very briefly below.

Engineering Conceptual Design

Several variations of the engineering conceptual design problem are encountered in the course of engineering design. The variations are slight and depend on whether the object being designed is

(1) a new product (usually an assembly),
(2) a subassembly within a product, or
(3) a part within a product or subassembly.

The desired state of information that is to result from the engineering conceptual design process is called the physical concept. It includes information about the physical principles by which the object will function. In addition, in the case of products and their subassemblies, the physical concept also includes identification of the principal functional subassemblies and components of which the product will be composed, including particular functions and couplings within the product. By "couplings" we mean their important interrelationships, such as physical connections, or the sharing of energy or other resources.

Engineering conceptual design problems, like all major engineering design problem types, are solved by guided iteration. The first step in the guided iteration process is problem formulation. In engineering conceptual design, this means preparation of an Engineering Design Specification. This Specification records the product's quantitative functional requirements as well as specific information on requirements for such factors as weight, cost, size, required reliability, and so on.

Generating alternatives in engineering conceptual design requires a creative process called *conceptual decomposition*, or just decomposition. We might, for example, decompose a wheelbarrow into a wheel subassembly, the tub, and the carrying handle subassembly. The subsidiary subassemblies and components that make up a product or subassembly are created during the decomposition process for a product or other subassembly. There are many options open to designers in this creative decomposition process, and many opportunities for innovation. The decomposition process is very important because the physical concepts chosen have a tremendous impact on the final cost and quality of the designed object.

For evaluating competing conceptual solutions, a number of methods are available, as discussed in, among other places, Dixon and Poli. The conceptual design of parts is a bit different from the issues in the design of products and subassemblies, and it is discussed separately in Dixon and Poli.

An activity that permeates the engineering design process is the selection of the materials of which the parts are to be made, as well as the processes by which they are to be manufactured. Because there are literally thousand of materials and hundred of processes, these problems of choice are complex. Moreover, materials, processes, and the functional requirements of parts must all be compatible. Though most often it is only necessary at the conceptual design stage to select the broad class of material and process (e.g., plastic or thermoplastic injection molded, or aluminum die cast), we will cover the subject of materials and process selection rather fully in Chapter 13, "Selecting Materials and Processes for Special Purpose Parts."

Configuration Design of Parts and Components

During the conceptual design of products and their subassemblies, a number of components (that is, standard modules, standard parts, and special purpose parts) are created as concepts.

In the case of standard modules and standard parts, configuration design involves identifying and selecting their type or class. For example, if a standard module is a pump, then configuration design involves deciding whether it is to be a centrifugal pump, a reciprocating pump, a peristaltic pump, or some other type. Another example: if a standard part is to be a spring, then configuration design involves deciding whether it is to be a helical spring, a leaf spring, a beam spring, or another type.

For special purpose parts, configuration design includes determining the geometric features (e.g., walls, holes, ribs, intersections, etc.) and how these features are connected or related physically; that is, how the features are arranged or configured to make up the whole part. In the case of a beam, for example, configuration design makes the choice between I-beam or box beam and all the other possible beam cross-section configurations. The features in an I-beam are the walls (called *flanges* and *web*), and they are configured as in the letter "I."

At the configuration design stage, exact dimensions are not decided, though approximate sizes are generally quite obvious from the requirements of the Engineering Design Specification.

More information about material classes and manufacturing processes may be added at the configuration stage if the information is relevant to evaluating the configuration during the guided iteration process. For example, it may be necessary for evaluation to know that a high-strength engineering plastic is to be used, or that an aluminum extrusion is to be heat-treated to a high level of strength. However, this more detailed information should be generated only if it is really needed; least commitment is always the basic policy.

Configuration design of special purpose parts begins with problem formulation, usually in the form of a part requirements' sketch. Once the requirements are established in a sketch, then alternative configurations are sketched and evaluated with important help from qualitative reasoning based on physical principles and, it is hoped, DFM concepts.

Parametric Design

In the parametric design stage, the initial state of information is the configuration. In other words, just about everything is known about the designed object except its exact dimensions and tolerances. It may also be in some cases that the exact material choice is also unknown, though the basic class of material will usually be included in the configuration information.

The goal of parametric design, therefore, is to add the dimensions and any other specific information needed for functionality and manufacturability. The specific material is also selected if it has not previously been designated. In other words, parametric design supplies all the dimensions, tolerances, and detailed materials information critical to the design consistent with both the marketing concept and the Engineering Design Specification.

In the spirit of least commitment design, parametric design need not and should not specify information to any degree of precision that is not actually required by the Specifications, but most dimensions are in fact usually determined at the parametric stage.

Detail Design

Detail design supplies any remaining dimensions, tolerances, and material information needed to describe the designed object fully and accurately in preparation for manufacturing. As every manufacturing engineer knows, in a large complex product, the result seldom provides all the dimensions and tolerances that exist. The rest of the details, if needed, are established in the manufacturing design phase.

1.6 PRODUCTION DESIGN

Production design overlaps with detailed design. It involves, for example, finishing any of the design details left undone by detailed design, detailed design of the tooling, planning the manufacturing process, planning for quality control, "ramping up" the actual production, planning for quality and process control, and supporting the initial production runs. It is a substantial design task, though it is usually referred to as a part of the overall manufacturing or production process.

1.7 SCOPE OF THE BOOK

1.7.1 Mode of Operation in Industry

The sequential mode of operation depicted in Figure 1.2 is still the prevalent mode of operation found in industry today. Although changes are taking place, "old habits are hard to break," and the linear sequence displayed still seems to prevail.

As shown in Figure 1.2, the sequence begins with the conception of an idea for a new or revised product. These ideas for new and improved products commonly come from customers, employees, and new technology.

Customers

Competitive manufacturing businesses require constant feedback from the customers who buy, sell, repair, or use the company's products. Getting such feedback cannot be left to chance or to mail surveys, or even to formal complaint records, though such mechanisms are certainly a part of the process. There are marketing professionals who know how to get this information. But in addition, if a design engineer is looking for positive new ideas as well as for shortcomings

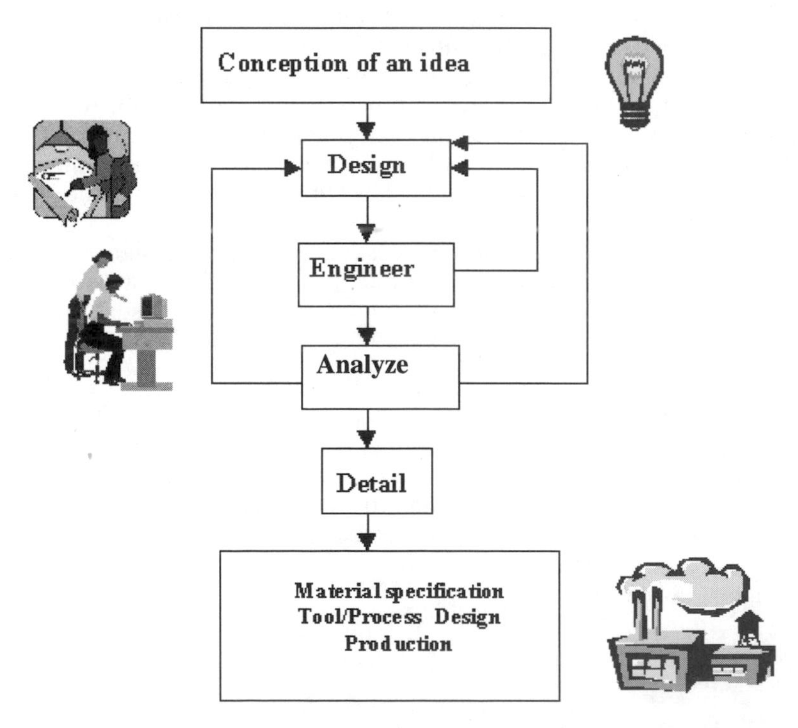

FIGURE 1.2 *Sequence of events prevalent in industry for the design and manufacture of products.*

of current products, then he or she must get out personally and talk to the customers.

A famous example is the Japanese automaker who sent design engineers to talk to a large number of American shoppers in supermarket parking lots discovering, among other things, that making it easy to get heavy grocery bags into and out of car trunks was a convenience valued by many potential customers.

Design and product development engineers must also get out and talk to their "customers" within the company. These include not only marketing people, but also manufacturing engineers, analysts, draftsmen, accountants, and others.

Employees

Employees in the factory, shops, and offices are also an extremely valuable source of new ideas for products and product improvements. Good practice requires that there must be a believable, financially rewarding, well understood, and low threshold (i.e., easy) mechanism for employees to get their new product, product improvement, and process improvement ideas heard and seriously considered.

Employees must also get rapid feedback on what happens to their ideas (good and bad), and why.

New Technology

Keeping abreast of new technologies and methodologies in materials, manufacturing, design, engineering, and management is another important source of ideas for new and improved products. Coupling new technological information with the search for new or improved product ideas is an essential part of the product development process that is not, strictly speaking, engineering design as we have defined it. But it is important for engineering designers to be able to contribute for best results within a company.

Upon approval of the idea, the new or improved product is then designed, engineered, and analyzed for function and performance. The design phases in this case consist of (a) an industrial or product design phase in which artistic renderings or nonfunctional models are produced, and (b) an engineering design phase. In the engineering phase of the design the decision of whether to use standard or special purpose parts is made. In the case of special purpose parts, the overall geometric configuration of the parts is first determined (box-shaped, flat, etc.), as well as the presence and location of various features such as ribs, holes, and bosses. Following this configuration design, a parametric design phase takes place in which more detailed dimensions and tolerances are added.

Next an analysis of the design from the point of view of function and performance takes place. Subsequently the design is detailed as the remaining dimensions and tolerances are added, the material is specified, and production drawings are produced. Finally, the product is turned over to manufacturing where both production design and process design takes place.

1.7.2 Goals of This Book

According to high-ranking representatives from industry, there are two main problems with the sequential approach depicted in Figure 1.2:

1. Decisions made during the early conceptual stages of design have a great effect on subsequent stages. In fact, quite often more than 70% of the manufacturing cost of a product is determined at this conceptual stage, yet manufacturing is not involved.

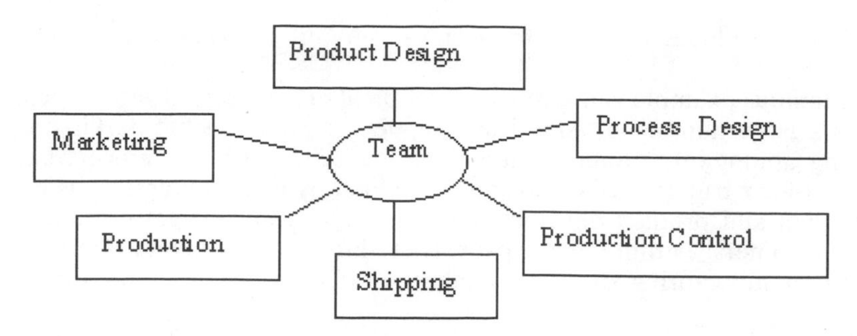

FIGURE 1.3 *The team approach often used in industry.*

2. No single person or group is in charge. Each group blames the other for problems. Subsequently, difficulties and delays occur and the design and manufacturing process goes out of control.

Two possible solutions that have been proposed are:

1. Form teams (Figure 1.3) that involve everyone from the beginning to the end of the entire life cycle. In the team approach, product design and process design take place concurrently. With the team approach, an approach often used with success in industry, the team is in charge of the product from cradle to grave.
2. Educate designers about manufacturing. In other words, make designers more manufacturing literate.

The overall goals of this book are centered on this latter solution, namely to educate students as to how parts are manufactured and to train them to recognize costly-to-produce features (cost-drivers) so that they can be avoided.

1.7.3 Manufacturing Processes Considered

The vast majority of consumer products today consist of both standard parts and special purpose parts. Many of the special purpose parts are thin-walled parts produced by injection molding, die casting, and stamping. Occasionally a forged component is present as well. For this reason, and in order to best meet the goal of making designers more manufacturing literate, the emphasis in this book is on the study of these four processes. In particular, we will study the effect of part geometry on the ease or difficulty of creating the tooling required to produce the part and the effect that geometry and production volume have on both processing costs and overall part costs.

There are obviously other important manufacturing processes. For example, *machining* is used to create the dies and molds needed to produce parts by the processes mentioned above and one-of-a-kind parts. *Rolling* is used to produce structural shapes, rails, sheets, large diameter tubes, strips, and plates, and drawing is used to produce bars, wires, and small diameter tubes. *Extrusion* plays an important role in the production of long metal objects whose cross-sectional shapes are constant. And, of course, there are also a whole host of other casting processes (*sand casting, investment casting,* etc.) and polymer processing methods (*extrusion, compression molding, transfer molding*, etc.) used to produce thin-walled parts. Since these processes are adequately treated elsewhere (*American Society of Metals Handbook*, Vols. 14, 15, and 16), their treatment here will be limited.

The emphasis, as indicated above, will be on those processes used to produce the vast majority of special purpose parts found in consumer products today.

1.8 SUMMARY

This chapter has described the product realization process and to the three main stages of engineering design, namely, engineering conceptual design, part configuration design, and parametric design. All of these design stages use guided iteration, a method discussed in greater detail in Dixon and Poli, 1995, as a problem solving methodology and all involve design for manufacturing (DFM).

In the chapters that follow, a number of common manufacturing processes and their associated qualitative DFM guidelines are described. Of course, whole books are written that describe each of these processes in much more detail. The purpose of this book is limited, however, thus, we need only present sufficient information for a reasonable understanding of the rationale for the DFM guidelines.

Before beginning our discussion of DFM, we review, in the next chapter, tolerances, mechanical properties, and physical properties.

REFERENCES

Ashley, S. "Rapid Prototyping Systems." *Mechanical Engineering*, April 1991.

American Society of Metals International. *American Society of Metals Handbook*, 9th edition, Vol. 15. "Casting." Metals Park, OH: ASM International, 1988.

———. *American Society of Metals Handbook*, 9th edition, Vol. 16. "Machining." Metals Park, OH: ASM International, 1989.

———. *American Society of Metals Handbook*, 9th edition, Vol. 14. "Forming and Forging." Metals Park, OH: ASM International, 1988.

Boothroyd, G., and Dewhurst, P. *Product Design for Assembly.* Wakefield, RI: Boothroyd Dewhurst, 1987.

Box, G., and Bisgaard, S. "Statistical Tools for Improving Designs." *Mechanical Engineering*, January 1988.

Box, G., Hunter, W. G., and Hunter, J. *Statistics for Experimenters: An Introduction to Design, Data Analysis, and Model Building.* New York: Wiley, 1978.

Charney, C. *Time to Market: Reducing Product Lead Time.* Dearborn, MI: Society of Manufacturing Engineers, 1991.

Deming, W. E. *Out of the Crisis*, M.I.T., Cambridge, MA: Center for Advanced Study, 1982.

Dixon, J. R. *Design Engineering: Inventiveness, Analysis, and Decision Making.* New York: McGraw-Hill, 1966.

———. "Information Infusion Is Strategic Management." *Information Strategy.* New York: Auerbach Publishers, Fall, 1992.

———, and Poli, C. *Engineering Design and Design for Manufacturing: A Structured Approach.* Conway, MA: Field Stone Publishers, 1965.

Gatenby, D. A. "Design for 'X' (DFX) and CAD/CAE." Proceedings of the 3rd International Conference on Design for Manufacturability and Assembly, Newport, RI, June 6–8, 1988.

Galezian, R. *Process Control: Statistical Principles and Tools.* New York: Quality Alert Institute, 1991.

Hauser, D. R., and Clausing, D. "The House of Quality." *Harvard Business Review*, May–June 1988.

Hayes, R. H., Wheelwright, S. C., and Clark, K. B. *Dynamic Manufacturing: Creating the Learning Organization.* New York: The Free Press, 1988.

Johnson, H. T., and Kaplan, R. *Relevance Lost: The Rise and Fall of Management Accounting.* Cambridge, MA: Harvard Business School Press, 1987.

National Research Council. *Improving Engineering Design: Designing for Competitive Advantage.* Washington, DC: National Academy Press, 1991.
Nevins, J. L., and Whitney, D. E. *Concurrent Design of Products and Processes*. New York: McGraw-Hill, 1989.
Pugh, S. *Total Design: Integrating Methods for Successful Product Engineering*. Reading, MA: Addison-Wesley, 1991.
Ross, R. S. *Small Groups in Organizational Settings*. Englewood Cliffs, NJ: Prentice-Hall, 1989. ASM International.
Simon, H. A. *The Sciences of the Artificial.* Cambridge, MA: M.I.T. Press, 1969.
Ver Planck, D. W., and Teare, B. R. *Engineering Analysis*. New York: Wiley, 1954.
Smith, P. G., and Reinertsen, D. G. *Developing Products in Half the Time.* New York: Van Nostrand Reinhold, 1991.
Taguchi, G., and Clausing, D. "Robust Quality." *Harvard Business Review*, Jan.–Feb. 1990.
Wall, M. B., Ulrich, K., and Flowers, W. C. "Making Sense of Prototyping Technologies for Product Design." DE vol. 31, *Design Theory and Methodology*, ASME, April 1991.
Wick, C., editor. *Tool and Manufacturing Engineers Handbook*, Vol. 2. "Forming, 9th edition." Dearborn, MI: Society of Manufacturing Engineers, 1984.

QUESTIONS AND PROBLEMS

1.1 What is the smallest functional designed object you have heard of? The largest?

1.2 In a bicycle (or thermostat or other product of your choice), give an example of a standard part, a special purpose part, a special purpose subassembly, a standard module, and a component.

1.3 Have you ever used the basic guided iteration process to solve a design problem? If so, describe your experience, noting especially how you implemented each of the steps.

1.4 Have you ever used the basic guided iteration process to solve a problem other than a design problem? If so, describe your experience, noting especially how you implemented each of the steps.

1.5 Go to your library and find at least one book on design for manufacturing and read its introductory chapter and the preface. Report back to the class on how it is the same and how it is different from this chapter.

1.6 In Figure 1.1, identify the stages that constitute "product development" as defined in this chapter.

1.7 What are the disadvantages to the sequential mode of operation depicted in Figure 1.2?

1.8 Can you think of any disadvantages to the mode of operation depicted in Figure 1.3?

Chapter 2

Tolerances, Mechanical Properties, Physical Properties—A Review

2.1 INTERCHANGEABILITY OF PARTS

At first glance it might appear that a car's fuel pump, a laser jet printer cartridge, and an ordinary lightbulb have nothing in common. This is because what they do have in common we simply take for granted. For example, when our fuel pump goes we assume that we can go to our local auto parts supplier, and that, upon supplying the make, model, and year of our car, we can pick up a pump and "easily" replace the old pump with the new one. The same is true of our printer cartridge and lightbulb. We assume that we can purchase a replacement part and that it will fit—we assume that the parts will be interchangeable.

It has not always been possible to interchange parts. In the early days of manufacturing, each assembly operator was responsible for producing every individual piece-part and then making adjustments to the parts in order to complete the final assembly. Components on one assembly could not be interchanged with those on another assembly. For example, during the Revolutionary War, components from one musket would not be interchangeable with "identical" components from another musket.

Modern manufacturing, be it the high-volume mass production of cars, pens, or watches, or the low-volume batch production of computers, copy machines, or printers, depends on the interchangeability of parts.

2.1.1 Size Control

Interchangeability depends on size control, and size control depends upon the existence of a standard unit of length and suitable measuring equipment such as *gage blocks*, *dial gages*, *calipers*, *steel rules*, and other measuring instruments (Figures 2.1 and 2.2). Size control also depends upon the inspection of parts and quality control.

2.2 TOLERANCES

2.2.1 Introduction

A tolerance is a designer-specified allowed variation in a dimension or other geometric characteristic of a part. Proper tolerances are crucial to the proper

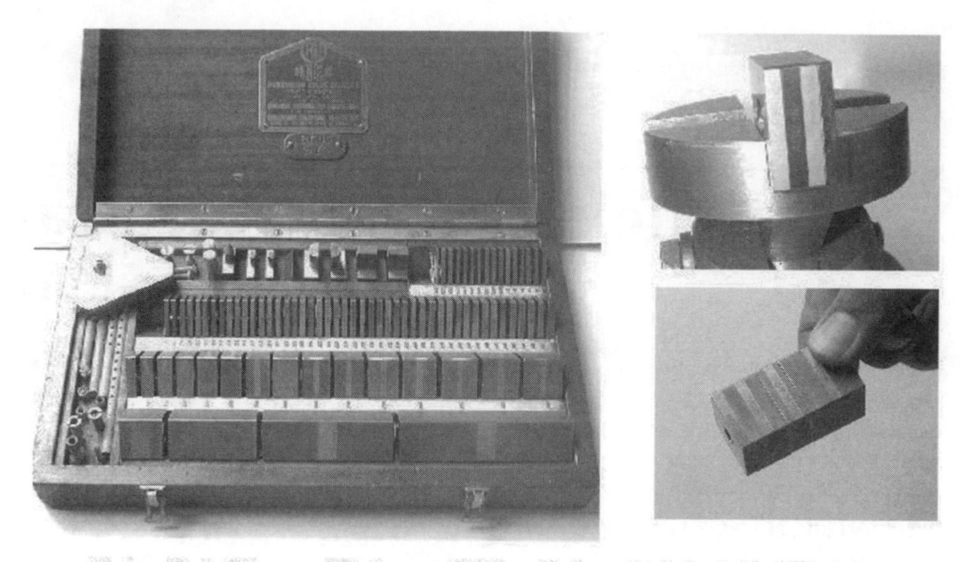

Developed by Carl Johansson of Sweden around 1900, gage blocks saw limited use in World War I. Henry Ford brought him to the U.S. after WWI and was responsible for the large scale production and use of gage blocks. Gage blocks are end-standards and can be used for direct measurement, as in measuring the width of a groove or keyway (top right), or for calibrating line standard measuring instruments (Fig. 2.2). The insert on the bottom right shows several gages blocks wrung together in order to produce a special purpose gage block dimension.

FIGURE 2.1 *Gage blocks.*

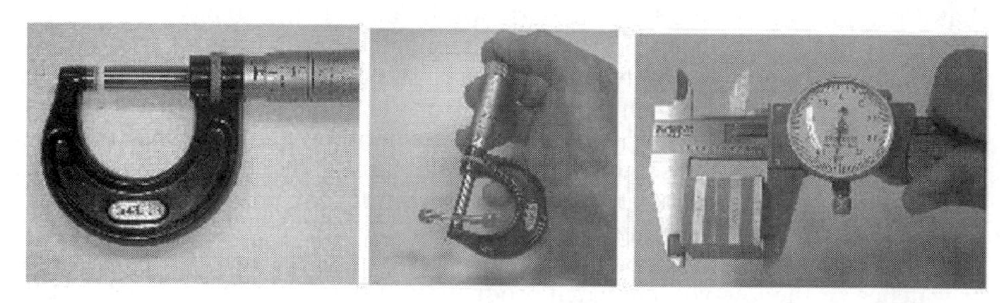

FIGURE 2.2 *Some line standard measuring instruments (micrometer on left, vernier caliper on the right) used for size control.*

functioning of products. However, the most common cause of excessive manufacturing cost is the specification by designers of too many tolerances or tolerances that are tighter than necessary.

On part drawings, simple dimensional tolerances are usually attached to dimensions, as shown in Figures 2.3 and 2.4.

The issue of tolerances may be found at the intersection between the requirements of functionality and the capabilities of manufacturing processes. Mass-produced parts cannot all be produced to any exact dimensions specified by designers. In a production run, regardless of the process used, there will always be variations in dimensions, from the nominal; that is, not all parts produced will have the same dimensions. Some of the reasons are: tools and dies wear; processing conditions change slightly during production; and raw materials vary in composition and purity. Modern methods for controlling processes are achieving

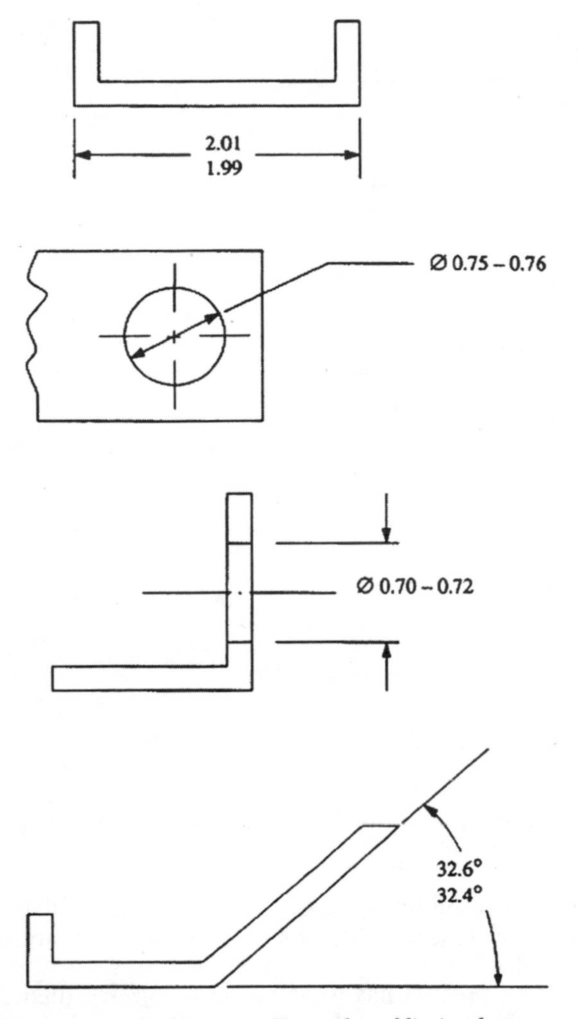

FIGURE 2.3 *Examples of limit tolerances.*

ever more consistent and accurate dimensions, but variations of some frequency and magnitude are inevitable. Different manufacturing processes are also inherently more capable of reducing these variations, and hence holding tighter tolerances, than others.

Tolerances can be specified by designers to limit the range of dimensions on parts that are to be considered acceptable. That is, designers can, in order to get the functionality required, limit the range of dimensional variations in those parts that reach the assembly line. However, the more strict these limitations are, and the more dimensions that are subject to special tolerance limitations, the more expensive the part will be to produce.

Fortunately, the functionality of parts seldom, if ever, requires that all or even most dimensions of parts be controlled tightly. To achieve the desired functionality (and other requirements) of a part, a few dimensions and other geometric characteristics (e.g., straightness or flatness) may require quite accurate control, while others can be allowed to vary to a greater degree. Thoughtful design can result in parts and assemblies configured so that the number of characteristics

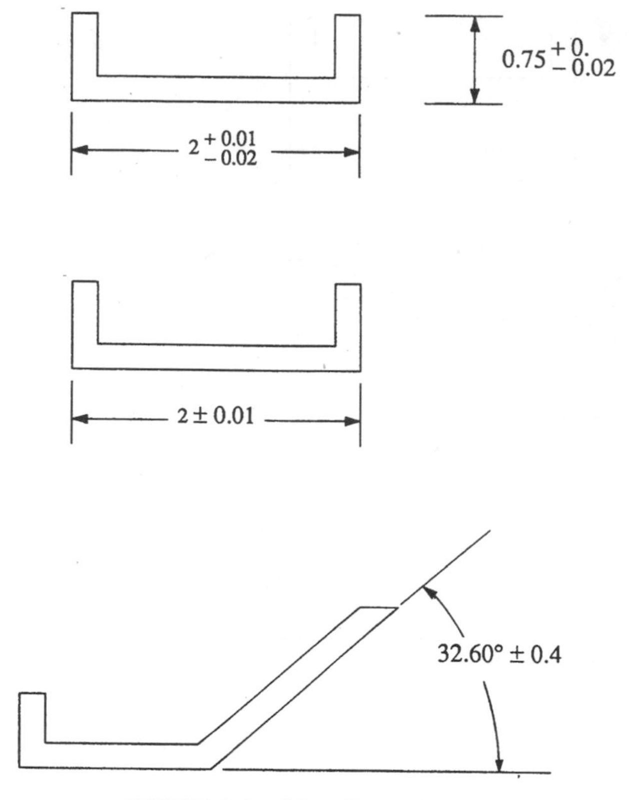

FIGURE 2.4 *More limit tolerances.*

requiring critical control is minimal. Also, those dimensions that do require critical control may be of a type or at a location where they are more easily controlled during manufacturing.

Every manufacturing process has what are most often called standard or commercial tolerances. These are the tolerance levels that can be produced with the normal attention paid to process control and inspection. Though standard tolerance values are not always completely and accurately defined—and available in print for designers—they are well known to manufacturing process engineers. There is no need to guess; designers can go visit their friendly manufacturing colleague.

Parts designed so that there are no tolerance requirements tighter than standard will be the least expensive to produce. Moreover, and this is an important point for designers, the number of tolerances that must be critically controlled, whether standard or tighter, is crucial to the ease and cost of manufacturing. Controlling one or two critical dimensions, unless they are of an especially difficult type or extremely tight, is often relatively easy to do if the other dimensions of the part do not need special control.

2.2.2 Some Definitions

Before providing a precise definition of tolerance we must first understand some preliminary definitions. The *limits of size* are the two extreme permissible sizes of a part between which the actual size must lie. For the part shown in Figure 2.3

the *maximum limit of size* is 2.01. The *minimum limit of size* for this same part is 1.99. *Tolerance* is the difference between the maximum limit of size and the minimum limit of size. For the part shown in Figure 2.3 the tolerance, T_a, is 0.02.

The size by reference to which the limits of size are fixed is called the *basic size*. For the parts shown in Figure 2.4 the basic size is 2.

Tolerances can be applied bilaterally, as shown in Figure 2.4, or unilaterally as indicated below.

$$2^{+0.02}_{-0.00}$$

When parts must fit together (a shaft inside a hole, for example), the tolerances on both parts must be considered simultaneously. The relationship resulting from the difference, before assembly, between the sizes of the two parts to be assembled is called the *fit*.

A *clearance* exists if the diameter of the hole is greater than the diameter of the shaft. This allows for relative movement (translation or rotation) between the two parts. An *interference* exists if the diameter of the hole is less than the diameter of the shaft. This is used when an alignment of parts is required or stiffness is needed.

As stated above, regardless of the process used, there will always be variations in dimensions from the nominal, and not all parts produced will have the same dimensions. Hence, in the case of a shaft and hole, a *clearance fit* exists if the minimum hole diameter is greater than the maximum shaft diameter. An *interference fit* is said to exist if the maximum hole diameter is less than the minimum shaft diameter. A *transitional fit* is said to exist if we sometimes get clearance and sometimes get interference.

At times there is a tendency to confuse tolerance with allowance. Allowance is the intentional desired difference between the dimensions of two mating parts. It is the difference between the dimensions of the largest interior-fitting part and the smallest exterior-fitting part. In the case of a clearance fit, allowance is the minimum clearance between the two parts. In the case of an interference fit, allowance is the maximum interference between two mating parts.

Example

If D_H represents the diameter of a hole and D_S represents the diameter of a shaft, then can you explain why the following hole-shaft combination results in a transitional fit?

$$D_H = 50.00^{+0.10}_{-0.00}$$

$$D_S = 50.00^{+0.10}_{-0.00}$$

2.3 MECHANICAL AND PHYSICAL PROPERTIES

2.3.1 Introduction

The design of special purpose parts involves the determination of a material class (steel, aluminum, thermoplastic, etc.) and the basic manufacturing process to be used (injection molding, forging, die casting, etc.). A part's size, shape, and geometry, and the material's mechanical and physical properties, affect the choice of material-process combination. Chapters 3 to 11 are devoted to a discussion of how a part's size, shape, and geometry affect the ease or difficulty of producing a part by various processes. Chapter 13, "Selecting Materials and Processes for Special Purpose Parts," presents a general methodology for selecting one or more material-process combinations for special purpose parts at the conceptual design stage. In the remainder of this chapter we will briefly review what is meant by the mechanical and physical properties of materials.

2.3.2 Mechanical Properties

Mechanical properties of materials are generally expressed in terms of the elastic behavior of the material, the *yield stress*, σ_y, and the *ultimate stress*, σ_u. The elastic behavior is characterized by *Young's modulus*, E, and the *engineering stress*, σ_e.

The most common method for determining these properties is via a uniaxial tension test. In this test a test specimen is placed in a tensile testing machine (Figure 2.5) and subjected to increasing loads in order to elongate it until it fractures.

Engineering stress, we know, is defined as

$$\sigma_e = \frac{P}{A_o}$$

and engineering strain is defined as

$$\varepsilon_e = \frac{l - l_o}{l_o}$$

In this case P is the tensile load applied to the specimen, A_o is the initial cross-sectional area of the specimen, l_o is the original gage length of the specimen, and l is the final length.

Also in common use are the *true stress*, defined as

$$\sigma = \frac{P}{A}$$

and the *natural strain*, defined as

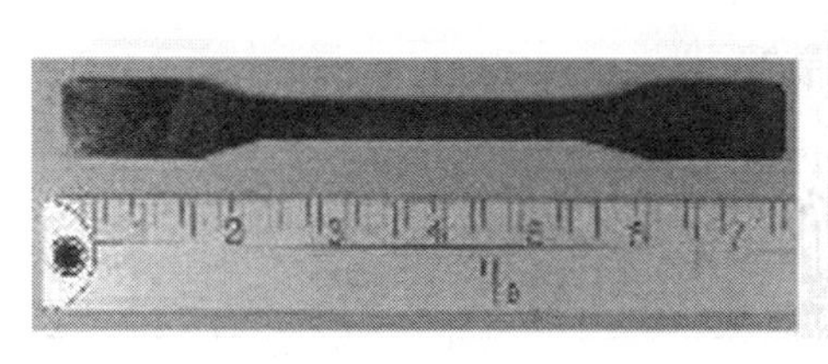

FIGURE 2.5 *A test specimen being subjected to uniaxial loads.*

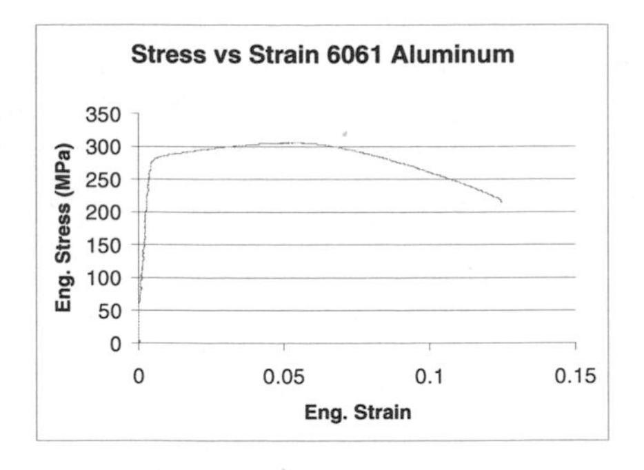

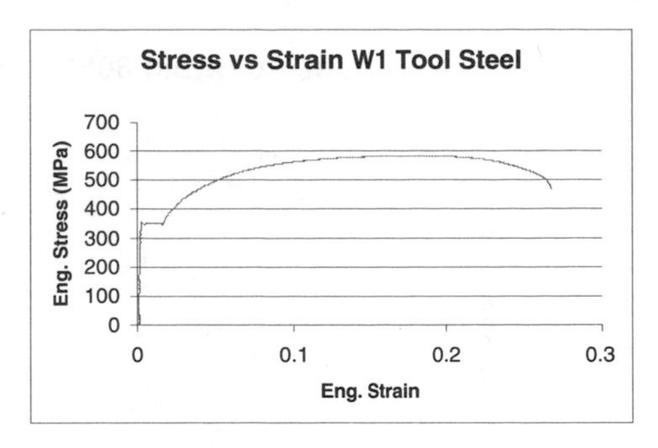

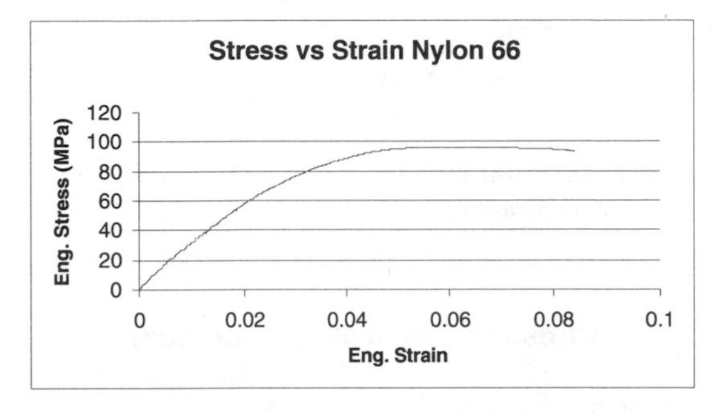

FIGURE 2.6 *Plots of engineering stress versus engineering strain for some specimens tested using the apparatus shown in Figure 2.5.*

$$\varepsilon = \ln\left(\frac{1}{1_o}\right)$$

In this case the true cross-sectional area is represented by A.

It is easy to show that the natural strain and engineering strain are related by the following expression, namely,

$$\varepsilon = \ln(1 + \varepsilon_e)$$

Figure 2.6 shows some test results obtained using the setup shown in Figure 2.5. The actual shape of the curve depends upon the specific material or alloy used. Figure 2.7 shows a plot of true stress versus natural strain superimposed upon a plot of the engineering stress versus engineering strain curve for the 6061 aluminum specimen shown in Figure 2.6.

Figure 2.7 shows that while the engineering stress reaches a maximum, referred to as the ultimate tensile stress (UTS), and then decreases until the specimen fractures, the true stress always increases. This is due to the fact that the actual cross-sectional area continually decreases.

A study of the curves shown in Figure 2.7 shows that both curves have linear and nonlinear regions and that initially the two curves coincide. In the linear region where E is the slope of the curve and is referred to as Young's modulus

$$\sigma_e = E\varepsilon_e$$

Young's modulus is a measure of the stiffness of the beam and is used in the calculation of the deflection of beams. In this region, if the load is released the specimen returns to its original length.

Beyond the elastic limit, the test specimen will not return to its original length upon removing the load. In this nonlinear region the specimen is said to deform

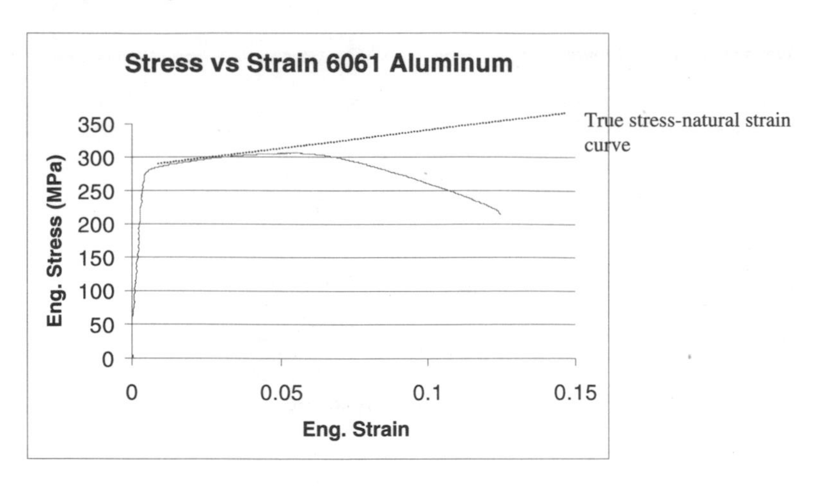

FIGURE 2.7 *A true stress–natural strain curve for 6061 aluminum superimposed on an engineering stress–engineering strain curve.*

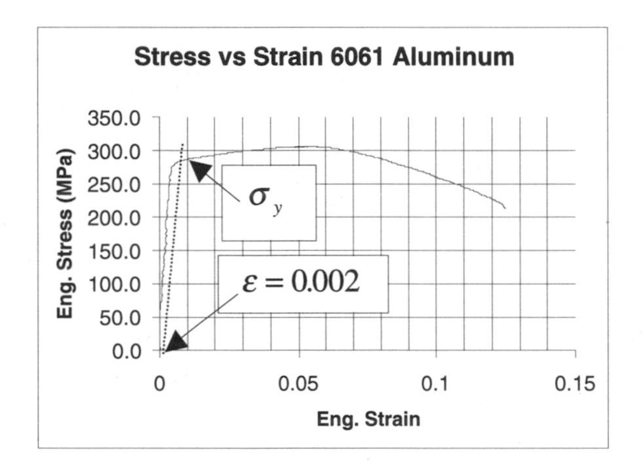

FIGURE 2.8 *Determination of the offset yield stress.*

plastically. In general it is difficult to determine the exact value of the stress when elastic deformation no longer takes place. The offset yield stress, σ_y, is used to indicate where plastic deformation begins. It is determined as indicated in Figure 2.8.

Knowing the value of the yield stress is important in design where in general we prefer not to obtain stresses that will result in permanent deformation of our product. Knowing the value of the yield stress is also important in metal forming where in general we do wish to obtain permanent deformation.

The true stress–natural strain curve is often represented by the following expression, namely,

$$\sigma = K\varepsilon^n$$

where K is a constant (= E in the elastic region) and n is referred to as the strain-hardening coefficient (= 1 in the linear region; varies between 0.1 and 0.5 in the nonlinear or plastic region). At necking, which begins when the engineering stress becomes equal to the ultimate tensile stress, it can be shown that the natural

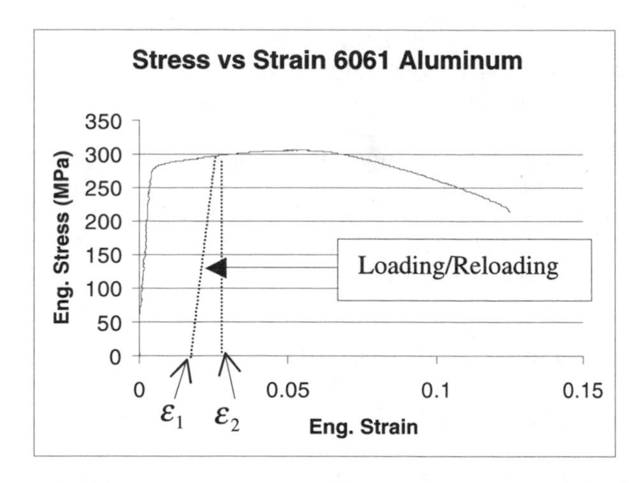

FIGURE 2.9 *Stress-strain curve showing the effects of loading and reloading in the non-linear regions of the curve.*

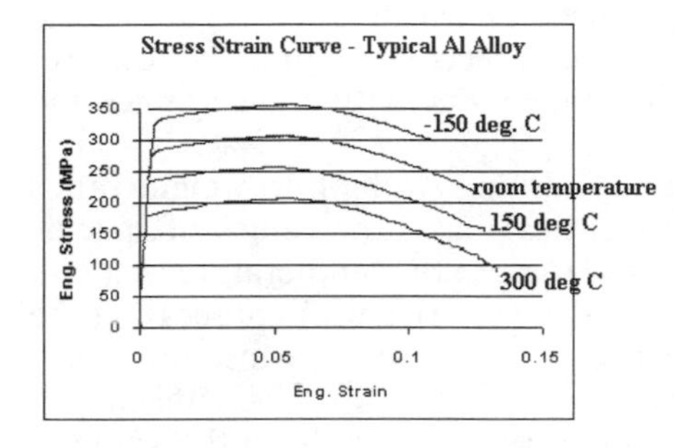

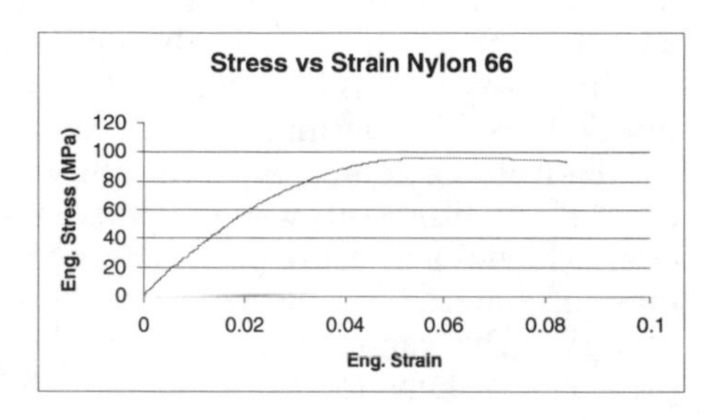

FIGURE 2.10 *Stress-strain curves for a typical aluminum alloy (plots on the left) at various temperatures and the stress-strain curve for nylon 66.*

strain, ε, is equal to n. Thus, the higher the value of n the greater the amount of elongation that can take place before the onset of necking—that is, the more ductile the material.

When the specimen is stretched beyond the yield point and then unloaded, it follows a path parallel to the linear region (Figure 2.9). Reloading the specimen will result in retracing our steps along the same curve we followed when unloading the specimen. As seen in Figure 2.9 the new yield stress is increased in value. This phenomena is referred to as strain hardening. *Springback*, which is the difference in the value of the strain when the specimen is loaded and the value of the strain when the specimen is unloaded, places an important role in the creation of stamped sheet-metal parts. For example, because of springback, a part that is to be bent, say 30°, must be over bent in order to accommodate springback.

2.3.3 Metals versus Plastics

In this section we briefly review some of the basic differences between metals and plastics. Figure 2.10 helps explain some of these differences.

Figure 2.10 shows that as the temperature changes the shape of the curve and the modulus of elasticity, E remains essentially the same. Figure 2.10 also shows that the yield stress, σ_y, the ultimate tensile stress, UTS, and strain hardening all decrease. In addition, it turns out that over a reasonably wide temperature range around room temperature, the value of E varies only slightly.

On the other hand, the modulus for plastics can change dramatically with both time and temperature, even for temperature changes of as little as 20°C. In addition, the shape of the curve changes with temperature and the material "creeps." Hard plastic generally refers to a plastic with a high modulus, a strong plastic is one with a high tensile yield stress.

2.4 PHYSICAL PROPERTIES OF MATERIALS

The physical properties of materials also play a significant role in the selection of a material-process combination for the production of special purpose parts. By physical properties we mean such characteristics as the *melting point, thermal conductivity, density, specific heat, thermal expansion, electrical conductivity, electrical resistivity*, and *corrosion resistance*.

The melting point of a material can affect energy costs for shaping parts via injection molding and casting. Thermal conductivity, which is a measure of the ease with which heat flows through a material, affects cycle time, hence, processing costs. Metals have high values of thermal conductivity, and ceramics and plastic have low thermal conductivity.

Density, of course, is important when considering strength-to-weight ratios as well as stiffness-to-weight ratios and weight savings. Low values of specific heat, which is a measure of the energy required to raise the temperature of a unit mass of material by one degree, cause temperatures to rise while processing the material. This can have a detrimental effect while machining because the workpiece can become too hot to handle and can result in a poor surface finish.

The amount of thermal expansion that takes place depends on the value of the coefficient of thermal expansion. The amount of thermal expansion that takes place is important when clearances and running fits are needed in an assembly.

Corrosion resistance or degradation is a measure of the ability of a metal or plastic to resist deterioration. This plays an extremely important role in the selection of a material for a given part and in turn affects which processes can or cannot be used to produce the part. For example, steel has poor resistance to corrosion and stainless steel has high resistance to corrosion.

In summary, selecting a material based on physical properties affects the choice of the process and the ease or difficulty of forming the material. For example, stainless steel resists corrosion but, as we will find out in Chapter 6, "Metal Casting Processses," it cannot be die cast. Also, stainless steel has a higher melting temperature than plastics, which implies that more energy is required to melt stainless steel. Although in some applications titanium may be a good choice based on its physical and mechanical properties, titanium is also more difficult to forge than aluminum due to its higher yield strength.

2.5 SUMMARY

This chapter has described the importance of the interchangeability of parts to modern manufacturing. In addition, this chapter has also discussed how the physical and mechanical properties of materials can effect the choice of process to be used to manufacture a part, and how these properties can affect the ease or difficulty of creating the part via a particular process.

REFERENCES

Degarmo, E. Paul, Black, J. T., and Kohser, Ronald A. *Materials and Processes in Manufacturing*, 8th edition. Upper Saddle River, NJ: Prentice Hall, 1997.

Flinn, Richard A., and Trojan, Paul K. *Engineering Materials and Their Applications*, 2nd edition. Boston: Houghton Mifflin, 1981.

Kalpakjian, Serope. *Manufacturing Engineering and Technology*, 2nd edition. Reading, MA: Addison-Wesley, 1992.

Schey, John A. *Introduction to Manufacturing Processes*, 3rd edition. New York: McGraw-Hill, 2000.

QUESTIONS AND PROBLEMS

2.1 Why is the interchangeability of parts important in the production of consumer products?

2.2 Explain the difference between tolerance and allowance.

2.3 What type of fit would best describe the following:
- a) Cork in a wine bottle.
- b) Floppy disk at the entrance to a floppy drive.
- c) Cover of a ballpoint pen.
- d) Laser printer cartridge and the printer.

2.4 Discuss the difference between true stress and engineering stress, and natural strain and engineering strain. Which stress-versus-strain curve makes more sense?

2.5 Imagine you are faced with the situation of choosing a metal that has high ductility. Can you define what is meant by ductility? How would you recognize which material has the higher ductility?

2.6 Name a product or component of a product for which physical properties are more important than mechanical properties. Explain!

2.7 Describe situations where it would be desirable to have a part that has:
- a) high density,
- b) low density,
- c) high melting point,
- d) low melting point,
- e) high thermal conductivity,
- f) low thermal conductivity.

2.8 Imagine you are faced with the decision of choosing between a plastic or a metal for a part you are designing. What are some of the reasons you would use for selecting plastic? What are some of the reasons you would use to select a metal?

Chapter 3

Polymer Processing

3.1 THE PROCESSES

A large number of polymer processing techniques exist; among the most common are injection molding, compression molding, transfer molding, extrusion, and extrusion blow molding.

Injection molding, compression molding, and transfer molding are capable of the economical production of complex parts (with significant levels of geometric detail) and simple parts (with little detail). Extrusion is limited to the production of long parts with a uniform cross-section. Extrusion blow molding is confined primarily to relatively simple hollow objects such as bottle containers. Each of these processes is described in more detail later in this section.

3.2 MATERIALS USED IN POLYMER PROCESSING

There are literally hundreds—maybe thousands—of polymeric materials available for processing, and more will continue to be developed. In general, these materials fall into two broad classes: thermoplastics and thermosets. Some polymers are available in both thermoplastic and thermoset formulations.

Thermoplastic materials, like water and wax, can be repeatedly softened by heating and hardened by cooling, and are formed into parts primarily by injection molding, extrusion, and extrusion blow molding.

For common product applications, most parts made by injection molding use thermoplastic materials. Examples are gears, cams, pistons, rollers, valves, fan blades, rotors, washing machine agitators, knobs, handles, camera cases, battery cases, telephone and flashlight cases, sports helmets, luggage shells, housings and components for business machines, power tools, and small appliances.

Thermoplastics are divided into two classes: crystalline and amorphous. Crystalline thermoplastics have a relatively narrow melting range. They are opaque, have good fatigue and wear resistance, high but predicable shrinkage, and relatively high melt temperatures and melt viscosities. Reinforcement of crystalline polymers with glass fibers or other materials improves their strength significantly. (Such reinforced plastics are often called composites.) Examples of crystalline plastics include acetal, nylon, polyethylene, and polypropylene (PP).

Amorphous thermoplastics melt over a broader temperature range, are transparent, and have less shrinkage, but they have relatively poor wear and fatigue resistance. The use of reinforcing fibers does not significantly improve the strength of amorphous thermoplastics at high temperatures. Examples of amorphous materials are ABS, polystyrene, and polycarbonate.

It should be noted that although amorphous polymers have no crystallinity, no polymer is more than about 90% crystalline. Thus, many thermoplastic polymers exhibit a mixture of amorphous as well as crystalline properties.

The small number of thermoplastics noted above include some that are called commodity or general purpose plastics (polystyrene, polyethylene, and polypropylene) as well as engineering thermoplastics (ABS, polycarbonate, acetal, and nylon-6). The largest proportion of thermoplastics used are commodity plastics used to produce film, sheet, tubes, toys, and such throwaway articles as bottles and food packaging. Compared to the commodity plastics, engineering thermoplastics are capable of supporting higher loads, for longer periods of time, and at higher temperatures.

Thermoset materials are polymeric materials that, similar to an egg, transform permanently on heating and cannot be remelted. Thermosets are formed primarily by compression molding and transfer molding. Parts made of thermoset materials can be subjected to higher temperatures without creeping, tend to have a harder surface, and are more rigid than thermoplastic parts made by injection molding. For this reason, parts used at higher temperatures (molded fryer pan housings, electrical connections, etc.), or parts that may be subjected to harsher environments (automotive carburetor spacers, automatic transmission thrust washers, etc.) are made of thermoset materials. Thus, such parts are formed by compression or transfer molding.

Typical thermosets include phenolics, ureaformaldehyde, epoxies, polyesters, and polyurethanes. Thermoplastics and thermosets can both be combined with one or more additives (colorants, flame retardants, lubricants, heat or light stabilizers), fillers or reinforcements (glass fibers, hollow glass spheres), or with other polymers to form a blend or alloy in order to increase dimensional stability and improve their mechanical properties. Some of the commodity plastics, such as polypropylene, are reclassified as engineering plastics when they are reinforced with glass fibers.

With all the basic polymer materials available, together with all the possible combinations of fillers and additives, there is a dizzying array of possibilities for designers to choose from. There are also pitfalls, as not all the properties of all these combinations are well known. Consultation with a polymer materials expert is well advised!

3.3 INJECTION MOLDING

In injection molding, thermoplastic pellets are melted, and the melt is injected under high pressure (approximately 10,000 psi or about 70 MPa) into a mold. There the molten plastic takes on the shape of the mold, cools, solidifies, shrinks, and is ejected. Figure 3.1 shows a stripped down version of an injection-molding machine along with a mold used to form a simple box-shaped part.

Molds are generally made in two parts: (1) the cavity half gives a concave part its external shape, and (2) the core half gives such a part its internal shape. As the geometry of a part becomes complex, molds of course increase in complexity—and hence in cost.

As a part cools, it shrinks onto the core. Therefore, an ejector system is needed to push the part off. Because the fixed (cavity) half of the mold contains the "plumbing system"—elements called runners, sprues, and gates—used to transfer the melt to the mold, the ejector system is usually in the core (moving) half of the mold. The ejector system generally consists of pins that are used to push the part off the core. Careful examination of most injection-molded parts will reveal the marks of the ejection pins—slight circular depressions about 3/16 of an inch in diameter. If a satisfactory flat part surface area does not exist to accommodate a pin, then a blade may be used to press against a narrow rib or part edge to eject the part.

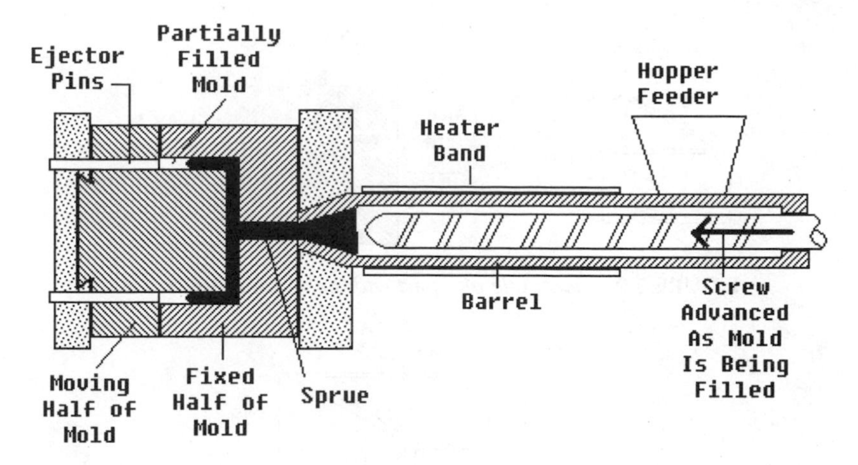

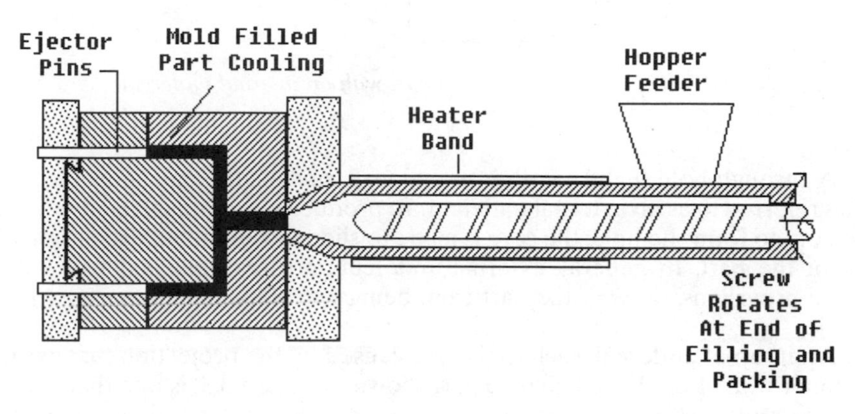

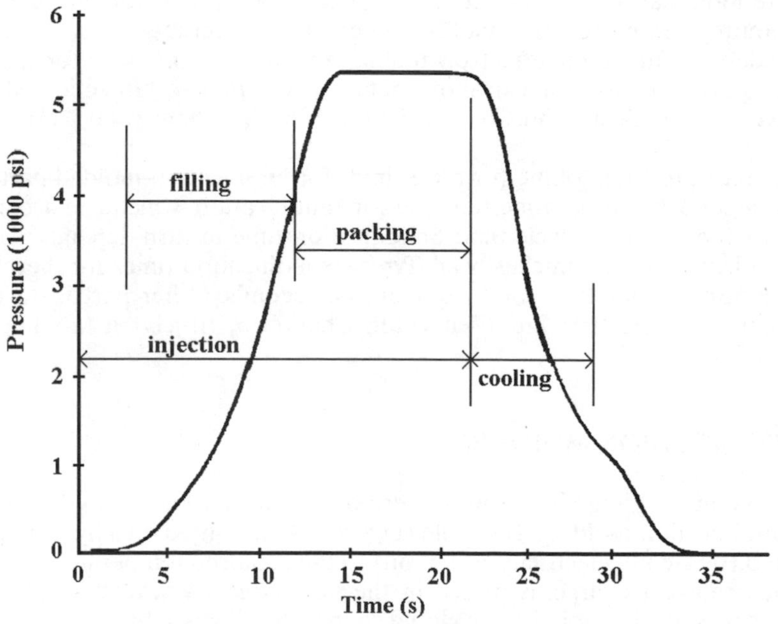

FIGURE 3.1 *Screw-type injection-molding machine.*

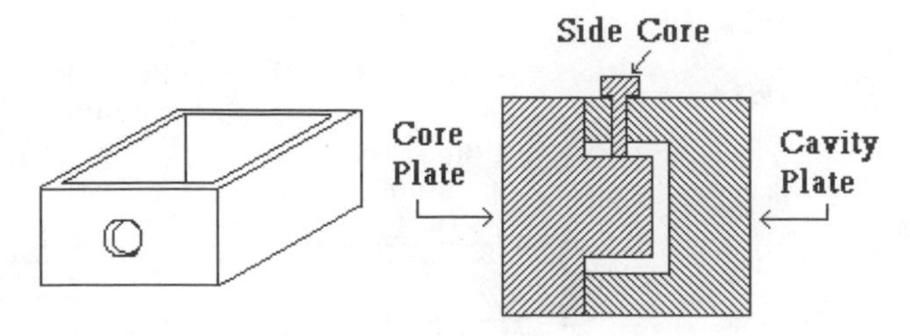

FIGURE 3.2 *Example of a part with an external undercut.*

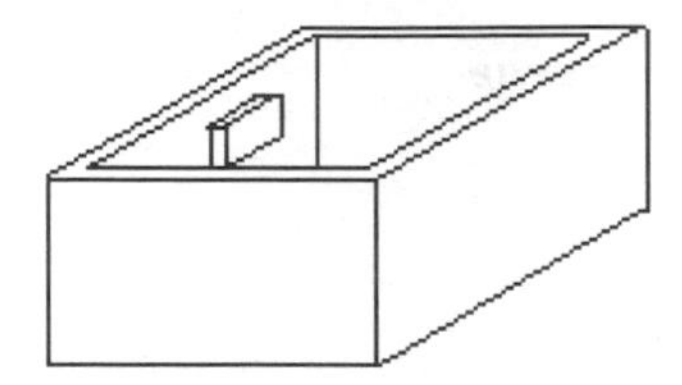

FIGURE 3.3 *Example of a part with an internal undercut.*

A through-hole feature in the vertical wall, such as the one shown in Figure 3.2, is referred to as an external undercut. To produce it requires a relatively costly side core to form the hole; the core is made to slide out of the way to permit ejection of the part. In general, external undercuts are features that will, without special provisions, prevent the part from being extracted from the cavity half of the die.

An internal undercut, such as the one caused by the projection that exists on the inner wall of the boxed-shaped part shown in Figure 3.3, is one that prevents the core mold half of the part from being extracted. In general, internal undercuts require even more costly molds than external undercuts.

Undercuts and their effect on tooling and processing costs for injection-molded parts are discussed in more detail in Chapter 4, "Injection Molding: Relative Tooling Cost," and Chapter 5, "Injection Molding: Total Relative Part Cost."

Per-part processing time (or cycle time) for an injection-molded part is primarily dependent on the time required for solidification, which can account for about 70% of the total cycle time. Solidification time in turn depends primarily on the thickness of the thickest wall. Typical solidification times for thermoplastic parts range from 15 seconds to about 60 seconds. Other part features that also influence cycle time are discussed in Chapter 5, "Injection Molding: Total Relative Part Cost."

3.4 COMPRESSION MOLDING

Compression molding for forming thermoset materials uses molds similar to those for injection molding. The mold (Figure 3.4), mounted on a hydraulic press, is heated (by steam, electricity, or hot oil) to the required temperature. A slug of material, called a charge, is placed in the heated cavity where it softens and becomes plastic. The mold is then closed so that the slug is subjected to pressures between 350 kPa (50 psi) to 80,000 kPa (12,000 psi) forcing the slug to take the shape of the mold.

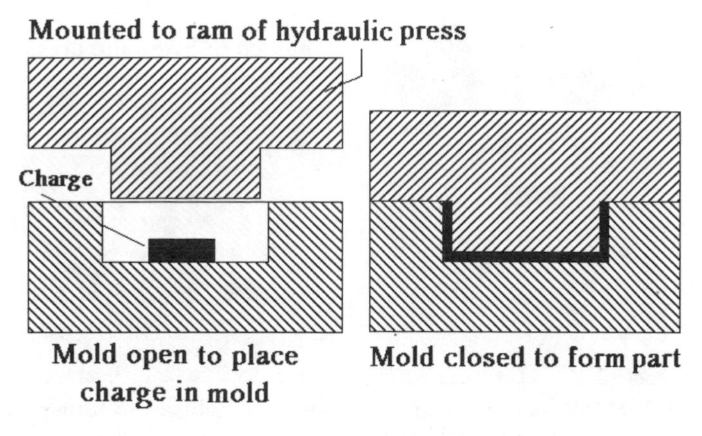

FIGURE 3.4 *Tooling for compression molding.*

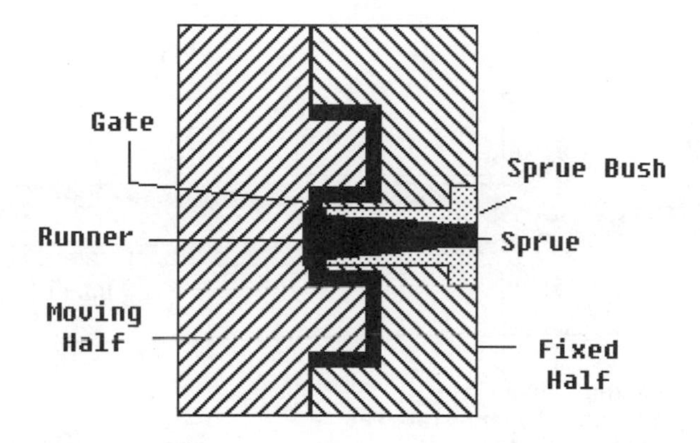

FIGURE 3.5 *Two-plate mold showing sprue, gate, and runner system.*

The mold remains closed under pressure until the part hardens (cures). The mold is then opened, the part removed, and the cycle repeated. The cure time for parts can be as low as 20 seconds for small, thin-walled parts, from 1 to 3 minutes for larger parts, and as long as 24 hours for massive, thick-walled parts such as an aerospace rocket nozzle.

Compression molded parts may have external undercuts, but as in injection molding, undercuts increase tooling cost and should be avoided if possible. Compression molds, however, are somewhat simpler than injection molds since the compression molds do not need a "plumbing system" (sprue, runner, and gates) to feed and distribute the melt. Figure 3.5 shows a two-plate injection molding-type mold with sprue, gate, and runner system.

3.5 TRANSFER MOLDING

The main difference between compression molding and transfer molding is in the mold. In a transfer mold (Figure 3.6), the upper portion of the mold contains a pot where a slug (or charge) is placed, heated, and melted. After the charge is melted the mold is closed, forcing the liquid resin through a sprue into the lower

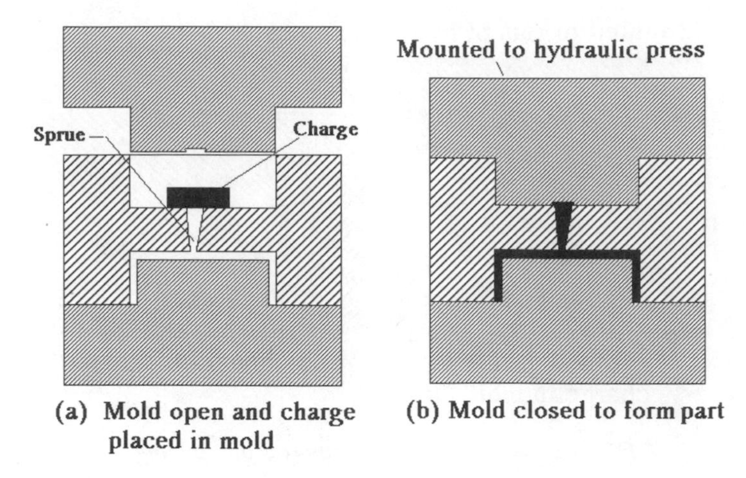

FIGURE 3.6 *Tooling for transfer molding.*

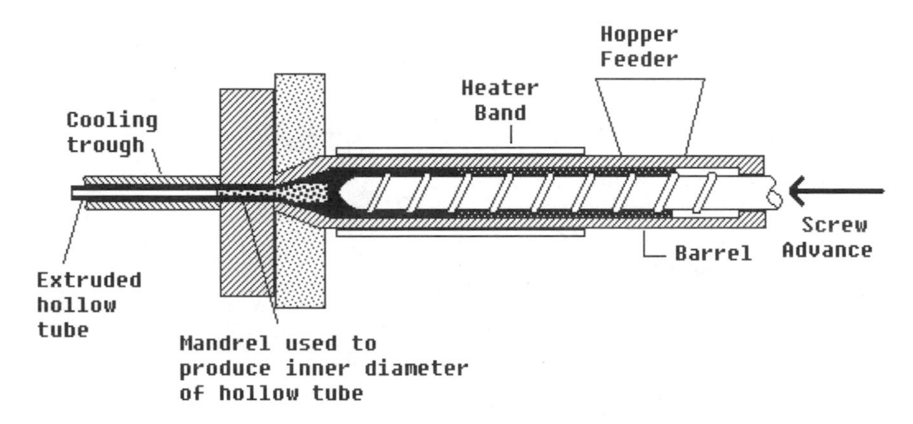

FIGURE 3.7 *A screw-type extruder.*

portion of the mold. The melt then takes the shape of the mold, hardens, and is removed.

3.6 EXTRUSION

Plastic extrusion is a process in which thermoplastic pellets are placed into a hopper that feeds into a long cylinder (called a barrel) that contains a rotating screw (Figure 3.7). The screw transports the pellets into a heated portion of the barrel where the pellets are melted and mixed to form a uniform melt. The resulting melt is then forced through a die hole of the desired shape to form long parts of uniform cross-section such as tubes, rods, molding, sheets, and other regular or irregular profiles. Figure 3.8 shows some common structural shapes produced by extrusion.

Extruded parts are generally long in comparison with their cross-section. Short parts with uniform cross-sectional shapes can also be produced by injection molding as well as extrusion, but the longer the part, the more advantageous the use of extrusion.

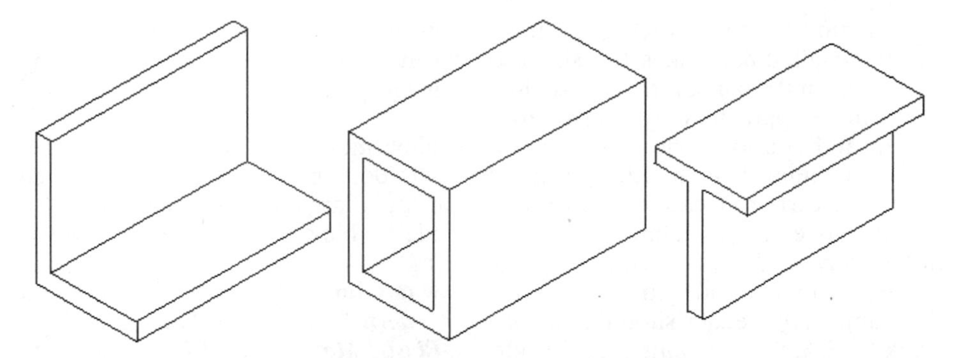

FIGURE 3.8 *Some common structural shapes produced by extrusion.*

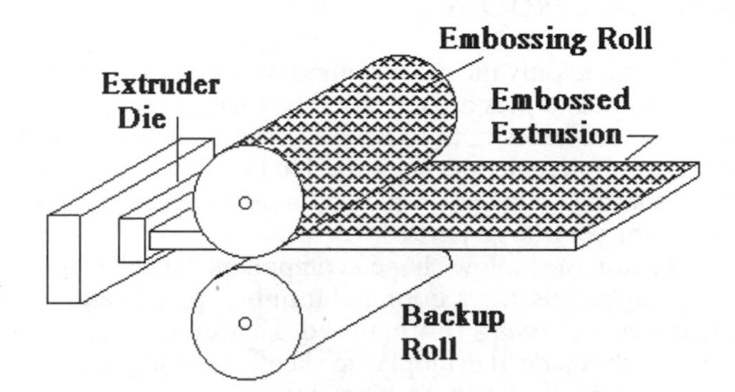

FIGURE 3.9 *Post-processing of extruded sheets.*

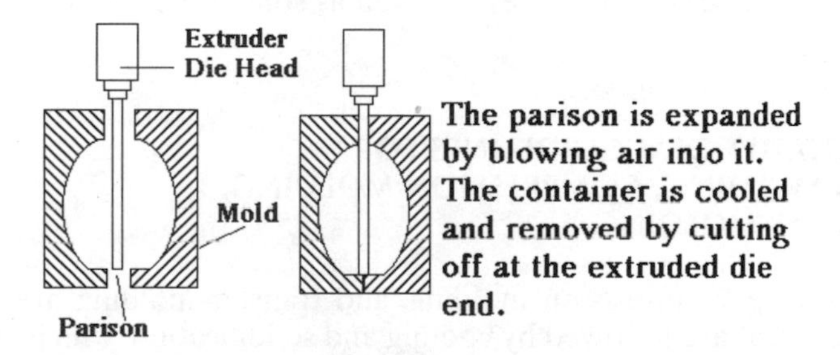

FIGURE 3.10 *Extrusion blow molding.*

Shapes formed by extrusion can be subjected to post-processing techniques by passing them through rollers (Figure 3.9) or stationary blades or formers that modify the shape of the (still hot and soft) extrusion. Sheets, for example, can be embossed to form patterns on them.

3.7 EXTRUSION BLOW MOLDING

Extrusion blow molding is a process used to form hollow thermoplastic objects (especially bottles and containers). The process (Figure 3.10) takes a thin-walled tube called a *parison* that has been formed by extrusion, entraps it between two halves of a larger diameter mold, and then expands it by blowing air (at about

100 psi) into the tube, forcing the parison out against the mold. The outside of the thin-walled part takes the shape of the inside of the mold. By controlling variations in the parison thickness along its length, the wall thickness of the final part can be approximately controlled.

In addition to bottles and containers, blow molding is used to form such shapes as simple balls, lightweight baseball bats, dolls, and animal toys. Although items like carrying cases for instruments and tools, large drums, ducts, and automobile glove compartments can also be made by blow molding, this process is not usually used to produce such "engineering" type parts.

No further consideration is given to blow molding in this book. For more on this subject, the reader should consult the *Modern Plastic Encyclopedia* and the book by S. S. Schwarz and S. H. Goodman, *Plastic Materials and Processes.*

3.8 OTHER POLYMER PROCESSES

We have so far described only the most commonly used polymer processing techniques. Others exist—examples are calendering, foam processing, and rotational molding—but these processes tend to be used for rather specialized or low production runs. For example, calendering is used to produce film and sheeting, foam processing for disposable cups and food containers, and rotational molding for battery cases and for very large parts.

For the production of shallow-shaped components, such as bus panels, boats, camper tops, lighting panels, trays, door and furniture panels, and other products, a process called *thermoforming* is often used. Thermoforming involves heating to soften a previously made thermoplastic sheet, clamping it over a mold, and then drawing and forcing it (via air pressure or vacuum) so that it takes the shape of the mold.

The books *Modern Plastic Encyclopedia* and *Plastic Materials and Processes* contain detailed description of these, as well as some other, polymer processing techniques.

3.9 QUALITATIVE DFM GUIDELINES FOR INJECTION MOLDING, COMPRESSION MOLDING, AND TRANSFER MOLDING

Injection molding, compression molding, and transfer molding are all internal flow processes that are followed by cooling and solidification, which are followed by ejection from the mold. That is, in each of these processes, a liquid (plastic resin) flows into and fills a die cavity. Then the liquid is cooled to form a solid, and finally the part is ejected. The physical nature of these processes—flow, cooling to solidify, and ejection—provides the basis for a number of the qualitative DFM guidelines or rules of thumb that have been established. Parts should ideally be designed so that: (1) the flow can be smooth and fill the cavity evenly; (2) cooling, and hence solidification, can be rapid to shorten cycle time and uniform to reduce warpage; and (3) ejection can be accomplished with as little tooling complexity as possible.

To design parts properly for these manufacturing processes, designers must at least understand the meaning of (1) mold closure direction and (2) parting surface. Dies are made in two parts forming a cavity that is very close to the shape of the part. (The cavity may be slightly different from the part to allow for inevitable shrinkage and warping.) Thus there is a closure direction for the die halves, and a parting "surface" (not necessarily planar) created where

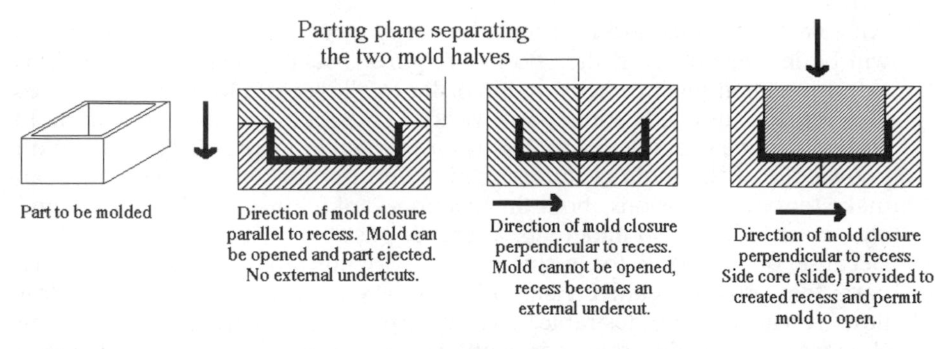

FIGURE 3.11 *External undercut created by choice in the direction of mold closure.*

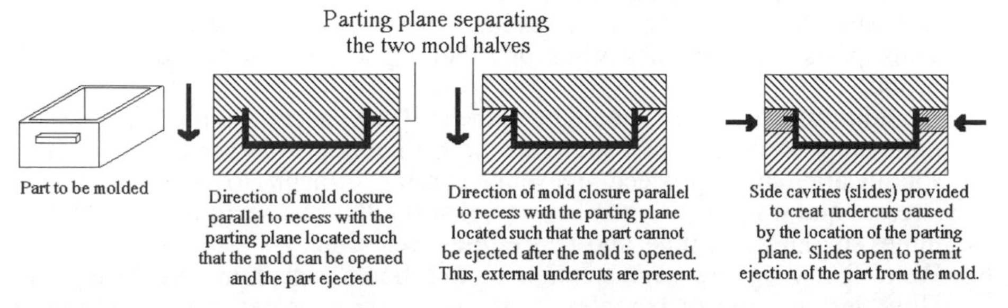

FIGURE 3.12 *External undercuts created by location of the parting plane between the die halves.*

the die sections meet when they are closed. The location of the parting surface, the direction of closure, and the design of the part must be considered simultaneously in order to provide for ejection of the part from the mold after solidification.

Knowing the mold closure direction enables designers to recognize and thus possibly avoid designing unnecessary undercuts. A fairly easy way to identify a potential undercut is to consider the shadows that would be created on the part from a light shining in the mold closure direction. If a part casts shadows onto itself, then the feature causing the shadow is an undercut. This is discussed again in the next chapter.

Figures 3.11 and 3.12 illustrate how the choice of mold closure direction and the location of the parting surface influence design and, in particular, tool design and tool cost.

With knowledge of mold closure direction and the location of the parting surface—and keeping in mind that the material should flow smoothly into and through the mold, solidify rapidly and uniformly, and then be easily ejected—designers can well understand and make good use of the following DFM guidelines:

1. In designing parts to be made by injection molding, compression molding, and transfer molding, designers must decide—as a part of their design—the direction of mold closure and the location of the parting surface. Though these decisions are tentative, and advice should be sought from a manufacturing expert, it is really impossible to do much design for manufacturing in these processes without considering the mold closure direction and parting surface location.

2. An easy to manufacture part must be easily ejected from the die, and dies will be less expensive if they do not require special moving parts (such as side cores) that must be activated in order to allow parts to be ejected. Since undercuts require side cores, parts without undercuts are less costly to mold and cast. Some examples of undercuts are shown in Figures 3.2 and 3.3. With knowledge of the mold closure direction and parting surface, designers can make tentative decisions about the location(s) of features (holes, projections, etc.) in order to avoid undercuts wherever possible.
3. Because of the need for resin or metal to flow through the die cavity, parts that provide relatively smooth and easy internal flow paths with low flow resistance are desirable. For example, sharp corners and sudden changes or large differences in wall thickness should be avoided because they both create flow problems. Such features also make uniform cooling difficult.
4. Thick walls or heavy sections will slow the cooling process. This is especially true with plastic molding processes since plastic is a poor thermal conductor. Thus, parts with no thick walls or other thick sections are less costly to produce.
5. In addition, every effort should be made to design parts of uniform, or nearly uniform, wall thickness. If there are both thick and thin sections in a part, solidification may proceed unevenly causing difficult to control internal stresses and warping. Remember, too, that the thickest section largely determines solidification time, and hence total cycle time.
6. We will not discuss gate location in this book except in this paragraph. However, in large or complex parts, two or more gates may be required through which resin will flow in two or more streams into the mold. There will, therefore, be fusion lines in the part where the streams meet inside the mold. The line of fusion may be a weak region, and it may also be visible. Therefore, designers who suspect that multiple gates may be needed for a part should discuss these issues with manufacturing experts as early as possible in the design process. With proper design and planning, the location of the fusion lines can usually be controlled as needed for appearance and functionality.

These DFM "rules" are not absolute, rigorous laws. Note, for example, how the molded-in, very thin hinge, referred to as a "living hinge" by custom molders, in a computer disk carrying case (Figure 3.13) violates the general thrust of the fifth rule above. If there are designs that have great advantages for function or marketing, then those designs can be given special consideration. Manufacturing engineers can sometimes solve the problems that may be associated with highly desirable functional but difficult to manufacture designs at a cost low enough to justify the benefit.

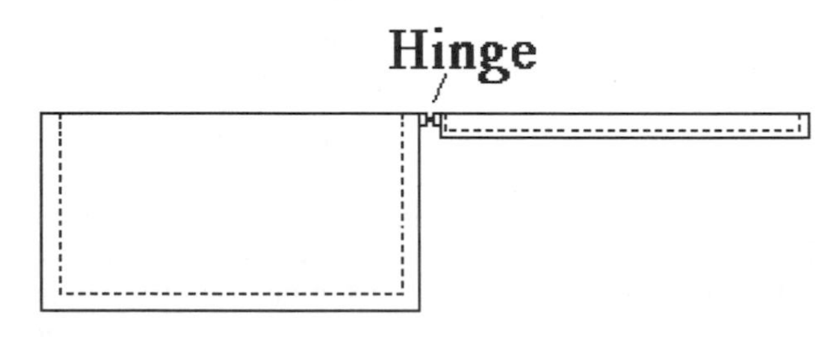

FIGURE 3.13 *Living hinge on computer disk carrying case.*

However, relatively easy to manufacture designs should always be sought. More often than not, a design can be found that will be both efficient from a functional viewpoint and relatively easy to manufacture.

3.10 SUMMARY

This chapter has described some of the most common polymer processing methods and materials used for the economical production of both complex parts (with significant levels of geometric detail) and simple parts (with little geometric detail). Included in this chapter was a discussion of design for manufacturing issues as they apply to the production of plastic parts. The chapter concluded with a set of qualitative DFM guidelines for injection molding, compression molding, and transfer molding.

REFERENCES

Bralla, J. G., editor. *Handbook of Product Design for Manufacturing*. New York: McGraw-Hill, 1986.

Kalpakjian, S. *Manufacturing Engineering and Technology*. Reading, MA: Addison-Wesley, 1989.

Modern Plastic Encyclopedia. Hightstown, NJ: McGraw-Hill/Modern Plastics, 1991.

Poli, C., Dastidar, P., and Graves, R. A. "Design Knowledge Acquisition for DFM Methodologies." *Research in Engineering Design*, Vol. 4, no. 3, 1992, pp. 131–145.

Schwarz, S. S., and Goodman, S. H. *Plastic Materials and Processes*. New York: Van Nostrand Reinhold, 1982.

Wick, C., editor. *Tool and Manufacturing Engineers Handbook*, Vol. 2. "Forming, 9th edition." Dearborn, MI: Society of Manufacturing Engineers, 1984.

QUESTIONS AND PROBLEMS

3.1 Figure P3.1 shows the sectional view of two proposed alternative designs for an injection-molded box-shaped part that is enclosed on four sides. From the point of view of tooling cost, which design is more costly to produce? Why? Assume that the wall thickness is the same in both designs.

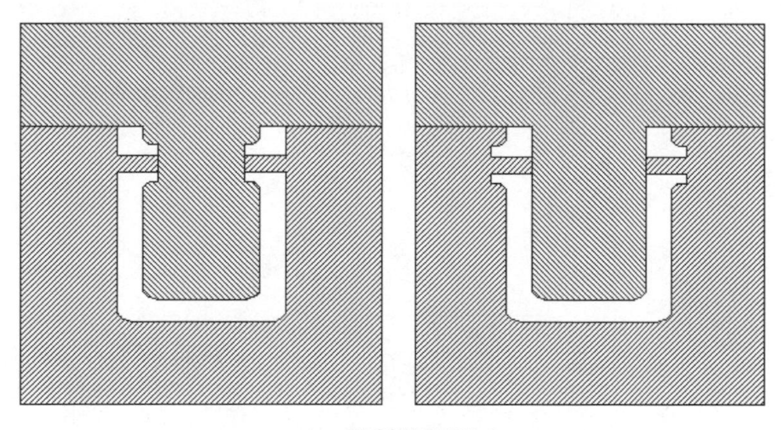

FIGURE P3.1

3.2 Figure P3.2 shows the sectional view of two proposed alternative designs for an injection-molded box-shaped part that is enclosed on four sides. From the point of view of tooling costs, which of the two designs is the most costly? Assume that the wall thickness is the same in all designs.

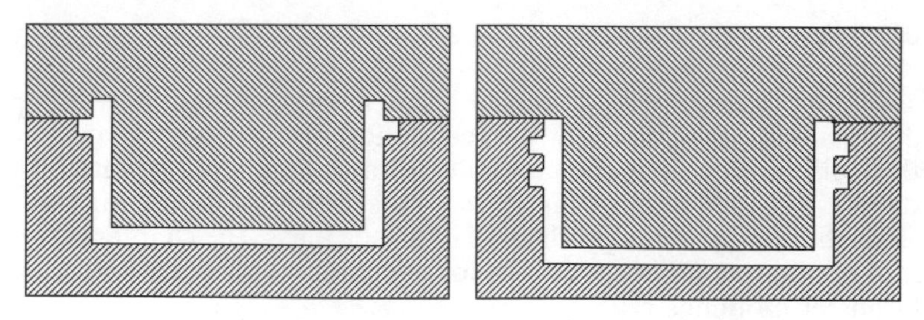

FIGURE P3.2

3.3 Figure P3.3 shows the preliminary sketch of two proposed designs. Assuming the part is to be injection molded, which of the two designs is less costly to produce? Why? Assume that the wall thickness is the same in both designs.

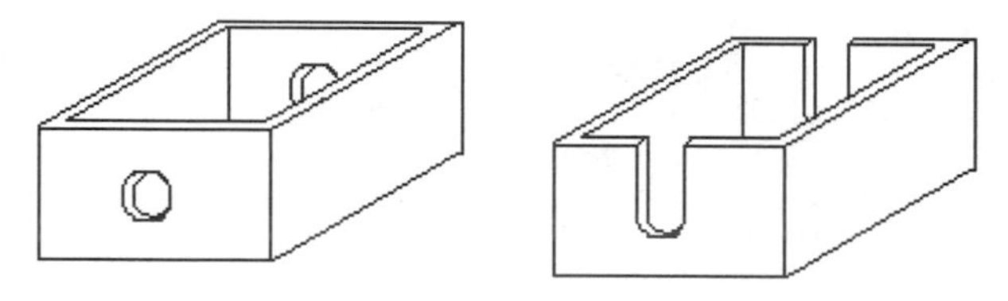

FIGURE P3.3

3.4 Figure P3.4 shows the preliminary sketch of two proposed injection-molded designs. From the point of view of tooling, which of the two designs is least costly to produce? Why? Assume both parts have the same wall thickness.

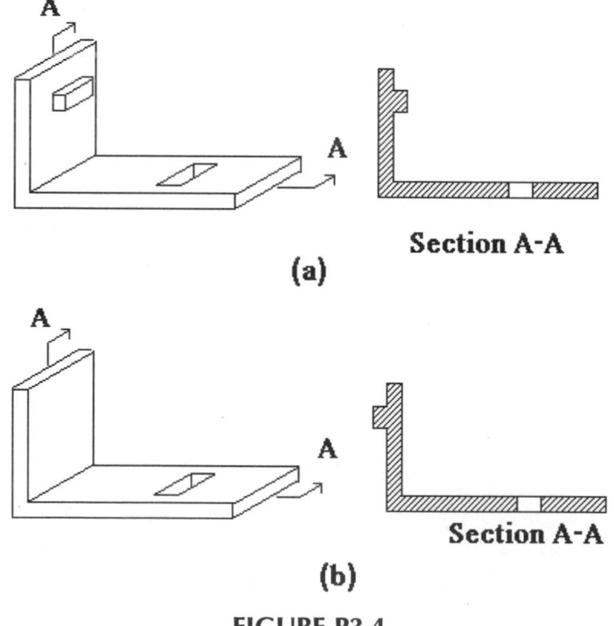

FIGURE P3.4

3.5 In an effort to become more competitive, a large automotive company has decided to expand its design for manufacturing group. Assume that you have applied for a position with that group. As part of the interview process you have been shown the proposed design of a compression-molded part similar to the part shown in Figure P3.5. What suggestions would you make in order to reduce the cost to mold the part? What suggestions would you make if the part were to be injection molded?

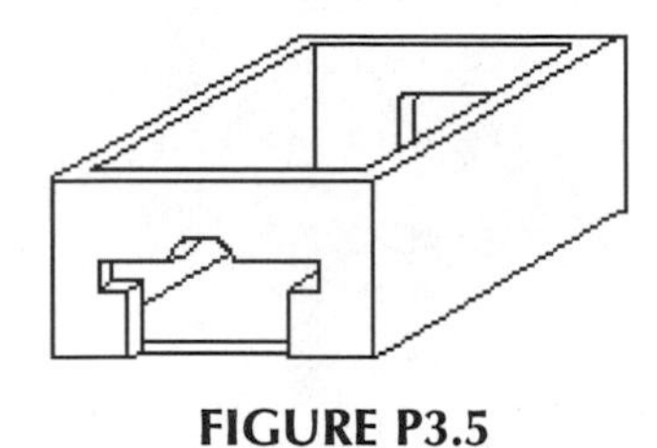

FIGURE P3.5

3.6 Repeat Problem 3.5 for the part shown in Figure P3.6.

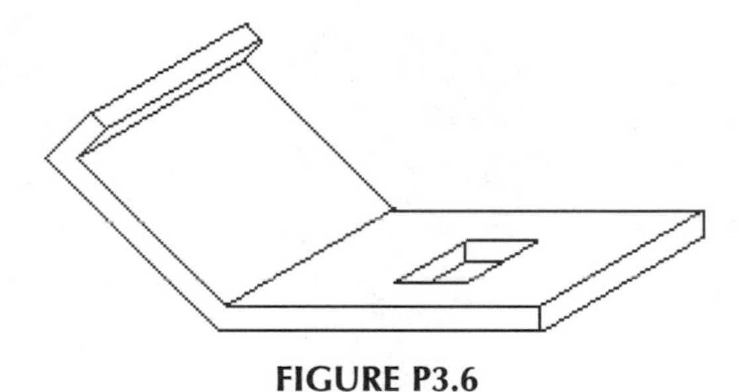

FIGURE P3.6

3.7 Name two or three "engineering" parts that are better suited for plastics than for metals.

3.8 How do thermoplastics differ from thermosets?

3.9 How do engineering plastics differ from commodity plastics?

Chapter 4

Injection Molding: Relative Tooling Cost

4.1 INTRODUCTION

Generally during the early stages of part design the procedures followed by a designer will result in several possible part configurations whose generation has been guided by qualitative reasoning related to both function and, it is hoped, manufacturability. In this chapter, we will discuss methods by which designers can perform more formal quantitative manufacturability evaluations of injection-molded parts at the configuration stage. In Chapter 5, "Injection Molding: Total Relative Part Cost," we will make use of the near final dimensions, locations, and orientation of features to evaluate the relative cost to process a part by injection molding.

The ability to evaluate part manufacturability at the configuration stage—before exact dimensions have been determined—is important because, in practice, design decisions made at this stage often become essentially irrevocable. Thus, before moving on to the sometimes computationally difficult and time-consuming task of assigning values to attributes (i.e., to parametric design), it is important for designers to be as certain as possible that the configuration selected is the best possible one considering both function and manufacturability.

In this book, we present detailed, quantitative manufacturability methods for three manufacturing domains: injection molding, die casting, and stamping. Design for manufacturability issues in other processes—such as assembly, forging, extrusion, and others—are also covered, but not in as much detail as these three. The essential reason for concentrating on these particular manufacturability domains is that they account for more than 70% of the special purpose parts found in consumer products.

It should always be remembered that the best method for reducing assembly costs is to reduce the number of parts in an assembly. This is often accomplished by combining several individual parts into one (sometimes more complex) part using either injection molding, die casting, or stamping. To do this by taking full advantage of the capabilities of the process, but without exceeding those capabilities, requires that designers be able to perform detailed DFM analyses.

This chapter is devoted to DFM methods for the ease of tool design for injection molding. Chapter 7, "Die Casting: Total Relative Part Cost," and Chapter 9, "Stamping: Relative Tooling Cost," deal with DFM for ease of tool designs for die casting and stamping.

4.2 ESTIMATING RELATIVE TOOLING COSTS FOR INJECTION-MOLDED PARTS

The cost of an injection-molded part consists of three subcosts: (1) tooling (or mold) cost, K_d/N; (2) processing cost (or equipment operating cost), K_e; and (3) part material cost, K_m.

$$\text{Total Cost of a Part} = K_d/N + K_e + K_m \qquad \text{(Equation 4.1)}$$

where K_d is the total tooling cost for a part and N is the number of parts to be produced with the mold—that is, the production volume.

At low production volumes (N less than, 20,000, e.g.), the proportion of the part cost due to tooling is often relatively high compared with processing costs, since the total cost of tooling, K_d, is divided by a small value of N. As the production volume increases, however, the total tooling cost (K_d) does not change; consequently, the tooling cost per part decreases, while the material and processing cost per part remains essentially the same. Thus, at high production volumes the proportion of total part cost due to tooling is relatively low, and the major costs are due to processing and materials.

With only configuration information available, we can do little to estimate processing costs (K_e) or part material costs (K_m). However, we can make a reasonably accurate estimate of tooling costs (K_d). In fact, if we restrict ourselves to relative tooling cost, then the analysis is accurate enough to permit a comparison between competing designs. As pointed out above, tooling costs are often important, and the fact that we can perform a DFM analysis of them at the configuration stage (before parametric design) is useful. The analysis helps designers identify the features of proposed configurations that contribute most significantly to tooling costs so that the features can be eliminated, or at least so that their negative impact on manufacturing cost can be reduced.

4.2.1 Relative Tooling Cost (C_d)

The DFM methodology to be presented here determines the tooling cost of injection-molded parts as a ratio of expected tooling costs to the tooling costs for a reference part. This ratio is called *relative cost.* Actual costs depend upon local practices and methods and can vary considerably from one location to another. To eliminate these effects costs relative to a reference part are used. Thus, relative to a reference part, total die costs are

$$C_d = \text{Cost of Tooling for Designed Part/Cost of Tooling for Reference Part}$$

$$C_d = (K_{dm} + K_{dc})/(K_{dmo} + K_{dco}) \qquad \text{(Equation 4.2)}$$

where K_{dmo} and K_{dco} refer to die material cost and die construction cost for the reference part. (In this case, the reference part is a flat 1 mm thick washer with OD = 72 mm and ID = 60 mm. The approximate tooling cost for this reference part—in the 1991–1992 time frame—is about \$7,000, including about \$1,000 in die material costs.)

Equation (4.2) can be written as

$$\begin{aligned} C_d &= K_{dm}/(K_{dmo} + K_{dco}) = K_{dc}/(K_{dmo} + K_{dco}) \\ &= A(K_{dm}/K_{dmo}) + B(K_{dc}/K_{dco}) \end{aligned} \qquad \text{(Equation 4.3)}$$

where

$$A = K_{dmo}/(K_{dmo} + K_{dco})$$

$$B = K_{dco}/(K_{dmo} + K_{dco})$$

Based on data collected from mold-makers, a reasonable value for A is between 0.15 and 0.20 and a reasonable value for B is between 0.80 and 0.85. For our purposes we will take A and B to be 0.2 and 0.8, respectively. Hence, Equation 4.3 becomes

$$C_d = 0.8C_{dc} + 0.2C_{dm} \qquad \text{(Equation 4.4)}$$

where C_d is the total die cost of a part relative to the die cost of the reference part, C_{dc} is the die construction cost for the part relative to the die construction cost of the reference part, and C_{dm} is the die material cost for the part relative to the die material cost of the reference part.

In this section, we show how to determine the relative tooling construction costs (C_{dc}). The following section deals with the relative tool material costs (C_{dm}).

4.2.2 Relative Tooling Construction Costs (C_{dc})

To estimate relative tool construction costs for a part, designers must understand in some detail the complex relationships between the part and its mold. Certain features and combinations of features result in more complex molds and, hence, higher tooling costs. It may be that, in order to meet a part's function, such features or their combinations cannot be changed or eliminated, but in many cases they can be—saving time and money. In any case, designers should know the tooling costs their designs are causing, and they should make every attempt to reduce them.

The time required for tooling to be designed, manufactured, and tested is also a factor. In general, however, the higher the cost of tooling, the longer the time required for making the tool.

Relative tooling construction cost, C_{dc}, is computed here as the product of three factors:

$$C_{dc} = C_b C_s C_t \qquad \text{(Equation 4.5)}$$

where C_b = The approximate relative tooling cost due to size and basic complexity;
C_s = A multiplier accounting for other complexity factors called subsidiary factors;
C_t = A multiplier accounting for tolerance and surface finish issues.

We will now discuss how to compute each of these factors, and look at examples of their use with actual parts. Readers of this chapter will be rewarded with an easy-to-use understanding of design for manufacturing principles and practices for injection-molded parts. Much of what is learned will be useful in understanding other manufacturing domains as well. In order to use the methodology, however, a reader must be familiar with a number of concepts related specifically to the manufacture of injection-molded parts. Though there appears to be a large number of them (they are explained in the next subsections), the concepts are individually relatively easy to understand. All are explained as they are introduced.

4.3 DETERMINING RELATIVE TOOLING CONSTRUCTION COSTS DUE TO BASIC PART COMPLEXITY (C_b)

4.3.1 Overview

Values for C_b—the relative tooling cost factor due to basic part complexity—are found in the interior boxes of the matrix in Figure 4.1.

The numbers above the slanted lines in the boxes in Figure 4.1 apply to flat parts; those below the slanted line apply to box-shaped parts. Note that the value for C_b in the upper left corner of the matrix for a flat part is 1.00—thus this box corresponds to the cost of the reference part.

Readers should note that, in general, values for C_b decrease significantly as one moves up and to the left in the matrix. This fact will help guide designers to redesigns that can reduce tooling costs.

4.3.2 The Basic Envelope

Figure 4.1 requires that the part to be evaluated be classified as either flat or box-shaped. (This is done because, in general, box-shaped parts require more mold machining time and, hence, result in higher tool construction costs, than flat parts.) In order to determine whether a part is flat or box-shaped, we determine the ratio of the sides of the basic envelope for the part. The basic envelope is the smallest rectangular prism that completely encloses the part.

The lengths of the sides of the basic envelope are denoted by L, B, and H, where $L \geq B \geq C$. (See Figure 4.2.) A part is considered flat if L/H is greater than about 4; otherwise it is considered box-shaped.

In order not to overestimate the amount of mold machining time required, in determining the basic envelope, small, isolated projections are ignored. Isolated projections are considered small if their greatest dimension parallel to the surface from which they project is less than about one-third times the envelope dimension in the same direction (as shown in Figure 4.3). This is done so that a part that is basically flat when the projection is ignored is not classified as a box-shaped part. If more than one projection exists, each should be examined separately.

4.3.3 The Mold Closure Direction

As noted briefly in Chapter 3, "Polymer Processing," designers of injection-molded parts must consider the direction of mold closure in order to be able to design for ease of manufacturability. The reason is that the orientation of certain part features and configurations in relation to the mold closure direction can have an important influence on tooling construction costs. This is also reflected in the fact that knowledge of the mold closure direction is essential in order for designers to use Figure 4.1, and hence to estimate tooling construction costs.

Knowledge of the mold closure direction is also essential in order to identify and possibly redesign the features that may be causing high tooling costs.

Recessed Features

In order to determine the best or most likely direction of mold closure, it is necessary to understand the meaning of a recessed feature. A recessed feature is

1 in = 25.4 mm; 100 mm/25.4 mm = 3.94 in

Key: Flat Parts (upper value) / Box-Shaped Parts (lower value) — e.g., 6: 5.37 / 6.28

BASIC COMPLEXITY			SECOND DIGIT: L ≤ 250 mm (4) — Number of External Undercuts (5): zero	one	two	More than two	250mm < L ≤ 480mm — Number of External Undercuts (5): zero	one	two	More than two	L > 480 mm — Number of External Undercuts (5): zero	one	More than one
FIRST DIGIT			0	1	2	3	4	5	6	7	8	9	10
Parts Without Internal Undercuts (1)	Parts whose peripheral height from a planar dividing surface is constant (2) — Part in one half(3)	0	1.00 / 1.64	1.23 / 1.87	1.38 / 2.02	1.52 / 2.16	1.42 / 2.89	1.65 / 3.12	1.79 / 3.27	1.94 / 3.41	1.83 / 4.28	2.07 / 4.51	2:33 / 4.77
Parts Without Internal Undercuts (1)	Parts whose peripheral height from a planar dividing surface is constant (2) — Part not in one half(3)	1	1.14 / 1.86	1.37 / 2.09	1.52 / 2.24	1.66 / 2.38	1.61 / 2.99	1.84 / 3.22	1.99 / 3.37	2.13 / 3.51	2.09 / 4.42	2.32 / 4.66	2.58 / 4.92
Parts Without Internal Undercuts (1)	Parts whose peripheral height from a planar Dividing Surface is not constant – or – parts with a non-planar Dividing Surface(2)	2	1.28 / 1.92	1.51 / 2.15	1.66 / 2.29	1.80 / 2.44	1.81 / 3.38	2.04 / 3.61	2.19 / 3.76	2.33 / 3.90	2.34 / 5.01	2.58 / 5.24	2.84 / 5.50
Parts With Internal Undercuts (1) — On Only One Face of the Part	Parts whose ONLY Dividing Surface (2) is planar, or parts whose peripheral height from a planar dividing surface is constant	3	2.33 / 3.19	2.57 / 3.43	2.71 / 3.57	2.86 / 3.72	2.75 / 4.44	2.98 / 4.68	3.13 / 4.82	3.27 / 4.97	3.17 / 5.83	3.40 / 6.07	3.66 / 6.33
Parts With Internal Undercuts (1) — On Only One Face of the Part	Parts whose peripheral height from a planar Dividing Surface is not constant – or – parts with a non-planar Dividing Surface(2)	4	2.98 / 3.73	3.21 / 3.97	3.36 / 4.11	3.50 / 4.26	3.52 / 5.20	3.75 / 5.43	3.89 / 5.58	4.04 / 5.72	4.04 / 6.82	4.28 / 7.06	4.54 / 7.32
Parts With Internal Undercuts (1) — On More Than One Face of the Part	Parts whose ONLY Dividing Surface (2) is planar, or parts whose peripheral height from a planar dividing surface is constant	5	4.20 / 5.37	4.43 / 5.61	4.58 / 5.75	4.72 / 5.89	4.62 / 6.62	4.85 / 6.86	4.99 / 7.00	5.14 / 7.14	5.03 / 8.01	5.27 / 8.24	5.53 / 8.51
Parts With Internal Undercuts (1) — On More Than One Face of the Part	Parts whose peripheral height from a planar Dividing Surface is not constant – or – parts with a non-planar Dividing Surface(2)	6	5.37 / 6.28	5.60 / 6.52	5.74 / 6.66	5.89 / 6.81	5.90 / 7.74	6.13 / 7.98	6.28 / 8.12	6.42 / 8.27	6.43 / 9.37	6.67 / 9.60	6.93 / 9.86

FIGURE 4.1 *Classification system for basic tool complexity, C_b. (The numbers in parentheses refer to notes found in Appendix 4.A.)*

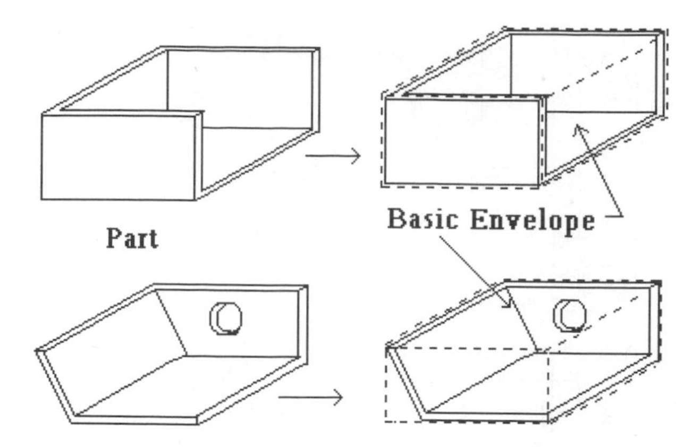

FIGURE 4.2 *Basic envelope for a part.*

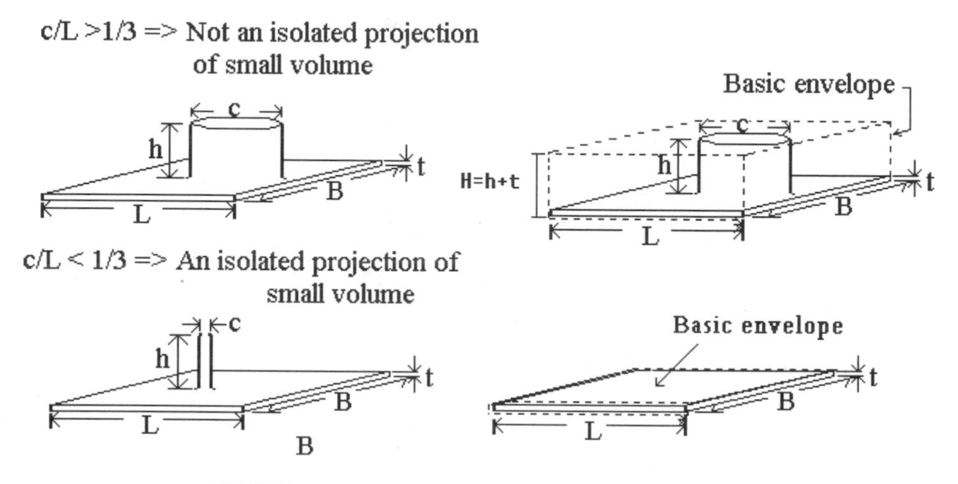

FIGURE 4.3 *Isolated projections of small volume.*

any depression or hole in a part, including also depressed features that come about due to closely spaced projecting walls. Figure 4.4 shows some examples of recessed features. Also shown in Figure 4.4 are sectional views of the molds that can be used to produce these features. The part shown in Figure 4.4d is not considered to have a recess because the direction of mold closure does not affect basic tool construction difficulty.

Holes and Depressions

Depressions are pockets, recesses, or indentations of regular or irregular contour that are molded into a portion of an injection-molded part. Holes are the prolongation of depressions that completely penetrate some portion of the molding.

Circular holes and depressions can be formed either by an integer mold in which the projections required to create the two holes are machined directly into the core half of the mold as shown in Figure 4.4f, or by an insert mold. In the case of an insert mold, as in Figure 4.4g, the projections shown in the first version of the mold are replaced by a core pin that is inserted into the core.

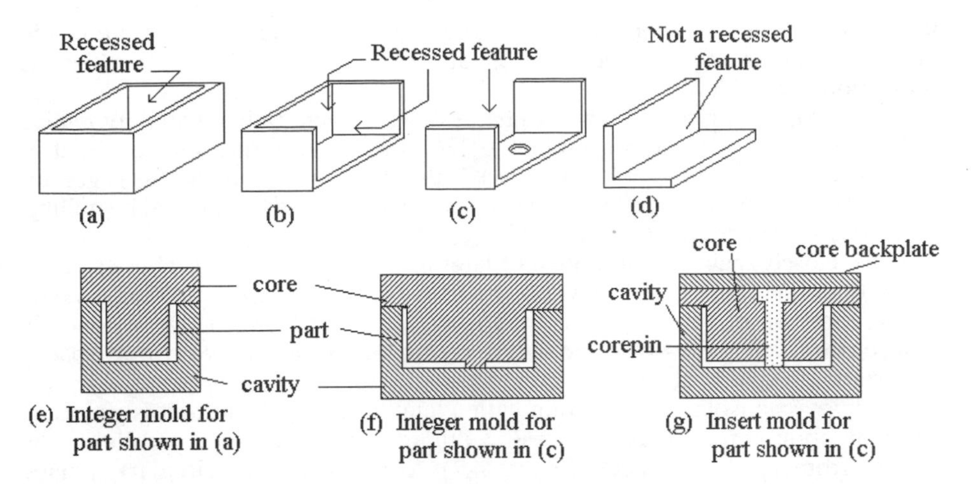

FIGURE 4.4 *Examples of parts with recessed features and section views of the molds used to produce them.*

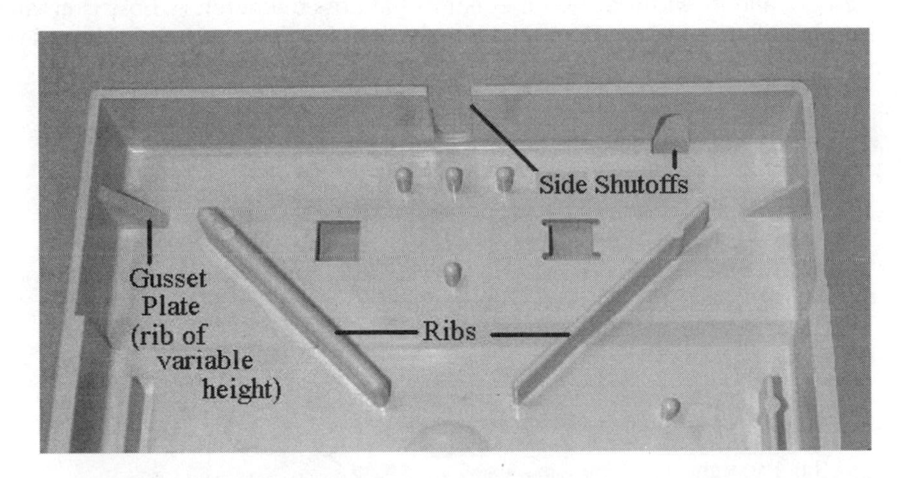

FIGURE 4.5 *Photograph of an injection-molded part showing ribs and shutoffs.*

Rectangular holes and irregularly shaped holes can also be formed by either the use of integer molds or insert molds, but the cost to create the tooling to form these holes is more costly (See "Questions and Problems" in Chapter 11, "Other Metal Shaping Processes.")

Projections

A feature that protrudes from the surface of a part is considered a projection. The most common examples are ribs and bosses.

A *rib* is a narrow elongated projection with a length generally greater than about three times its width (thickness), both measured parallel to the surface from which the feature projects (see Figure 4.5) and a height less than six times its width. Ribs may be located at the periphery or on the interior of a part or plate. A rib may run parallel to the longest dimension of the part (a longitudinal rib), or it may run perpendicular to this dimension (a lateral rib). Radial ribs and concentric ribs are also common. A rib may be continuous or discontinuous, or it may be part of a network of other ribs and projecting elements. If the height of a narrow elongated projection is greater than six times its width, then the projection is considered a wall.

Nonperipheral ribs and nonperipheral walls are generally created by milling or by electrical discharge machining (EDM) a cavity in either the core half or cavity half of the mold (see Figure 4.6a). If the minimum rib thickness is greater than or equal to about 3 mm (0.125 inches), then the cavity is machined by milling; otherwise it is machined by the EDM process.

Two closely spaced longitudinal or lateral ribs, that is two ribs whose spacing is less than three times the rib width, are usually formed by first milling a cavity in the core half of the plate and then using an insert to form the two closely spaced ribs (see Figure 4.6b). The cost to create this cluster of two closely spaced ribs is about equal to the cost to create a single rib.

A *boss* is an isolated projection with a length of projection that is generally less than about three times its overall width, the latter measured parallel to the surface from which it projects (Figure 4.7). A boss is usually circular in shape but it can take a variety of other forms called knobs, hubs, lugs, buttons, pads, or "prolongs."

Bosses can be solid or hollow. In the case of a solid circular boss, the length of the boss and its width are both equal to the boss diameter. A boss is created by simply milling a hole in the core half of the mold (Figure 4.6c). In the case of

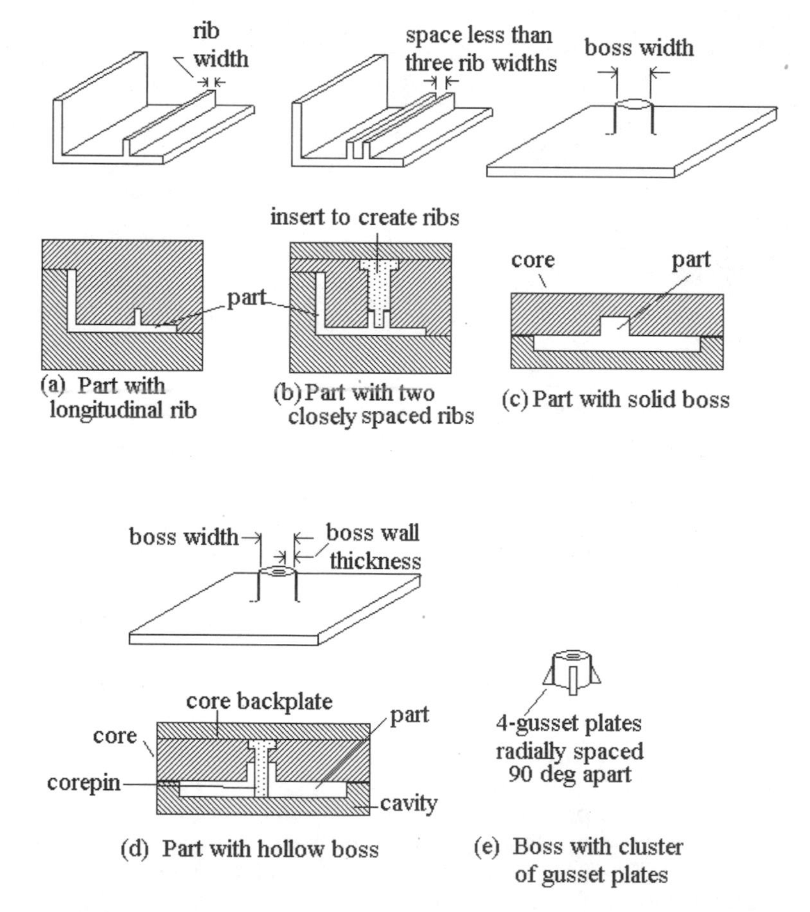

FIGURE 4.6 *Example of parts with ribs and bosses and sectional views of the molds used to produce them.*

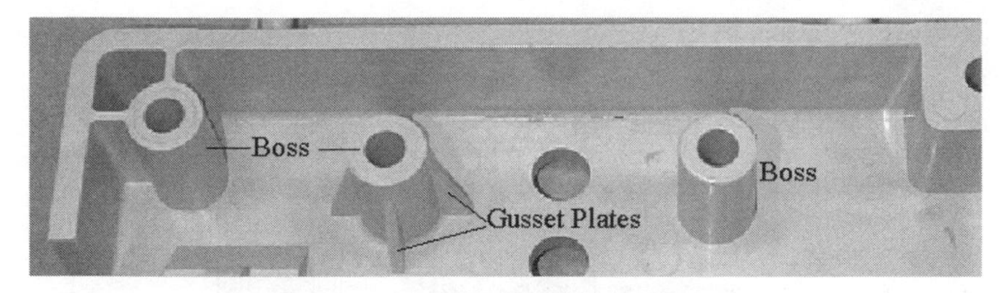

FIGURE 4.7 *Photograph of a part with bosses.*

a circular hollow boss, a pin is used to create the hole (Figure 4.6d). Although the width of the boss is still equal to the outside diameter, the boss thickness is equal to the difference between the outside and inside radii.

Bosses, and sometimes ribs, are supported by ribs of variable height called *gusset plates*. In the case of bosses, these supports are radially located, as shown in Figure 4.6e. These ribs are machined in simultaneously using the EDM process. For this reason this cluster of ribs costs about the same to create as a single rib.

Dividing and Parting Surfaces

One reason the determination of the direction of mold closure is so crucial to tooling cost evaluation is that it affects the location of the parting surface between the mold halves. The mold closure direction and the parting surface location together establish which recessed features can be molded in the direction of mold closure, and which (because they are not parallel to the mold closure direction) will require special tooling in the form of side action units or lifters in order to permit ejection of the part. (These subsidiary features, i.e., holes, projections, etc., are often referred to as add-ons.)

In order to determine where a parting surface should be located, we will introduce the concept of dividing surface. Given a direction of mold closure, the dividing surface (Figure 4.8) is defined as an imaginary surface, in one or more planes, through the part for which the portion of the part on either side of the surface can be extracted from a cavity conforming to the form of the outer shape of the portion in a direction parallel to the direction of mold closure.

If the dividing surface is in one plane only, it is called a *planar dividing surface*. In general, the dividing surface that results in the least costly tooling is the one that should be used as the parting surface in the actual construction of the tooling. Figure 4.9 shows the parting surface that was used in the tooling for the box-shaped part shown in Figure 4.8. Figure 4.10 shows the tooling used to produce an L-bracket similar to the one shown in Figure 4.8.

A dividing surface is a potential parting surface. A part may have several dividing surfaces, but of course a mold when constructed has only one parting surface.

Designers who understand the process of injection molding can usually plan for a convenient mold closure direction quite readily with just a little thought and study of the part.

4.3.4 Undercuts

In general, undercuts are combinations of part features created by recesses or by projections whose directions are not parallel to the mold closure direction. Undercuts are classified as either internal or external.

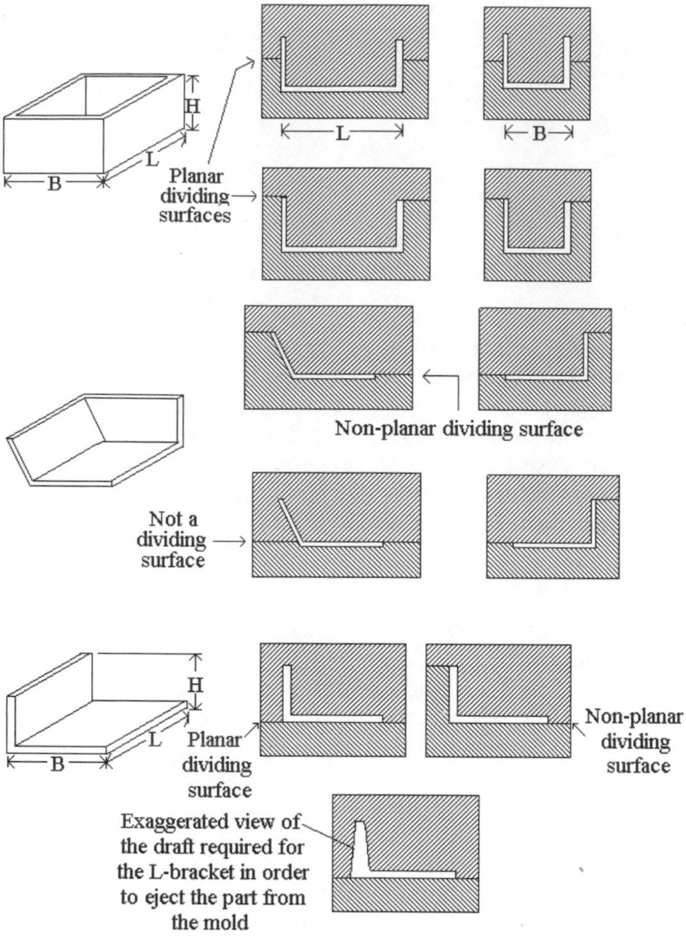

FIGURE 4.8 *Dividing surface of a part.*

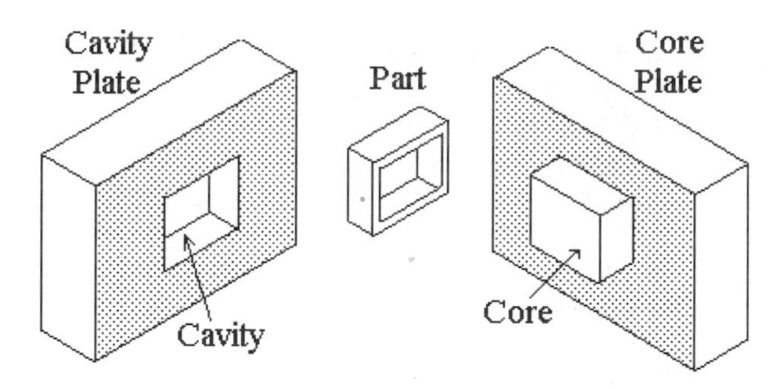

FIGURE 4.9 *Tooling for box-shaped part—shows parting surface between the core and cavity halves of the mold.*

By reference to Figure 4.1, readers should note how the number of external undercuts increases the part's tooling cost by moving the part's location to the right in the Figure. Similarly, note how the number of internal undercuts moves a part's place in the matrix downward, thus also increasing tooling cost.

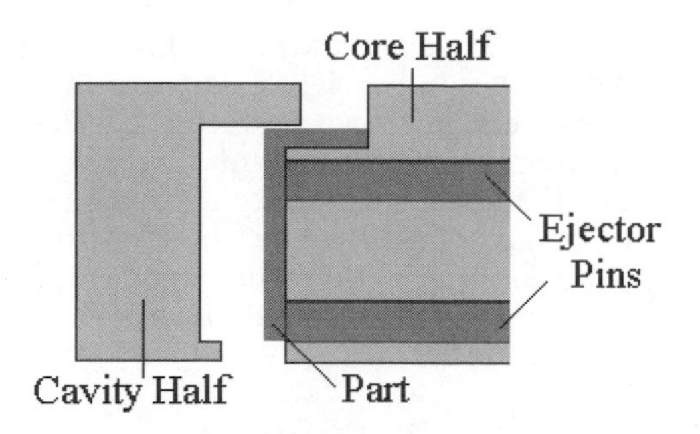

FIGURE 4.10 *Tooling for L-bracket.*

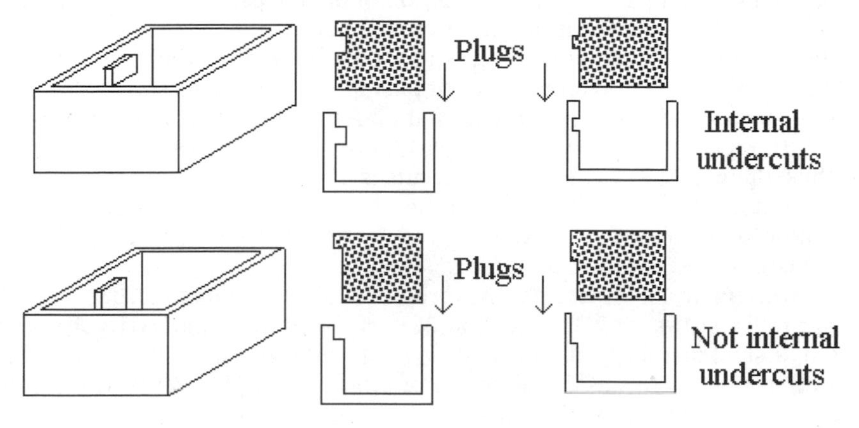

FIGURE 4.11 *Examples of internal undercuts.*

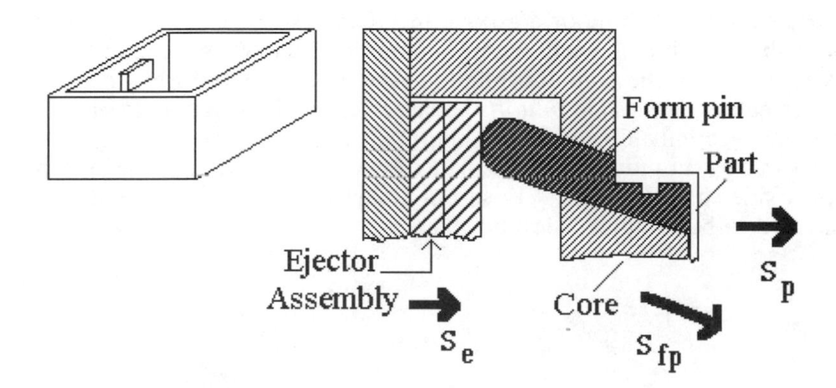

FIGURE 4.12 *Form pin used to form internal undercut.*

Internal Undercuts

Internal undercuts are recesses or projections on the inner surface of a part which, without special provisions, would prevent the mold cores from being withdrawn in the line of closure (often called the *line of the draw*). (See Figure 4.11.)

To permit withdrawal of the core when there is an internal undercut, hardened steel pins (called *form pins*) must be built into the cores. Figure 4.12 shows an illustration of a form pin. When the part solidifies, it shrinks onto the core.

The mold then opens and the core and part are withdrawn from the cavity. As the ejector assembly plate moves to the right, relative to the core (see vector S_e), the form pin slides in the direction shown by vector S_{fp}. The part then moves in a direction parallel to the motion of the ejector plate (see vector S_p) and the internal projection is lifted from the core pin cavity and the part removed or blown off. Alternatively, cores called *split cores* must be constructed in two or more parts. Both split cores and form pins add to the complexity and hence the cost of the tooling. Figure 4.13 shows a screen capture from an animation used to illustrate how internal undercuts are formed.

External Undercuts

In general, external undercuts are holes or depressions on the external surface of a part that are not parallel to the direction of mold closure (Figure 4.14). Some exceptions to this generalization are discussed below.

The number of external undercuts in a part is equal to the number of surfaces that contain external undercuts. For example, Figure 4.14a shows a part with only one surface that contains an external undercut; thus, the number of undercuts is 1. In Figure 4.14b, however, the part has two surfaces with external undercuts; thus, the number of external undercuts is 2.

In addition, projections located on the external surface of a part such that a single mold-dividing surface (planar or nonplanar) cannot pass through them all, are also considered external undercuts.

As with internal undercuts, the presence of external undercuts requires special provisions to allow for ejection of parts from the mold cavity. To permit ejection, a steel member called a *side cavity* or *side core* must be mounted and operated at right angles to the direction of mold closure. However, this solution also adds to tooling complexity and cost. See Figure 4.15.

Side Shutoffs

In some situations, a hole or a groove in the side wall of a part can be molded without the need for side action cores. Figure 4.16 shows two examples. In these cases, a portion of the core abutting the face of the cavity forms the hole. Such holes are called *simple side shutoffs* because contact between mold halves occurs on one surface only. *Complex side shutoffs* occur when contact between mold surfaces occurs on more than one plane. A tab (Figure 4.16) is an example of a complex side shutoff. Figure 4.17 shows a part with other features that are also considered to be complex side shutoffs.

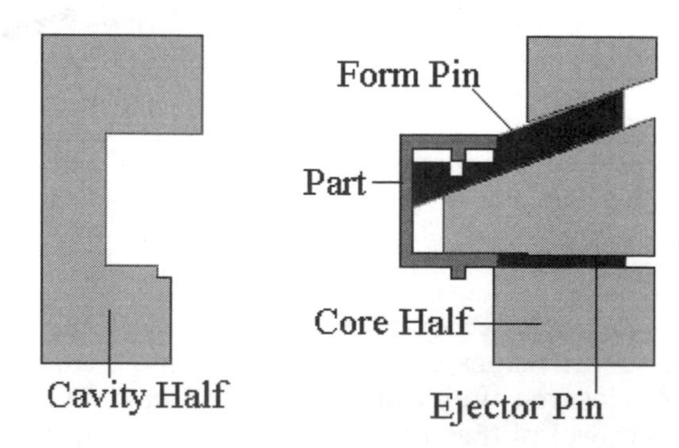

FIGURE 4.13 *A screen capture from an animation showing the formation of a part with an internal undercut.*

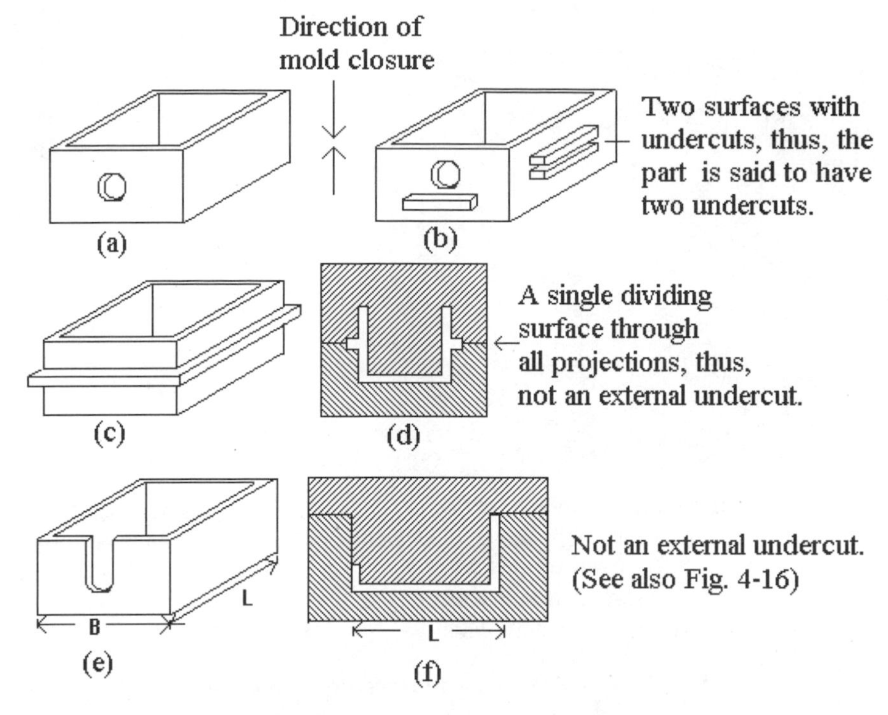

FIGURE 4.14 *External undercuts.*

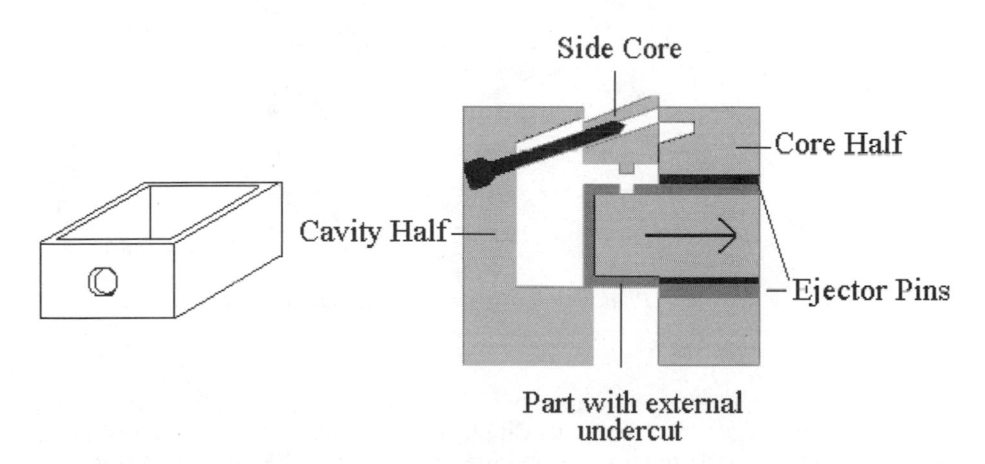

FIGURE 4.15 *Side core used to create an external undercut created by a circular hole.*

To determine whether a hole or depression is a side shutoff or an undercut, the following test can be applied: With a solid plug conforming to the exact shape of the inner surface of the part already inserted, imagine the part inserted into a plug conforming to the exact shape of the outer surface of the part. If the outer plug can now be removed, by the use of straight-line motion parallel to the direction of mold closure, the hole is considered a side shutoff. If the outer plug cannot be removed, the hole is considered an undercut.

A part with isolated grooves and cutouts on the external surface of a part can also sometimes be considered a part with side shutoffs and constant periph-

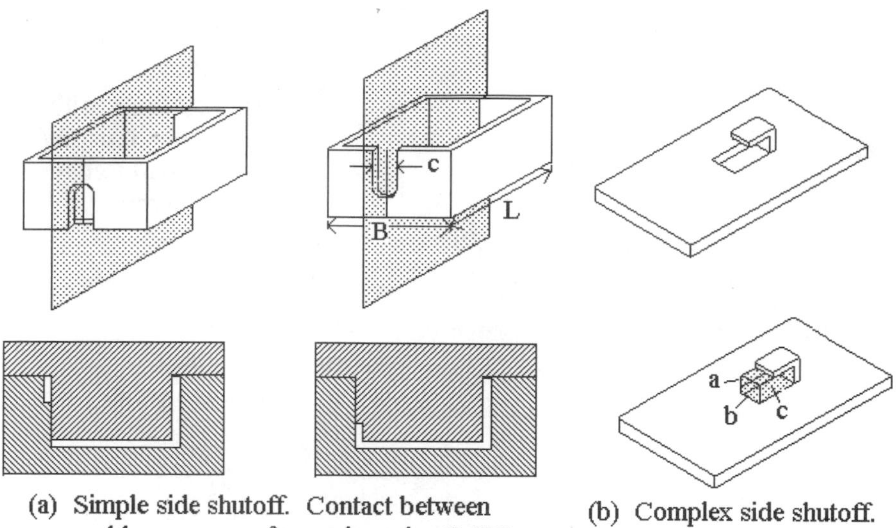

(a) Simple side shutoff. Contact between molds on one surface only and $c<0.33B$. Note: If $c \geq 0.33B$, peripheral height is not treated as constant and side shutoff is ignored in determining cavity detail.

(b) Complex side shutoff. Contact between molds along three side surfaces, a, b, and c.

FIGURE 4.16 *Simple and complex side shutoffs.*

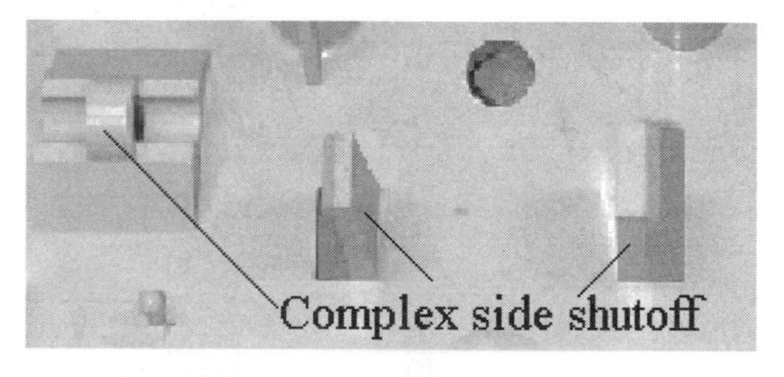

FIGURE 4.17 *Photograph of part with complex side shutoffs.*

eral height, rather than a part whose peripheral height is not constant. A groove or cutout is considered isolated if its dimension normal to the direction of mold closure is less than 0.33 times the envelope dimension in the same direction (see Figure 4.16).

4.3.5 Other Factors Influencing C_b

Parts Molded in One-Half the Mold

Mold costs are influenced to some extent by the amount of machining that must be done to create the core and cavity hollow sections. If the part cavity can be molded entirely in one-half of the mold, then the other part of the mold needs no special machining. A part is said to be in one-half the mold (Figure 4.18) when the entire part is on one side of a planar dividing surface.

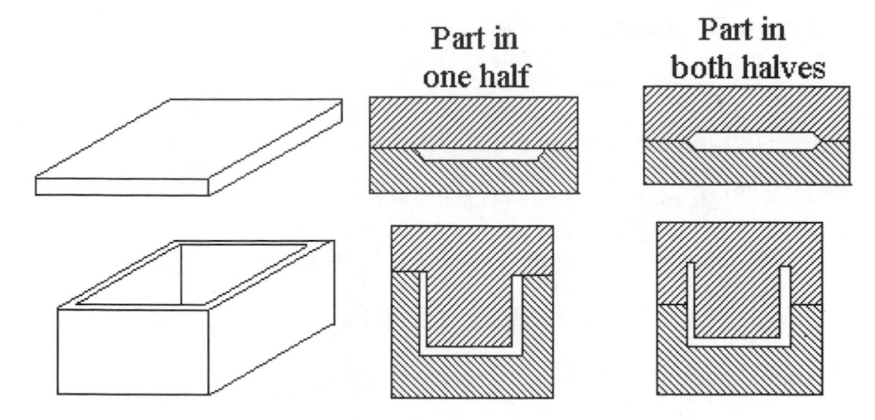

FIGURE 4.18 *Part with cavity in one-half of die.*

Peripheral Height

Although the L-shaped part shown in Figure 4.8 does have a planar dividing surface and could be molded using a planar parting surface, in general this would not be done. To use a planar parting surface would require a taper (draft) on the vertical wall, the wall parallel to the direction of mold closure, so that the part can be easily removed from the mold (see Figure 4.8).

To avoid the need for a taper, a nonplanar parting surface is generally used. To indicate situations where a nonplanar parting surface would probably be used, even if a planar dividing surface exists, the concept of a constant peripheral height is introduced. For the box-shaped part shown in Figure 4.8, a planar dividing surface exists and the peripheral height as measured from a planar dividing surface is constant (i.e. the wall height is constant); thus, a planar parting surface is used to produce the part (Figure 4.9). However, for the L-shaped part, the peripheral height from the planar dividing surface is not constant (i.e. the wall height is zero on three of the four peripheral surfaces) and, thus, a nonplanar parting surface would be used to construct the tooling.

4.3.6 Entering and Using Figure 4.1

The value of C_b can readily be determined from Figure 4.1 given the following information, all of which can be found easily by methods explained above in Sections 4.3.2 to 4.3.6:

1. the longest dimension of the basic envelope, L;
2. the number of external undercuts;
3. the number and location (on one or more faces of the part) of internal undercuts;
4. whether or not the part will be made in one-half of the mold;
5. whether the dividing surface will be planar or not;
6. whether or not the part's peripheral height from a planar dividing surface is constant or not; and
7. whether the part is flat or box-shaped.

For example, refer to Figure 4.1 and consider two parts with the following characteristics:

		PART A	PART B
1.	Longest dimension (mm)	400	200
2.	External undercuts	0	3
3.	Internal undercuts (faces)	0	1
4.	Dividing surface	Planar	Planar
5.	Peripheral height	Constant	Constant
6.	Part in one-half?	Yes	—
7.	Flat or box-shaped	Flat	Box

Readers should verify that the relative cost is 1.42 for Part A and 3.72 for Part B.

4.4 DETERMINING C_S

As noted previously, the relative tooling construction cost for a part is found from

$$C_{dc} = C_b \, C_s \, C_t \qquad \text{(Equation 4.5)}$$

where C_b = The approximate relative tooling cost due to size and basic complexity;
C_s = A multiplier accounting for other complexity factors called subsidiary factors;
C_t = A multiplier accounting for tolerance and surface finish issues.

In the preceding section, we showed how C_b is determined. In this section, we show how to obtain an appropriate value for C_s.

Features like ribs, bosses, holes, lettering, and other elements that are aligned with the mold closure direction contribute to mold complexity. We refer to the

Feature		Number of Features (n)	Penalty per Features	Penalty
Holes or Depressions	Circular		2n	
	Rectangular		4n	
	Irregular		7n	
Bosses	Solid (8)		n	
	Hollow (8)		3n	
Non-peripheral ribs and/or walls and/or rib clusters (8)			3n	
Side Shutoffs	Simple (9)		2.5n	
	Complex (9)		4.5n	
Lettering (10)			n	
			Total Penalty	

SMALL PARTS (L ≤ 250 mm)

Total Penalty ≤10 => Low cavity detail
10 < Total Penalty ≤20 => Moderate cavity detail
20 < Total Penalty ≤40 => High cavity detail
Total Penalty >40 => Very high cavity detail

MEDIUM PARTS (250 < L ≤ 480 mm)

Total Penalty ≤15 => Low cavity detail
15 < Total Penalty ≤30 => Moderate cavity detail
30 < Total Penalty ≤60 => High cavity detail
Total Penalty >60 => Very high cavity detail

LARGE PARTS (L > 480 mm)

Total Penalty ≤20 => Low cavity detail
20 < Total Penalty ≤40 => Moderate cavity detail
40 < Total Penalty ≤80 => High cavity detail
Total Penalty >80 => Very high cavity detail

1 in = 25.4 mm; 100 mm/25.4mm = 3.94 in

FIGURE 4.19 *Determination of cavity detail. (The numbers in parentheses refer to notes found in Appendix 4.A.)*

number and complexity of such features as cavity detail. Figure 4.19 shows the method for rating the cavity detail as low, moderate, high, or very high. Figure 4.20 shows photographs of two parts, one with low cavity detail (on the left) and one with high cavity detail (on the right).

In addition to the level of cavity detail, C_s is influenced by the complexity and number of external undercuts. Table 4.1 requires only that a judgment be made about whether extensive undercut complexity exists or does not exist. External undercuts other than unidirectional holes or depressions are considered extensive since the creation of such tooling is more costly. Figure 4.21 provides an example of a part that clearly has extensive external undercut complexity.

4.5 DETERMINING C_t

The effects of surface finish requirements, sometimes referred to as surface quality, and the strictness of required tolerances on relative tool construction

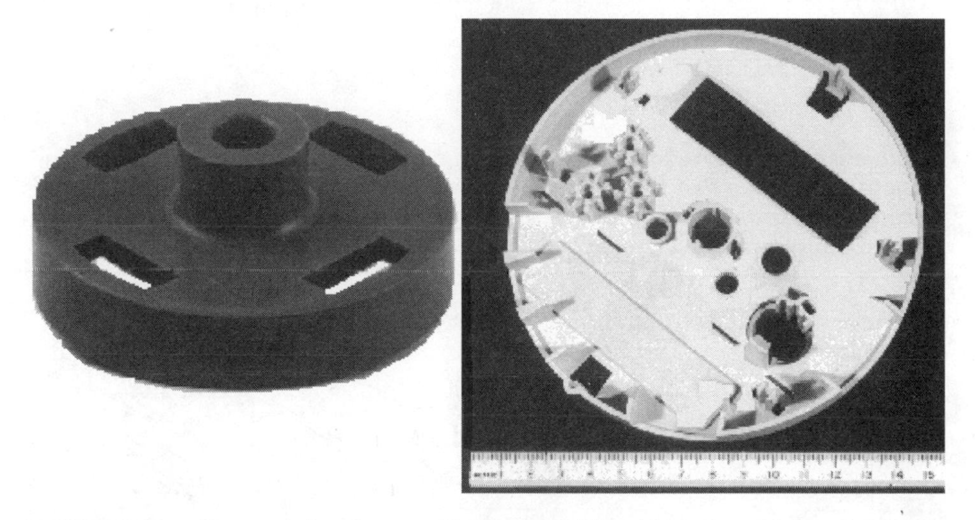

FIGURE 4.20 *Photographs of parts with low (on left) and high (on right) cavity detail.*

Table 4.1 Subsidiary complexity rating, C_s. (The numbers in parentheses refer to notes found in Appendix 4.A.)

				Fourth Digit	
				Without Extensive (7) External Undercuts (5)	With Extensive (7) External Undercuts (5)
				0	1
Third Digit	Cavity Detail (6)	Low	0	1.00	1.25
		Moderate	1	1.25	1.45
		High	2	1.60	1.75
		Very High	3	2.05	2.15

costs are accounted for by the factor C_t, which is obtained from Table 4.2. Guidelines relating the surface quality of the part and the various SPI (Society of the Plastics Industry) finishes of the mold are given below:

SPI-SPE 1: Used on transparent moldings requiring minimum distortions and surface blemishes. Good for most optical lenses.
SPI-SPE 2: Near optical. Used when require good transparent clarity and high gloss. Also good for bearing surface due to minimum of surface scratches.
SPI-SPE 3: Finely abraded surface. Resembles very lightly brushed stainless steel. Used when high gloss not required.
SPI-SPE 4: Medium, abraded surface resembling brushed steel. Used in nonaesthetic areas not usually seen. Inexpensive surface, yet provides easy ejection from the mold.
SPI-SPE 5: 40 micro-inch textured surface that has the appearance of frosted glass. Good for areas needing adhesive bonding or products requiring smooth, nonglass surface that absorbs light.
SPI-SPE 6: Medium-textured surface similar to 400-to-600-grit emery paper. Good for bonding and absorbing light. Inexpensive, appealing finish for industrial products and some consumer products.

4.6 USING THE PART CODING SYSTEM TO DETERMINE C_b, C_s, AND C_t

When analyzing a part for entry into Figure 4.1 and Tables 4.1 and 4.2, it is convenient to make use of the part coding system that has been developed for this purpose. The coding system involves six digits that, in effect, describe the part in the fashion of group technology.

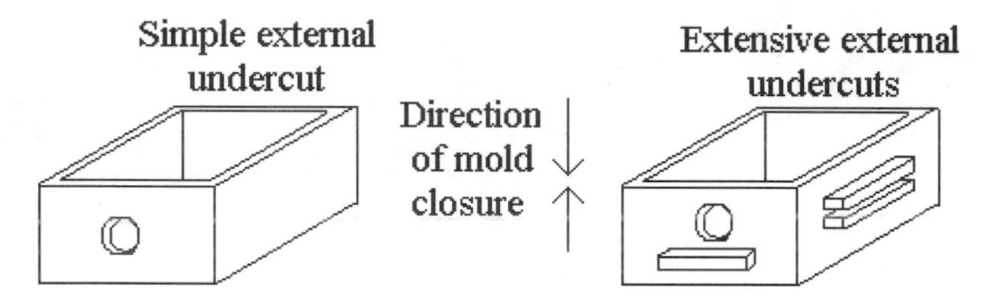

FIGURE 4.21 *External undercuts caused by features other than circular, unidirectional holes are considered extensive external undercuts because the tooling is more costly to create.*

Table 4.2 Tolerance and surface finish rating, C_t. (The numbers in parentheses refer to notes found in Appendix 4.A.)

				Sixth Digit	
				Commercial Tolerance, T_a	Tight Tolerance, T_a
				0	1
Fifth Digit	Surface Finish, R_a	SPI 5–6	0	—	—
		SPI 3–4	1	1.00	1.05
		Texture	2	1.05	1.10
		SPI 1–2	3	1.10	1.15

In group technology, as can be seen in Figure 4.1, data is organized in the form of a matrix that gives one a qualitative feel for the impact of the part attributes on the ease or difficulty of producing, in this case, the tooling for the part. In addition, it provides a comprehensive checklist of cost factors and presents the user with a consistent and systematic method for analyzing part designs for manufacturability. It also facilitates implementation on a computer.

Here are the descriptions of the meaning of the digits in the coding system and their interpretation.

For Figure 4.1

First Digit (0–6): The first digit in the coding system identifies the row in Figure 4.1 (for C_b) that describes the part. It is fixed by (1) the number of faces with internal undercuts, (2) whether the part is in one-half the mold or not, (3) whether the dividing surface is planar or nonplanar, and (4) whether the peripheral height is constant from the dividing surface.

Second Digit (0–9): The second digit identifies the column in Figure 4.1 that describes the part. It is thus fixed by (1) the part size (L), and (2) the number of external undercuts.

Together, the first and second digits locate the place in Figure 4.1 where the value of C_b is found. (Remember: the values above the slanted line in that Figure refer to flat parts; values below the line refer to box shaped parts.)

Readers should verify that the first two digits of the code for parts A and B just described are, respectively, 0–4 and 3–3.

For Table 4.1

C_s is determined from Table 4.1 by thc third and fourth digits of the coding system as follows:

Third Digit (0–3): The third digit in the coding system identifies the row in Table 4.1 (for C_s) that describes the part. It is determined by the level of cavity detail as determined from Figure 4.19.

Fourth Digit (0–1): The fourth digit identifies the column in Table 4.1 that describes the part. It is determined by the extent of external undercut complexity.

As an example, readers should verify from Figure 4.19 that the penalty factor for a part with five radial ribs, three hollow bosses, three simple side shutoffs, and localized lettering is 32.5—resulting in a level of cavity detail for a large part (L >480 mm) of Moderate. Also verify from Table 4.1 that C_s for such a part with extensive external undercuts and moderate cavity detail is 1.45. (The third and fourth digits in the coding system for this part are 1 and 1, respectively.)

For Table 4.2

The coding system for entry into Table 4.2 is as follows:

Fifth Digit (0–3): The fifth digit identifies the row in Table 11.2 (for C_t) that describes the part. It is fixed by the nature of the required surface finish.

Sixth Digit (0–1): The sixth digit identifies the column in Table 4.2 that describes the part. It is fixed by whether the tolerances required are commercial or tight.

4.7 TOTAL RELATIVE TOOLING CONSTRUCTION COST

As defined earlier, the total relative mold construction cost is:

$$C_{dc} = C_b C_s C_t \qquad \text{(Equation 4.5)}$$

and it is determined, as shown above in section 4.6, at the configuration stage of part design, that is, prior to any detailed knowledge concerning part dimensions, rib sizes, wall thickness, and so on. The results show very clearly and simply the aspects of a part's design that contribute most heavily to tooling construction costs. In Figure 4.1, for example, designers can see clearly that parts should be redesigned if possible to move the rating up and to the left in the matrix, and they can compute approximately how much can be saved. Removing undercuts accomplishes this goal, as do other simplifying changes that reduce detail or eliminate the need for special finishes or tight tolerances.

4.8 RELATIVE MOLD MATERIAL COST

In order to compute total relative tooling costs, we must be able to estimate the mold material cost as well as its construction costs. This is relatively easy to do from a knowledge of the approximate size of a part—which in turn dictates the required size of the mold. Referring to Figure 4.22, we define the following mold dimensions:

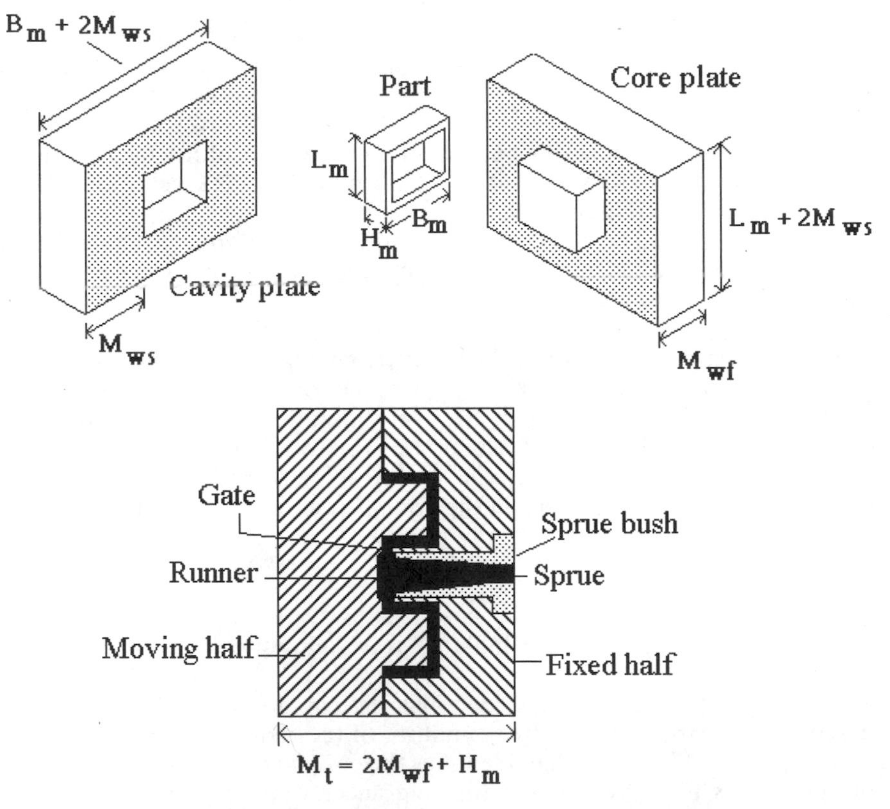

FIGURE 4.22 *Mold dimensions for two-plate mold.*

M_{ws} = Thickness of the mold's side walls (mm)

M_{wf} = Thickness of core plate (mm)

L_m and B_m = The length and width of the part in a direction normal to the mold closure direction (mm)

H_m = The height of the part in the direction of mold closure (mm) (H_m not necessarily equal to H)

M_t = The required thickness of the mold base (mm)

With these definitions, the following equations can be used sequentially to determine the projected area of the mold base, M_a, and the required thickness of the mold base, M_t, which in turn are used to obtain the relative mold material cost (C_{dm}) from Figure 4.24.

C = value obtained from Figure 4.23

$$M_{ws} = [0.006CH_m^4]^{1/3} \quad \text{(Equation 4.6)}$$

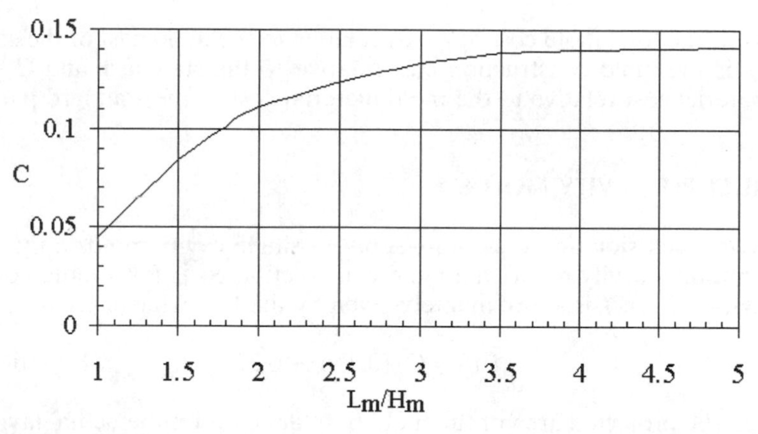

FIGURE 4.23 *Value of C for use in Equation 4.6. (If L_m/H_m <1, then use the value of H_m/L_m to determine C.)*

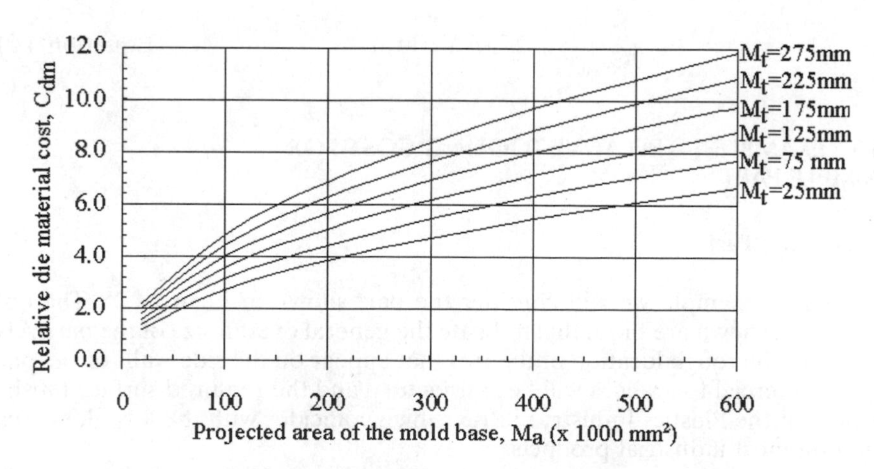

FIGURE 4.24 *Relative die material cost.*

$$M_{wf} = 0.04L_m^{4/3} \quad \text{(Equation 4.7)}$$

$$M_a = (2M_{ws} + L_m)(2M_{ws} + B_m) \quad \text{(Equation 4.8)}$$

$$M_t = (H_m + 2M_{wf}) \quad \text{(Equation 4.9)}$$

$$C_{dm} = \text{value obtained from Figure 4.24}$$

The data for the plot shown in Figure 4.24 is based on the assumption that a standard two-plate die block unit is being used. This corresponds to DME's A-series. It is also assumed that the highest quality steel, namely, P-20 and S-7, is used as the mold material. Using DME cost data, a parabolic curve was fit to the data and the plot shown in Figure 4.24 was obtained (see Juan R. Escudero, 1991).

4.8.1 Total Relative Mold Cost

The total relative mold cost is determined from Equation 4.4, namely,

$$C_d = 0.8C_{dc} + 0.2C_{dm} \quad \text{(Equation 4.4)}$$

where C_d is the total mold cost of a part relative to the mold cost of the standard part, C_{dc} is the mold construction cost relative to the standard, and C_{dm} is the mold material cost relative to the mold material cost of the standard part.

4.9 MULTIPLE CAVITY MOLDS

The above discussion and equations apply to single cavity molds only. For the case of multiple cavity molds, the mold construction costs for a mold consisting of n_c cavities, $C_{dc}(n_c)$, is approximately given by the following expression:

$$C_{dc}(n_c) = C_{dc}(0.73n_c + 0.27) \quad \text{(Equation 4.10)}$$

Although the projected area of the mold base depends on the actual layout of a multiple cavity mold, it is assumed here that the projected area is roughly given by the product of the projected area for a single cavity mold, M_a, times the number of cavities n_c, that is,

$$M_a(n_c) = M_a n_c \quad \text{(Equation 4.11)}$$

4.10 EXAMPLE 1—RELATIVE TOOLING COST FOR A SIMPLE PART

4.10.1 The Part

As our first example we will consider the part shown in Figure 4.25. The only dimensions shown are those that indicate the general overall size of the part. Also shown are the rough location of the ribs that appear on the side walls of the part.

Commercial tolerances will be satisfactory, and the required surface finish is (Society of the Plastics Industry) SPI-3, which coincides with the low-gloss finish found on most industrial products.

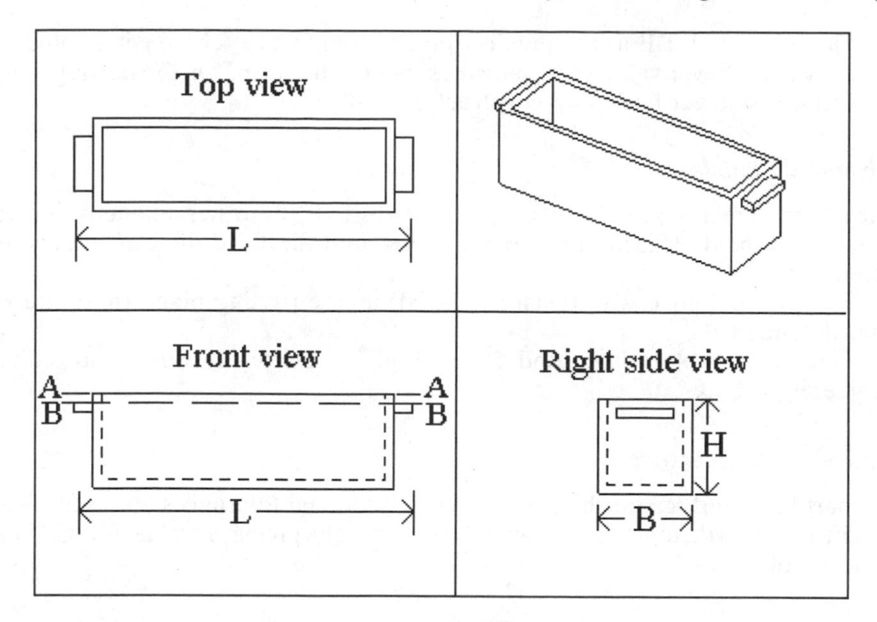

FIGURE 4.25 *Original Design—Example 4.1. (L = 180 mm, B = H = 50 mm.)*

4.10.2 Relative Tooling (Mold) Construction Cost

Basic Complexity

The dimensions of the basic envelope of this part are:

$$L = 180\,\text{mm}, \quad B = 50\,\text{mm}, \quad H = 50\,\text{mm}$$

Since L/H is less than 4, the part is box-shaped.

If the direction of mold closure is assumed to be in the direction of the recess, then a planar dividing surface exists for the part. Planes AA and BB are just two of the many planar dividing surfaces that exist for this part (i.e., just two of the many surfaces that could be used to separate or part the two halves of the mold). Initially, dividing surface AA is taken as the parting surface. In this case there are no internal undercuts, the peripheral height from the planar dividing surface (AA) is constant, and the part is in one-half of the mold. Thus, the first digit of the coding system is 0.

With dividing surface AA, there are two surfaces that contain external undercuts, hence there are two undercuts. Since L is less than 250 mm, the second digit is 2.

With the first two digits being 0 and 2, Figure 4.1 indicates a value for C_b of 2.02.

Since one of the major methods available for reducing mold manufacturability costs is to reduce the number of external and internal undercuts, the tooling cost for the part is reexamined using BB as the planar dividing surface.

In this case, the peripheral height from BB is still constant; however, the part is no longer in one-half the mold. Thus, the first digit is 1.

Since BB passes through both external projections, they are no longer considered undercuts. Thus, the second digit is 0.

Therefore, with BB as the dividing plane, from Figure 4.1 we get a value for C_b of 1.86. This lower value of C_b indicates that the use of BB as the parting plane will result in a lower basic tool construction cost.

Subsidiary Complexity

Since there are no ribs, bosses, holes, depressions, or other elements in the direction of mold closure, cavity detail is low and the third digit of the coding system is 0.

The fourth digit is also 0 since with BB as the parting plane there are no external undercuts.

Thus from Table 4.1 we find the multiplying factor, C_s, due to subsidiary complexity is 1.00.

Surface Finish/Tolerance

The part has a surface finish of SPI-3 and commercial tolerances are used. Thus, the fifth and sixth digits are 1 and 0, respectively giving a value for C_t from Table 4.2 of 1.00.

Total Relative Mold Construction Cost C_{dc}

$$C_{dc} = C_b C_s C_t = 1.86(1)(1) = 1.86$$

4.10.3 Relative Mold Material Cost

From Figure 4.23, for L_m/H_m of 3.6, C is 0.138. Thus, the thickness of the mold wall is given by:

$$M_{ws} = [0.006CH_m{}^4]^{1/3} = [0.006(0.138)(50)^4]^{1/3} = 17.3\,\text{mm}$$

and the thickness of the base is:

$$M_{wf} = 0.04\,L_m{}^{4/3} = 0.04(180)^{4/3} = 40.7\,\text{mm}.$$

Consequently, the projected area of the mold base is

$$M_a = [2(17.3) + 180][2(17.3) + 50] = 18155\,\text{mm}^2$$

and the required plate height is

$$M_t = [50 + 2(40.7)] = 131.4\,\text{mm}.$$

From Figure 4.24, the relative mold material cost, C_{dm}, for this part is approximately 1.6, and the total relative mold cost is

$$C_d = 0.8C_{dc} + 0.2C_{dm} = 0.8(1.86) + 0.2(1.6) = 1.81$$

4.10.4 Redesign Suggestions

The mold manufacturability costs for this part can be reduced slightly if the part is in one-half the mold. This can be done by moving the two side projections to

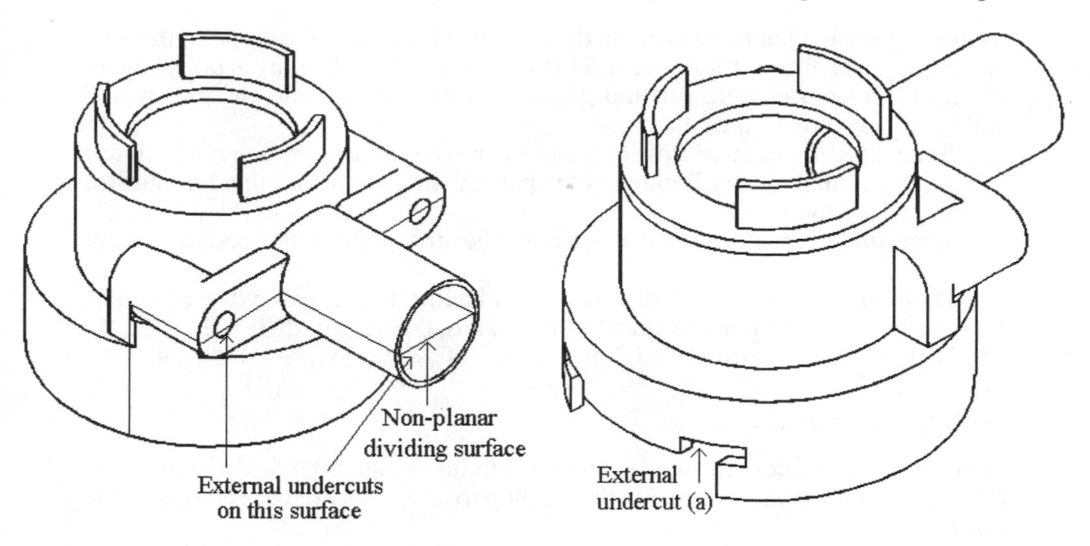

FIGURE 4.26 *Original Design—Example 4.2. (L = 55mm, B = 40mm, H = 20mm.)*

the top; that is, so that the tops of the projections are tangent to plane AA. In addition, relocating the side projections in this manner avoids the need for "reverse" taper or draft.

4.11 EXAMPLE 2—RELATIVE TOOLING COST FOR A COMPLEX PART

4.11.1 The Part

As a second example we will consider the part shown in Figure 4.26. We will assume that we are at the configuration stage of the design and that detailed dimensions concerning wall thickness, holes sizes, and other elements are not available. Thus, the only dimensions given are those that indicate the overall size and shape of the part.

4.11.2 Relative Mold Construction Cost

Basic Complexity

The dimensions of the basic envelope of this part are:

$$L = 55\,\text{mm}, \quad B = 40\,\text{mm}, \quad H = 20\,\text{mm}$$

It is possible that the projections indicated in Figure 4.26 are such that the largest dimensions parallel to the surface from which they project are less than 0.33 times the envelope dimension in the same direction. If so, they are isolated projections of small volume, and would consequently be ignored in determining the basic envelope. Since we are at the configuration stage of the design and the detailed dimensions are not yet known, it will be assumed that these are not isolated projections of small volume. Hence, the dimensions of the basic envelope of the part are those given above. Therefore, L/H is less than 4, and the part is box-shaped. (Even if the features were isolated projections of small volume, the part would still have been box-shaped.)

If the direction of mold closure is assumed to be in the direction of the major recess (i.e., normal to the LB plane of the part), then a nonplanar dividing surface is required for the part.

Since there are no internal undercuts, the first digit of the coding system is 2.

There are two surfaces with external undercuts, hence, two external undercuts are present, and L is less than 250 mm. Thus, the second digit is 2.

With the first digits of 2 and 2, the value of C_b from Figure 4.1 is 2.29.

Subsidiary Complexity

There is one set of concentric ribs, and one circular hole in the direction of mold closure. Thus the total penalty for this small part is 5, cavity detail is low, and the third digit is 0.

The fourth digit is 1 because the external undercuts are extensive. Therefore, from Table 4.1, the factor C_s, due to subsidiary complexity, is 1.25.

Surface Finish/Tolerance

The part has a surface finish of SPI-3 and commercial tolerances are used. Thus the fifth and sixth digits are 1 and 0, respectively. Thus, from Table 4.2, C_t is 1.00.

Total Relative Mold Construction Cost C_{dc}

$$C_{dc} = C_b C_s C_t = 2.29(1.25)(1) = 2.86$$

4.11.3 Relative Mold Material Cost

Since L_m = 55 mm and H_m = 20 mm, then L_m/H_m = 2.75; and from Figure 4.23, C is 0.13. Thus, the thickness of the mold wall is

$$M_{ws} = [0.006CH_m^4]^{1/3} = [0.006(0.13)(20)^4]^{1/3} = 5.0\,\text{mm}$$

and the thickness of the base is

$$M_{wf} = 0.04\,L_m^{4/3} = 0.04(55)^{4/3} = 8.4\,\text{mm}$$

Consequently, the projected area of the mold base is

$$M_a = [2(5.0) + 55][2(5.0) + 40] = 3250\,\text{mm}^2$$

and the required plate height is

$$M_t = [20 + 2(8.4)] = 36.8\,\text{mm}$$

Hence, from Figure 4.24, the relative mold material cost, C_{dm}, for this part is approximately 1.2.

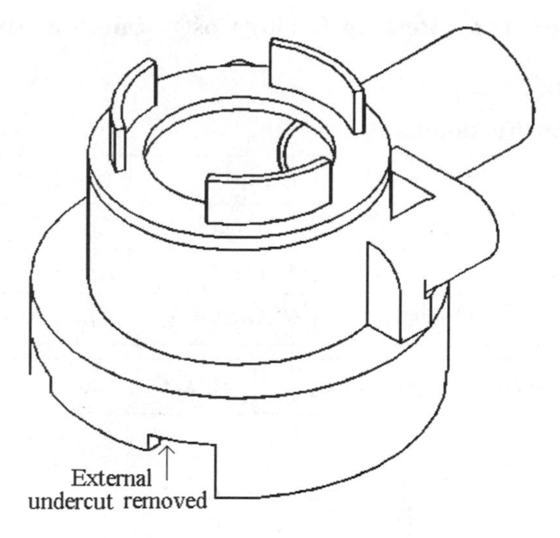

FIGURE 4.27 *Redesigned part.*

4.11.4 Total Relative Mold Cost

$$C_d = 0.8C_{dc} + 0.2C_{dm} = 2.53$$

4.11.5 Redesign Suggestions

The two features causing tooling complexity cost in this case are the two surfaces containing external undercuts. If external undercut (a) can be eliminated, as shown in Figure 4.27, then the new basic complexity code becomes B21, and C_b becomes 2.15.

With this redesign, the cavity detail remains low, and the third digit is still 0. The remaining external undercut does not constitute an extensive undercut. Hence, the fourth digit is 0, and C_s is 1.00.

With these values, the new total mold construction cost C_{dc} becomes

$$C_{dc} = (2.15)(1)(1) = 2.15$$

which is a 25% reduction in mold construction costs, and about a 23% reduction in total mold cost.

4.12 WORKSHEET FOR RELATIVE TOOLING COST

The determination of the relative die construction costs, the relative die material costs, and the overall relative die costs is a straightforward, though sometimes cumbersome, procedure. The following worksheet can be used to simplify the calculations. To illustrate the use of the worksheet, it has been filled out for the part shown in Figure 4.26 (Example 4.2).

A blank version of the worksheet is shown in Appendix 4.B at the end of this chapter. The worksheet shown in Appendix 4.B may be copied for use with this book.

Worksheet for Relative Tooling Costs—Injection Molding

Original Design

Relative Die Construction Cost

Basic Shape	$L = 55$	$B = 40$	$H = 16$	Box/Flat Box
Basic Complexity	1st Digit = 2	2nd Digit = 2	$C_b = 2.29$	
Sub. Complexity	3rd Digit = 0	4th Digit = 1	$C_s = 1.25$	
T_a/R_a	5th Digit = 1	6th Digit = 0	$C_t = 1.00$	

Total relative die construction cost C_{dc}	$= C_bC_sC_t = 2.86$

Relative Die Material Cost

$L_m = 55$	$B_m = 40$	$H_m = 20$
Die closure parallel to H	$L_m/H_m = 2.75$	Thus, $C = 0.13$

$M_{ws} = [0.006CH_m{}^4]^{1/3} = 5\,mm$
$M_{wf} = 0.04L_m{}^{4/3} = 8.4\,mm$
$M_a = (2M_{ws} + L)(2M_{ws} + B) = 3250\,mm$
$M_t = (H_m + 2M_{wf}) = 36.8\,mm$

Thus,

$C_{dm} = 1.2$	$C_d = 0.8C_{dc} + 0.2C_{dm} = 2.53$

Redesign Suggestions

Eliminate external undercut (a) as shown in Figure 4.27.

Basic Shape	$L = 55$	$B = 40$	$H = 16$	Box/Flat Box
Basic Complexity	1st Digit = 2	2nd Digit = 1	$C_b = 2.15$	
Sub. Complexity	3rd Digit = 0	4th Digit = 1	$C_s = 1.0$	
T_a/R_a	5th Digit = 1	6th Digit = 0	$C_t = 1$	

Total relative die construction cost C_{dc}	$= C_bC_sC_t = 2.15$
$C_d = 0.8C_{dc} + 0.2C_{dm} = 1.96$	
% Savings = (2.49 – 1.96)/2.49 = 0.21 => 21%	

4.13 SUMMARY

This chapter has described a systematic approach for calling designers' attention to those features of injection molding that tend to increase the tooling cost to manufacture parts—and for estimating the relative costs of tooling. The system employs a six-digit coding system for determining total relative tooling cost, which groups parts according to their similarity in tool construction difficulty. The system highlights those features that significantly increase cost so that designers can minimize difficult-to-produce features.

Using the methodology presented, designers can perform a tooling cost evaluation of a proposed part using only the information available at the configuration stage of part design. That is, the evaluation can be performed from only the knowledge of whether certain features are present or absent and, if present, their approximate location and orientation. Detailed dimensions are not needed. The methodology points out what features or arrangements of features contribute to the cost so that the direction of improved redesign is made apparent.

REFERENCES

Dym, J. B. *Product Design with Plastics.* New York: Industrial Press, 1983.

Escudero, Juan R. "Two Methods to Assess the Effect of Part Design on Tooling Costs in Injection Molding." Mechanical Engineering Department, M.S. Thesis, University of Massachusetts at Amherst, Amherst, MA, 1988.

Fredette, Lee. "A Design Aid for Increasing the Producibility of Mold Cast Parts." M.S. Final Project Report, Mechanical Engineering Department, University of Massachusetts at Amherst, Amherst, MA, 1989.

Kuo, Sheng-Ming. "A Knowledge-Based System for Economical Injection Molding." Ph.D. Dissertation, University of Massachusetts at Amherst, Amherst, MA, Feb. 1990.

Poli, C., Escudero, J., and Fernandez, R. "How Part Design Affects Injection Molding Tool Costs." *Machine Design* 60 (Nov. 24, 1988): 101–104.

Poli, C., Fredette, L., and Sunderland, J. E. "Trimming the Cost of Die Castings." *Machine Design* 62 (March 8, 1990): 99–102.

Poli, C., Kuo, Sheng-Ming, and Sunderland, J. E. "Keeping a Lid on Mold Processing Costs." *Machine Design* 61 (Oct. 26, 1989): 119–122.

Rajagopalan, Swaminath. "Design for Injection Molding and Die Casting: A Knowledge Based Approach." Mechanical Engineering Department, M.S. Thesis, University of Massachusetts at Amherst, Amherst, MA, 1991.

Shanmugasundaram, S. K. "An Integrated Economic Model for the Analysis of Mold Cast and Injection Molded Parts." M.S. Final Project Report, Mechanical Engineering Department, University of Massachusetts at Amherst, Amherst, MA, August 1990.

QUESTIONS AND PROBLEMS

4.1 As part of a training program that you are participating in at Plastics.com you are explaining to some new hires that the tooling cost of transfer molded parts is a function of both *basic complexity* as well as *subsidiary complexity*. Explain in greater detail exactly which features of a part affect basic complexity and which features of a part influence subsidiary complexity.

4.2 As part of the same training program described in Problem 4.1 you have decided to use the parts shown in Figures P3.1 to P3.4 of Chapter 3 as examples to illustrate both basic complexity and subsidiary complexity. Explain which features of these parts, if any, affect basic complexity, and which features, if any, affect subsidiary complexity.

4.3 Assume that you work for DFM.com, a large cap manufacturing company whose product line consists of a family of widgets of various sizes and shapes. In an effort to become more competitive, the company has decided to completely redesign Widget A. In addition, it has decided to organize the team responsible for redesigning Widget A along the lines described in Figure 1.3 in Chapter 1. As one of the lead designers on the team, you are responsible for commenting on the proposed designs of all Widget A components. What comments would you make concerning the tooling costs for the component part shown in Figure P4.3? In other words, what suggestions would you make to reduce mold costs?

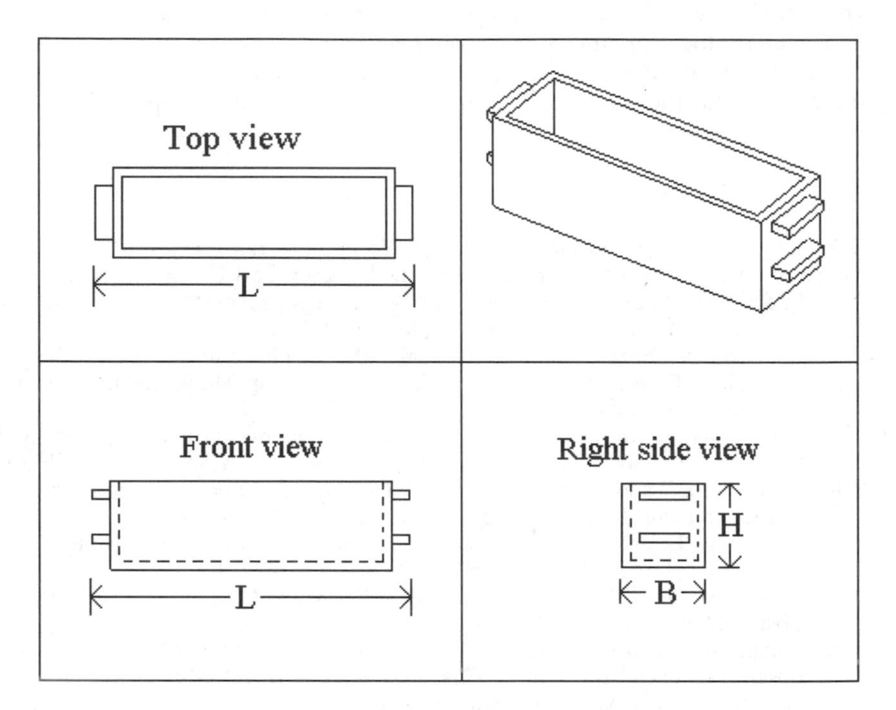

FIGURE P4.3

4.4 Some members of the integrated product and process design (IPPD) team formed in Problem 4.3 are unconvinced that your suggested redesign of the part shown in Figure P4.3 will yield significant savings in tooling cost. In an effort to convince them, calculate the savings in tool costs that you can achieve by redesigning the part. Assume that L = 180 mm, B = H = 50 mm, the part has a surface finish of SPI-3, and that commercial tolerances are to be used.

4.5 As another alternative to your redesign suggestion for the part shown in Figure P4.3 (see Problem 4.3), another member of the team has suggested redesigning the part as shown in Figure P4.5. What are the savings in tool costs between the design shown in Figure P4.5 and your proposed redesign of the part shown in Figure P4.3? Again, assume that the part has a surface finish of SPI-3, that commercial tolerances are to be used, and that L = 180 mm, B = H = 50 mm.

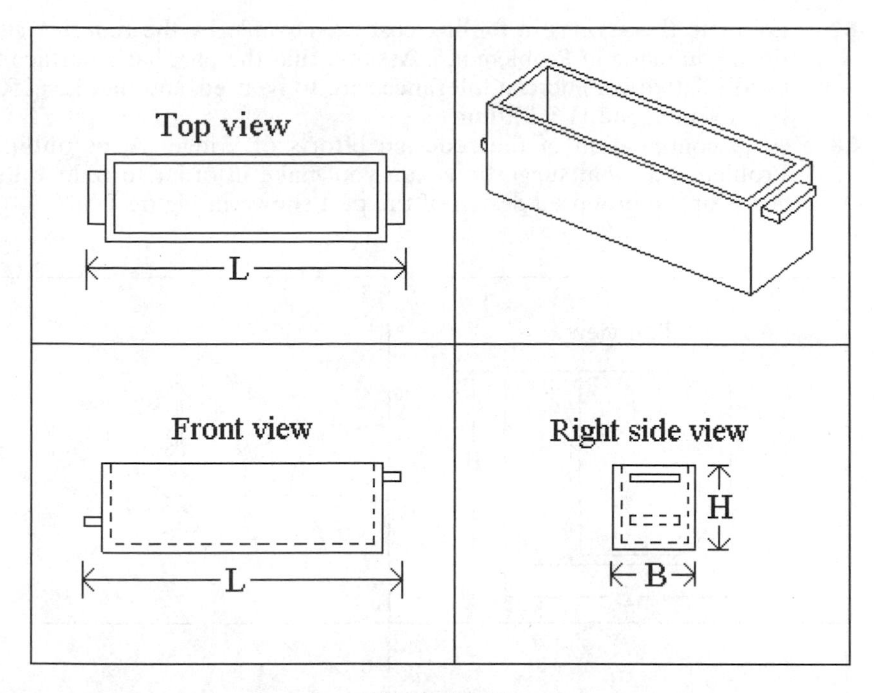

FIGURE P4.5

4.6 Although the total tooling costs for a part can be distributed over the entire production volume of the product, tooling costs are up-front costs and must be paid for before the mold is delivered to the vendor and any injection-molded parts can be produced. Therefore, as a lead member of the design team described in Problem 4.3, you are responsible for suggesting changes to the proposed design shown in Figure P4.6 so as to reduce tooling costs. What suggestions would you make?

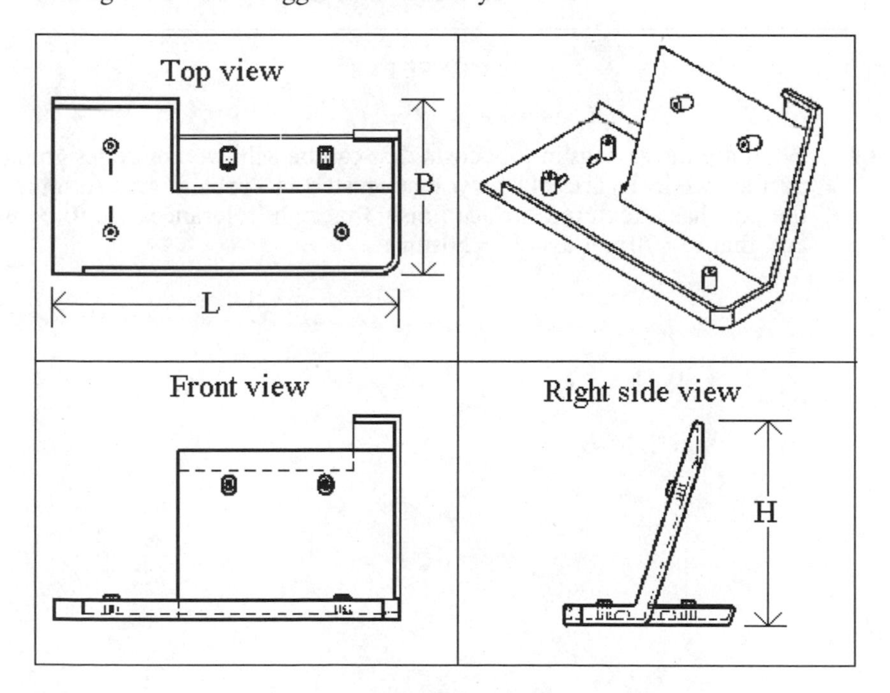

FIGURE P4.6

4.7 Estimate the savings in tooling costs achievable by the redesign suggestions you made in Problem 4.6. Assume that the part has a surface finish of SPI-3, that commercial tolerances are to be used, and that L = 250 mm, B = 130 mm, and H = 120 mm.

4.8 As a continuation of the redesign efforts of Widget A, as outlined in Problem 4.3, what suggestions can you make in order to reduce tooling costs for the proposed design of the part shown in Figure P4.8?

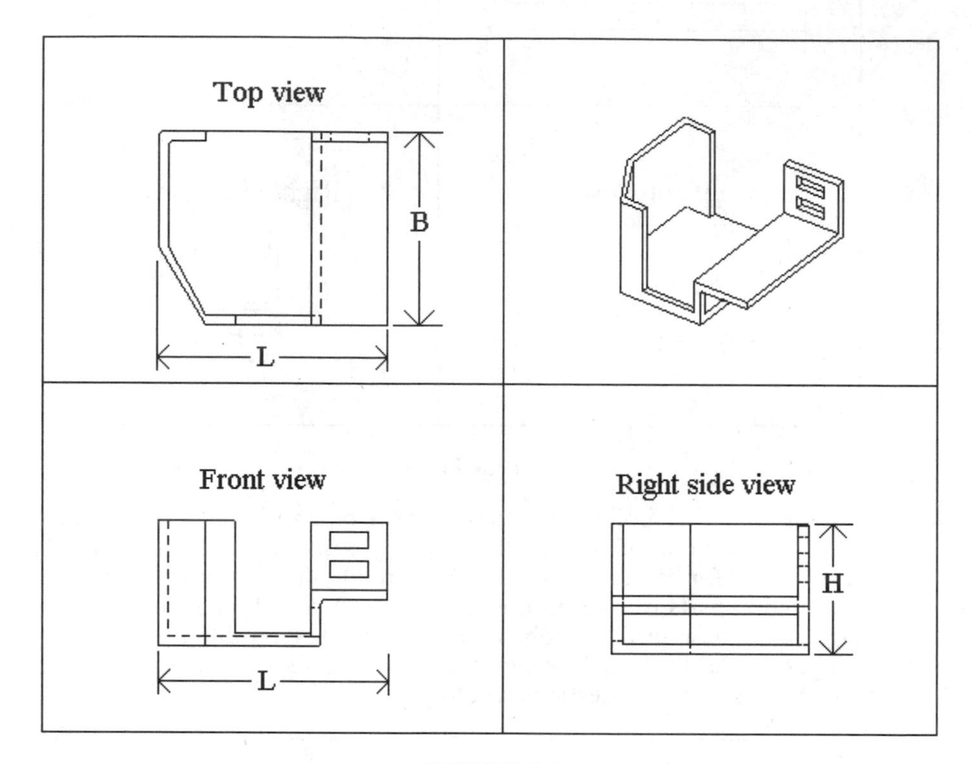

FIGURE P4.8

4.9 What are the savings in tool costs that can be achieved by redesigning the part shown in Figure P4.8 as you suggested in Problem 4.8? Assume that the part has a textured surface finish, that tight tolerances are to be used, and that L = 70 mm, B = H = 50 mm.

4.10 Determine the relative die cost for the part shown in Figure P4.10. Assume that the part has a surface finish of SPI-3, that commercial tolerances are to be used, and that L = 160 mm, B = 130 mm, and H = 13 mm.

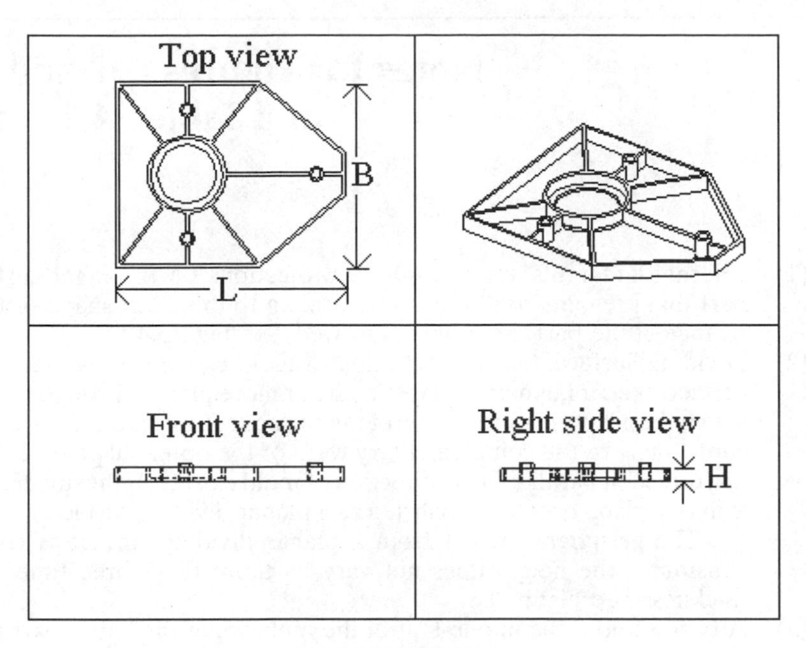

FIGURE P4.10

APPENDIX 4.A

Notes for Figures 4.1 and 4.19, and Tables 4.1 and 4.2

(1) Internal undercuts are recesses or projections on the inner surface of a part that prevents solid plugs, conforming to the exact shape of the inner surface of the part, from being inserted. See Figure 4.11.

(2) Dividing Surface. Given a direction of mold closure, a dividing surface is defined as an imaginary surface, in one or more planes, through the part, for which the portion on either side of the surface can be extracted from a cavity, conforming to the complementary form of the outer shape of the portion, in a direction parallel to the direction of mold closure. If the dividing surface is in one plane only, it is regarded as a planar dividing surface.

The peripheral height from a planar dividing surface is considered constant if the height does not vary by more than three times the wall thickness. See Figure 4.8.

(3) A part is said to be in one-half of the mold when the entire part is on one side of a planar dividing surface. See Figure 4.18.

(4) L is the longest dimension of the basic envelope of the part. If H is the smallest dimension of the basic envelope, then when L/H is greater than 4 the part is considered flat; otherwise it is considered box-shaped.

(5) External undercuts are holes or depressions on the external surface of a part that are not parallel to the direction of mold closure. Projections that are on the external surface of a part and are such that a single dividing surface, planar or nonplanar, cannot pass through all of them are also considered external undercuts.

The number of external undercuts is equal to the number of surfaces bearing unidirectional holes or depressions not in the direction of mold closure, and projections that prevent a single dividing surface from passing through all of the projections. See Figure 4.14.

(6) Cavity detail is a measure of the concentration of features parallel to the direction of mold closure. Typical features that increase cavity detail are ribs, bosses, and holes. See Figure 4.19.

(7) External undercuts other than unidirectional circular holes or depressions are considered extensive external undercuts. See Figure 4.21.

(8) A rib is a narrow elongated projection with a length generally greater than about three times its width (thickness), both measured parallel to the surface from which the feature projects, and a height less than six times its width. Ribs may be located at the periphery or on the interior of a part or plate. Peripheral ribs are not included in the rib count.

A narrow elongated projection with a height greater than six times its width is considered a wall.

A cluster of two closely spaced longitudinal ribs or lateral ribs, that is two ribs whose spacing is less than three times the rib width, are counted as one rib.

A boss is an isolated projection with a length of projection that is generally less than about three times its overall width, the latter measured parallel to the surface from which it projects. A boss is usually circular in shape, but it can take a variety of other forms called knobs, hubs, lugs, buttons, pads, or "prolongs."

Bosses can be solid or hollow. In the case of a solid circular boss, the length of the boss and its width are both equal to the boss diameter.

Bosses, and sometimes ribs, are supported by a cluster of ribs of variable height called gusset plates. This cluster of ribs is treated as one rib or one cluster of ribs. See Figures 4.5, 4.6, and 4.7.

(9) Holes in a component that do not need to be classified as undercuts are considered side shutoffs. Side shutoffs can be simple or complex. Isolated grooves or cutouts on the external surface of a part are also considered side shutoffs. A groove or cutout is considered isolated if the dimension of the cutout normal to the direction of mold closure is less than 0.33 times the envelope dimension in the same direction. Penalties due to simple side shutoffs are not considered for parts whose first digit is 2, 4, or 6. See Figures 4.16 and 4.17.

(10) All words and symbols at one location on the part are classified as a single lettering entity since the entire lettering pattern on the tooling will be made using one electrode.

APPENDIX 4.B

Worksheet for Relative Tooling Costs—Injection Molding

Original Design

Relative Die Construction Cost

Basic Shape	L =	B =	H =	Box/Flat
Basic Complexity	1st Digit =	2nd Digit =	C_b =	
Sub. Complexity	3rd Digit =	4th Digit =	C_s =	
T_a/R_a	5th Digit =	6th Digit =	C_t =	

Total relative die construction cost C_{dc}	$= C_b C_s C_t =$

Relative Die Material Cost

L_m =	B_m =	H_m =
Die closure parallel to	L_m/H_m =	Thus, C =

$M_{ws} = [0.006CH_m^4]^{1/3} =$
$M_{wf} = 0.04L_m^{4/3} =$
$M_a = (2M_{ws} + L_m)(2M_{ws} + B_m) =$
$M_t = (H_m + 2M_{wf}) =$

Thus,

C_{dm}=	$C_d = 0.8C_{dc} + 0.2C_{dm} =$

Redesign Suggestions

Basic Shape	L =	B =	H =	Box/Flat
Basic Complexity	1st Digit =	2nd Digit =	C_b =	
Sub. Complexity	3rd Digit =	4th Digit =	C_s =	
T_a/R_a	5th Digit =	6th Digit =	C_t =	

Total relative die construction cost C_{dc}	$= C_b C_s C_t =$
$C_d = 0.8C_{dc} + 0.2C_{dm} =$	
% Savings =	

Chapter 5

Injection Molding: Total Relative Part Cost

5.1 INJECTION-MOLDED PART COSTS

5.1.1 Introduction

As we learned in the previous chapter, the first stage of a manufacturability evaluation for injection-molded parts is an evaluation of tooling costs. It can be done at the configuration design stage where only approximate dimensions, locations, and orientations of features are known. At the parametric stage, making use of the near final dimensions, locations, and orientation of features, a manufacturing evaluation of the relative cost to process a part can be made. Then the total cost of a part can be computed as the sum of the per part tooling costs, processing costs, and material costs.

As in the previous chapter, throughout this chapter we will be dealing with the concept of relative part cost. Relative cost, you will recall, was defined as the cost of your current part compared with the cost of some standard part. The standard or reference part used in the previous chapter was a 1 mm thick flat washer whose outer and inner diameters are 72 mm and 60 mm, respectively.

As was pointed out before, actual part costs depend upon local practices and methods and can vary considerably from one plant or location to another. Since the main objective here is to develop a methodology for making design decisions among various competing alternative designs, actual costs aren't necessarily required. In general, it suffices to have an appreciation for the cost drivers associated with the particular process under consideration so that the relative costs of competing designs can be compared. In this way informed design decisions can be made, better original designs will be proposed, and ultimately unnecessary redesigns will be avoided.

5.1.2 Processing Costs

Processing costs (sometimes called operating costs) are the charges for use of the injection molding machine used to produce the part. They depend on the machine hourly rate, C_h (\$/hr), and the effective cycle time of the process, t_{eff}. The effective cycle time is the machine cycle time, t, divided by the production yield, Y. Production yield, or just yield, is the fraction of the total parts produced that are satisfactory and, hence, usable. Thus,

$$\text{Processing cost per part, } K_e = C_h t_{eff} = C_h(t/Y) \quad \text{(Equation 5.1)}$$

where

Table 5.1 Data for the reference part.

Material	*Polystyrene*
Material Cost (K_{po})	1.46×10^{-4} cents/mm^{3}(1)
Vol (V_o)	1244 mm^3
Die Material Cost (K_{dmo})	\$980(2)
Die Construction Time (Includes design and build hours)	200 hours(2)
Labor Rate (Die Construction)	\$30/hr(2)
Cycle time (t_o)	16 s(2)
Mold Machine Hourly Rate (C_{ho})	\$27.53(3)

(1) *Plastic Technology*, June 1989; (2) Data from collaborating companies; (3) *Plastic Technology*, July 1989.

Y = Production Yield (usable parts/total parts produced)

Part surface "quality" requirements and tolerances are the main causes for variations in production yield. A low yield reduces the number of acceptable parts that are produced in a given time, and thus increases the "effective cycle time" to a value higher than the actual machine cycle time, t. Increases of 10% to 30% in the effective cycle time for a given part are typical. The reasons for this increase are discussed in greater detail in Section 5.9.

The relative processing cost is the cost of producing a part relative to the cost of producing a reference part. Relative processing cost, C_e, can be expressed as:

$$C_e = \frac{tC_h}{t_o C_{ho}} = t_r C_{hr} \qquad \text{(Equation 5.2)}$$

where t_o and C_{ho} represent the cycle time and the machine hourly rate for the reference part, C_{hr} represents the ratio C_h/C_{ho}, and t_r is the total relative cycle time for the part compared with the reference part; that is:

$$t_r = \frac{t}{t_o} \qquad \text{(Equation 5.3)}$$

The reference part in this case is the same flat washer used as a reference part in Chapter 4: a 1-mm-thick flat washer whose outer and inner diameters are 72 mm and 60 mm, respectively. Some additional data (part material, material cost, tooling cost, etc.) for the reference part are given in Table 5.1.

5.1.3 Material Costs

The material cost for a part, K_m, is given by

$$K_m = VK_p \qquad \text{(Equation 5.4)}$$

where V is the part volume and K_p is the material cost per unit volume. Thus, if the subscript "o" is used to indicate the reference part, then the relative material cost can be expressed as

$$C_m = \frac{K_m}{K_{mo}} = \left(\frac{V}{V_o}\right)\left(\frac{K_p}{K_{po}}\right) = \left(\frac{V}{V_o}\right)C_{mr} \qquad \text{(Equation 5.5)}$$

Table 5.2 Relative material prices, C_{mr}, for engineering thermoplastics. (Based on material prices in *Plastics Technology*, June 1990.)

Material	C_{mr}
ABS	1.71
Acetal	2.92
Acrylic	1.54
Nylon 6	2.79
Polycarbonate	2.96
Polyethylene	0.71
Polypropylene	0.62
Polystyrene	1.00
PPO	2.33
PVC	0.62

Table 5.2 contains the relative material prices for the most often used engineering thermoplastics. The prices are all relative to polystyrene.

5.1.4 Total Cost

The total production cost of a part, K_t, can be expressed as the sum of the material cost of the part, K_m, the tooling cost, K_d/N, and processing cost, K_e, where K_d represents the total cost of the tool and N represents the production volume or total number of parts produced with the tool or mold. Thus,

$$K_t = K_m + \frac{K_d}{N} + K_e \qquad \text{(Equation 5.6)}$$

If the manufacturing cost of a reference part is denoted by K_o, then the rclative total cost of the part, C_r, can be expressed as:

$$C_r = \frac{K_m + K_d/N + K_e}{K_o} \qquad \text{(Equation 5.7)}$$

In the remainder of this chapter, we present methods for computing the relative costs for injection-molded parts. We have already described, in Chapter 4, "Injection Molding: Relative Tooling Cost," how to estimate relative tooling costs for injection-molded parts.

A prerequisite step to determining total relative processing cost is the determination of the total relative cycle time. Thus we will begin in the next section with relative cycle time.

5.2 DETERMINING TOTAL RELATIVE CYCLE TIME (t_r) FOR INJECTION-MOLDED PARTS—OVERVIEW

As noted in the previous chapter, statistical studies of tooling cost as a function of part geometries have shown that tooling cost is more a function of

1. Overall part size (small, medium, or large) and shape (flat or box);
2. Presence and location of holes and projections that, depending upon their location, can lead to undercuts;

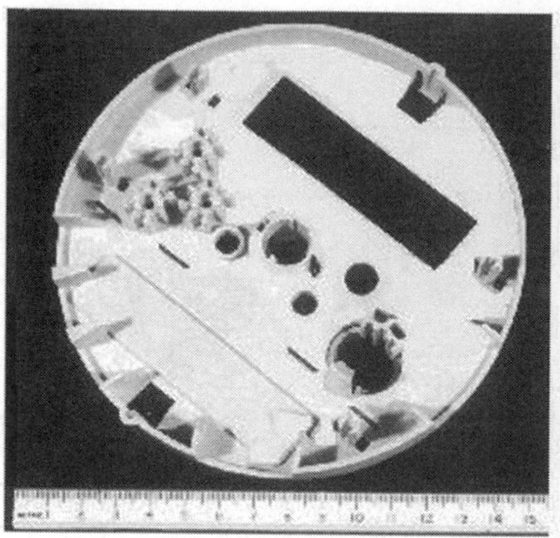

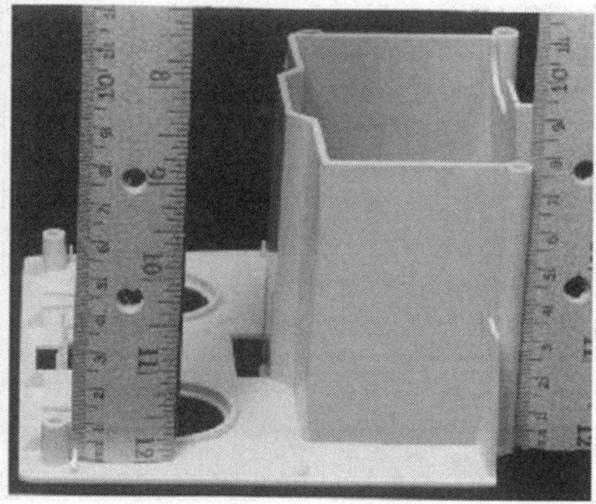

FIGURE 5.1 *Photographs of two injection-molded parts.*

3. Mold closure direction and parting surface location that can also, if poorly selected, lead to undercuts;

and less a function of localized features and details. On the other hand, similar statistical studies of processing costs as a function of part geometries have shown that cycle times are more a function of the localized features and details of parts. For example, in the case of parts such as those shown in Figure 5.1, the question becomes, is there a particular part feature (rib, boss, wall, etc.) whose wall thickness, height, layout, and other features result in a solidification time for that feature to be greater than all other part features? For it is the feature that takes longest to solidify that determines the cycle time of the part.

Efforts to emulate the conditioned knowledge possessed by molders, based upon years of experience, has led to a part-coding system for determining injection-molding processing costs similar to the coding system used for tooling costs. The mechanics of using the system are explained in the next several subsections. The overall result, however, is that the total cycle time relative to the standard or reference part, t_r, is obtained as a function of three parameters:

1. The *basic relative cycle time*, t_b
2. An *additional relative cycle time*, t_e, due to the presence of inserts and internal threads, and
3. A *multiplying penalty factor*, t_p, to account for the effects of part surface quality and tolerances.

In terms of these parameters, the *total relative cycle time*, that is the cycle time relative to the reference part, t_r, is given by

$$t_r = (t_b + t_e)t_p \qquad \text{(Equation 5.8)}$$

The basic relative cycle time, t_b, is given by values found in one of the three matrices shown in Figure 5.2, which define the first and second digits of the coding system. Note that to use this figure, the meaning of a number of basic terms must be understood, including: partitionable and non-partitionable parts; slender (S), non-slender (N), and frame-like parts; elemental plates; part thickness (w); grilles and slots; ribs and types of ribs; gussets; significant ribs and bosses; and easy-versus difficult-to-cool parts. We also must recall the definition of the basic enve-

SECOND DIGIT

Wall Thickness

FIRST DIGIT

(a) Slender Partitionable Parts, S

				$w < 1$ mm	$1\text{mm} \le w \le 2\text{mm}$	$2\text{mm} < w \le 3\text{mm}$	$3\text{mm} < w \le 4\text{mm}$	$4\text{mm} < w \le 5\text{mm}$	$w > 5\text{mm}$
				0	1	2	3	4	5
Parts with $L_u/B_u \ge 10$ (1) or Frames (2)	Plates with $L_u/2w < 100$ (3) without lateral projections (4)	Without ribs	0	Difficult to fill or eject	1.00	1.35	1.70	2.55	Use foamed materials (9)
		With ribs	1		1.10	1.45	1.85	2.70	
	Plates with $L_u/2w \ge 100$ and/or plates with lateral projections (4)	Without ribs	2		1.15	1.55	2.00	2.85	
		With ribs	3		1.25	1.82	2.20	3.85	

(b) Non-Slender Partitonable Parts, N

					$w < 1$ mm	$1\text{mm} \le w \le 2\text{mm}$	$2\text{mm} < w \le 3\text{mm}$	$3\text{mm} < w \le 4\text{mm}$	$4\text{mm} < w \le 5\text{mm}$	$w > 5\text{mm}$
					0	1	2	3	4	5
Parts with $\frac{L_u}{B_u} < 10$ (1)	Plates without significant rib (5) or significant bosses (6) with or without non-peripheral ribs or bosses	Plates which are grilled or slotted (7)	Without non-peripheral ribs	0	Difficult to fill or eject	1.68	2.39	3.11	3.82	Use foamed materials (9)
			With non-peripheral ribs	1		1.82	2.53	3.25	3.96	
		Plates which are not grilled or slotted (7)	Without non-peripheral ribs	2		1.96	2.67	3.39	4.10	
			With concentric or cross ribbing	3		2.10	2.81	3.53	4.24	
			With radial or unidirectional ribbing	4		2.24	2.96	3.67	4.39	
	Plates with significant ribs (5) and/or significant bosses (6)	Plates with rib and/or boss thickness less than the wall thickness (8)	Ribs/bosses supported by gusset plates	5		2.38	3.10	3.81	4.53	
			Ribs/bosses not supported by gusset plates	6		2.52	3.24	3.95	4.67	
		Plates with rib and/or boss thicknes (8) greater than or equal to the wall thickness		7		2.66	3.38	4.09	4.81	

(c) Non-Partitionable Parts

			$w < 1$ mm	$1\text{mm} \le w \le 2\text{mm}$	$2\text{mm} < w \le 3\text{mm}$	$3\text{mm} < w \le 4\text{mm}$	$4\text{mm} < w \le 5\text{mm}$	$w > 5\text{mm}$
			0	1	2	3	4	5
Parts which are not partitionable	Easy to cool (10)	0	Difficult to fill or eject	2.66	3.38	4.09	4.81	Use Foamed Materials
	Difficult to cool (10)	1		3,56	4.50	5.47	6.40	

1 in = 25.4 mm; 1 mm/25.4 mm = 0.04 in

FIGURE 5.2 *Classification system for basic relative cycle time, t_b. (The numbers in parentheses refer to notes found in Appendix 5.A.)*

Table 5.3 Additional relative time, t_e, due to inserts and internal threads. (The numbers in parentheses refer to notes found in Appendix 5.A.)

Third Digit	Parts without internal threads (11)	Without molded-in inserts (12)	0	0.0
		With molded-in inserts (12)	1	0.5*
	Parts with internal threads (11)	Without molded-in inserts (12)	2	0.1*
		With molded-in inserts (12)	3	0.1*/0.5*

Table 5.4 Time penalty, t_p, due to surface requirements and tolerances. (The numbers in parentheses refer to notes found in Appendix 5.A.)

					Fifth Digit	
					Tolerances not difficult to hold (14)	Tolerances difficult to hold (14)
					0	1
Fourth Digit	Plate surface requirements (13)	Low		0	1.00	1.20
		H	$1\,mm \leq w \leq 2\,mm$	1	1.30	1.43
		i	$2\,mm < w \leq 3\,mm$	2	1.22	1.41
		g	$3\,mm < w \leq 4\,mm$	3	1.16	1.37
		h	$4\,mm < w \leq 5\,mm$	5	1.10	1.32

lope of a part that was introduced in Chapter 4, "Injection Molding: Relative Tooling Cost."

The additional relative time due to inserts and internal threads, t_e, is found from Table 5.3. Table 5.3 defines the third digit of the coding system.

The time penalty factor, t_p, to account for surface requirements and tolerances, is found in Table 5.4, which defines the fourth and fifth digits in the coding system. To use Table 5.4, we must be able to distinguish between tolerances that are "easy" to hold and those that are more "difficult" to hold. We must also be able to distinguish surface finish requirements that are "low" from those that are "high." These issues are discussed and explained in Section 5.9.

As in the tooling evaluation system, the value of the basic relative cycle time, t_b, for the reference part is 1.00; it can be found in the upper-left-hand corner for slender or frame-like parts. (The washer chosen as the reference part is a frame-like part.) Note that the values for t_b increase significantly as one moves down and to the right of the matrix. This information, again, helps guide designers to redesigns that can reduce relative cycle time, hence, processing costs.

In the next few subsections, we will discuss the meaning of these terms, and then illustrate the use of the coding system in the evaluation of several example parts.

5.3 DETERMINING THE BASIC PART TYPE: THE FIRST DIGIT

In order to focus our attention on that particular feature or detail of a part that controls cycle time, when possible, parts are decomposed or partitioned into a series of elemental plates (Figure 5.3). The concept here is that every elemental plate has its own corresponding cooling time, hence, the plate with the longest

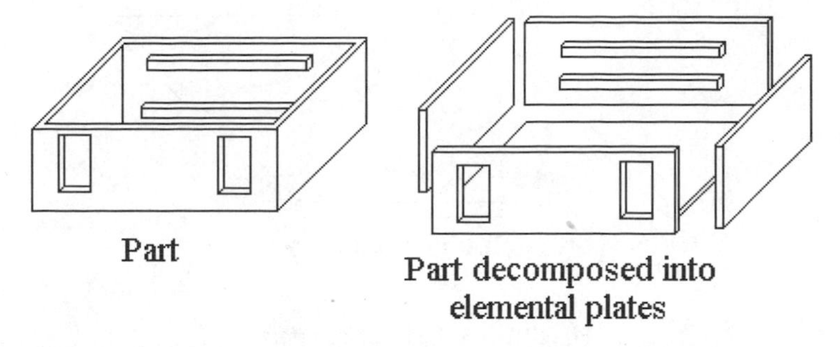

FIGURE 5.3 *Example of a part decomposed into a series of elemental plates.*

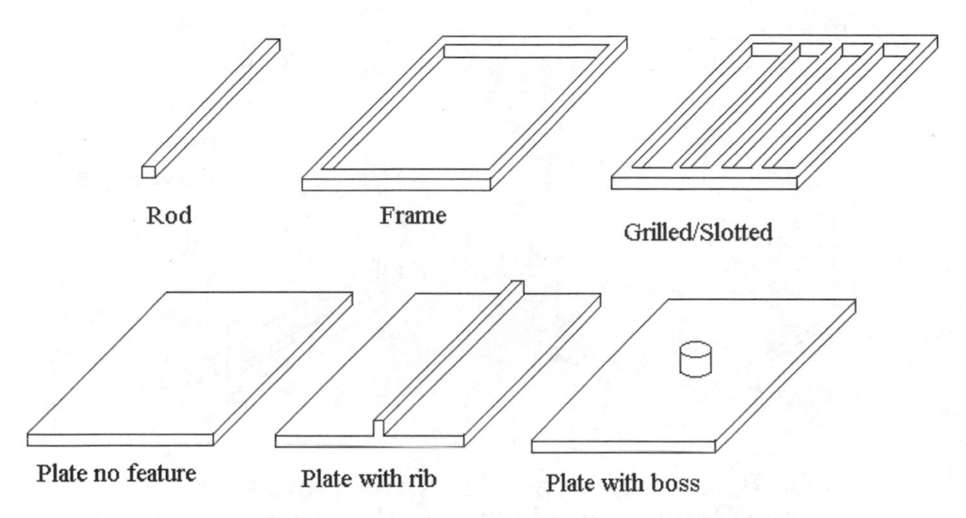

FIGURE 5.4 *Examples of elemental plates.*

cooling time controls the cycle time for the part. Thus, although cooling time for a plate is "local" its effect on the cycle time is "global."

Not all parts can be easily separated or partitioned into elemental plates, hence, partitionable parts are defined as those parts that can be easily and completely (except for add-ons like bosses, ribs, etc.) divided into a series of elemental plates. An elemental plate is a contiguous thin flat wall section whose edges are either not connected to other plates, or are connected via distinct intersections (e.g., corners). An elemental plate may have add-on features like holes, bosses, or ribs.

Examples of several types of elemental plates are shown in Figure 5.4. A method for partitioning a part into its elemental plates is described in Section 5.4.

For each elemental plate in a part, we will determine a cooling or solidification time. The plate with the longest cooling time controls the machine cycle time of the entire part. Thus, every elemental plate should be carefully designed so that its individual solidification time is minimized.

Partitionable parts are further classified as either slender or frame-like (S), or non-slender (N).

To distinguish quantitatively between slender and non-slender parts (see Figure 5.5 for a qualitative distinction), we consider the basic envelope of the part as shown in Figure 5.6. Given a basic envelope of dimensions (if you've forgotten what the definition of the basic envelope is you may want to reread

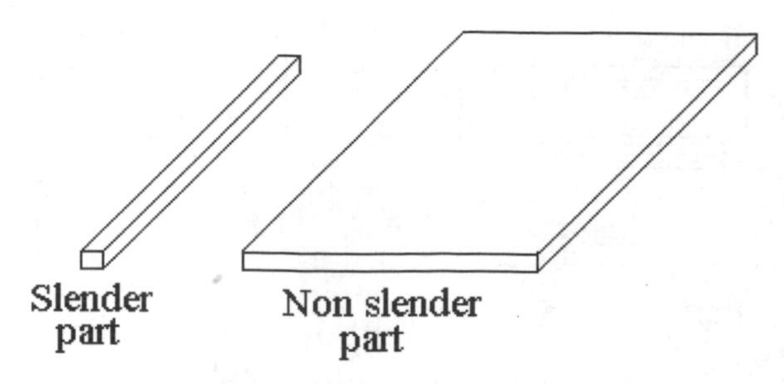

FIGURE 5.5 *A slender part.*

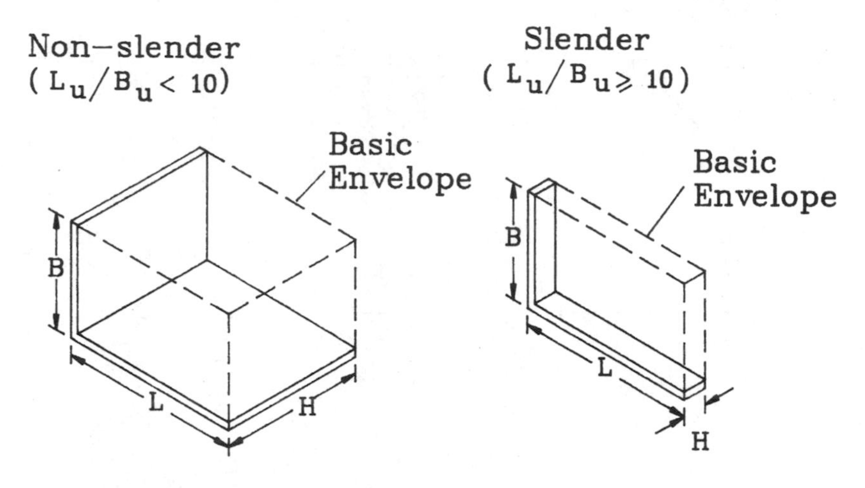

FIGURE 5.6 *Definition of a slender part.*

Section 4.3.2 of Chapter 4) L, H, and B, a slender partitionable part is one for which

$$L_u/B_u \geq 10$$

where

$$L_u = L + B$$

and

$$B_u = H$$

For parts with a bent or curved longitudinal axis, the unbent length, L_u, is the maximum length of the part with the axis straight (Figure 5.6). The width of this unbent part is referred to as B_u.

Frames are parts or elemental plates that have a through hole greater than 0.7 times the projected area of the part/plate envelope and whose height is equal to its wall thickness (Figures 5.7 and 5.8).

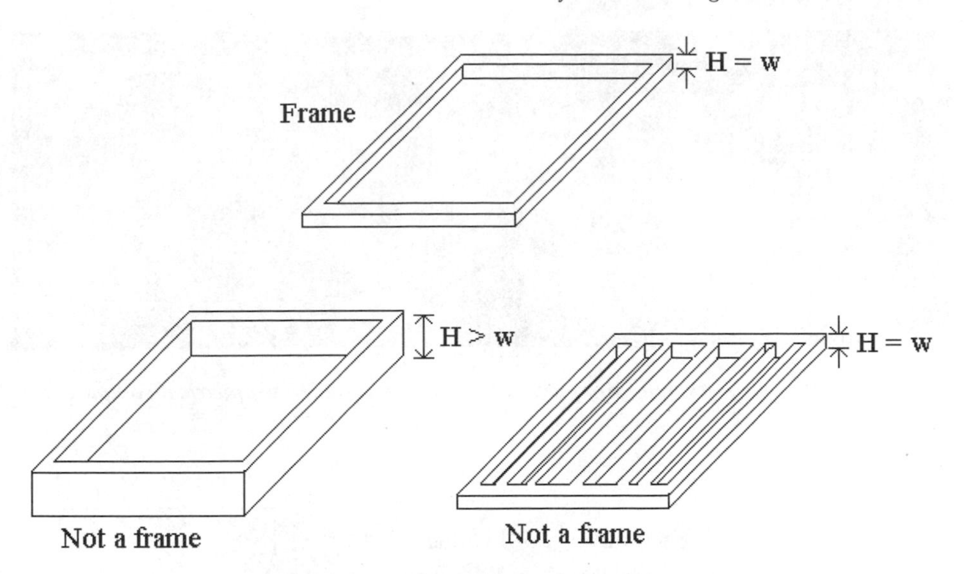

FIGURE 5.7 *Example of a frame-like part.*

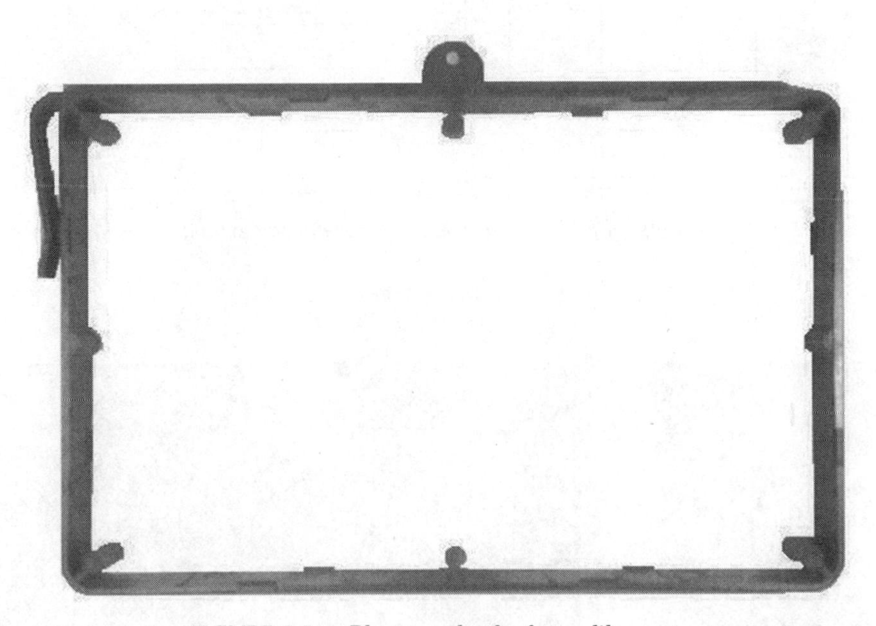

FIGURE 5.8 *Photograph of a frame-like part.*

Molds for slender parts and frames are generally more difficult to fill than molds for non-slender parts; however, slender parts are generally easier to cool.

Non-partitionable parts include parts with complex geometries, or parts with extensive subsidiary features such that they cannot be easily partitioned into elemental plates. We will also consider parts that have simple geometric shapes but contain certain difficult-to-cool features as non-partitionable. Examples of non-partitionable parts are shown in Figures 5.9, 5.10, and 5.11. A more complete discussion of non-partitionable parts is given in Section 5.5.

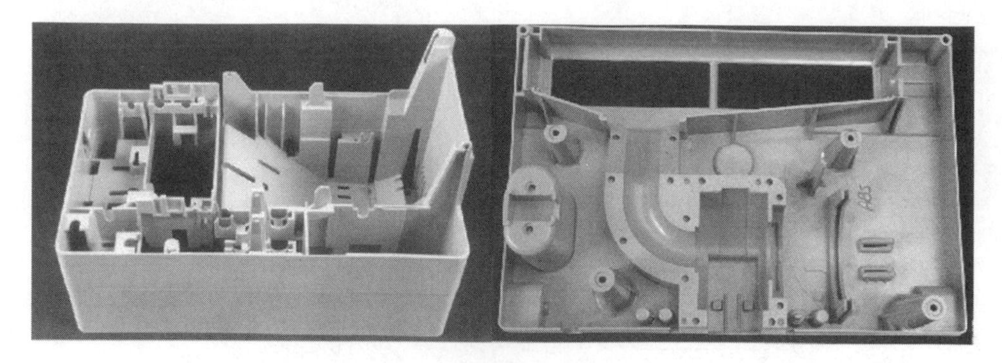

FIGURE 5.9 *Two examples of non-partitionable parts due to geometrical complexity.*

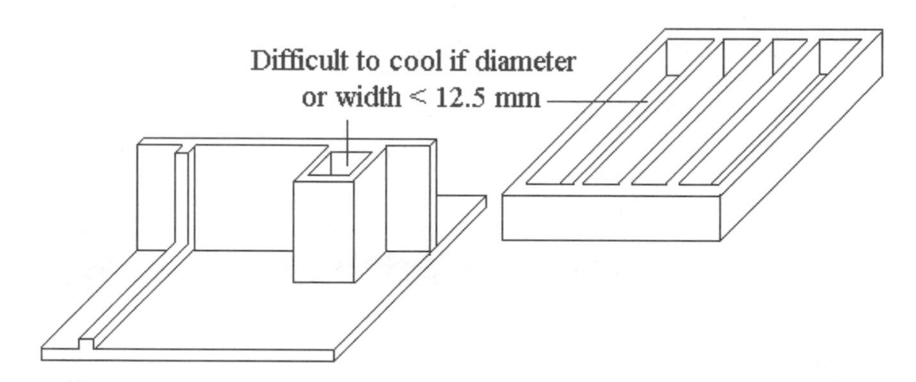

FIGURE 5.10 *Examples of non-partitionable parts due to extremely difficult to cool features.*

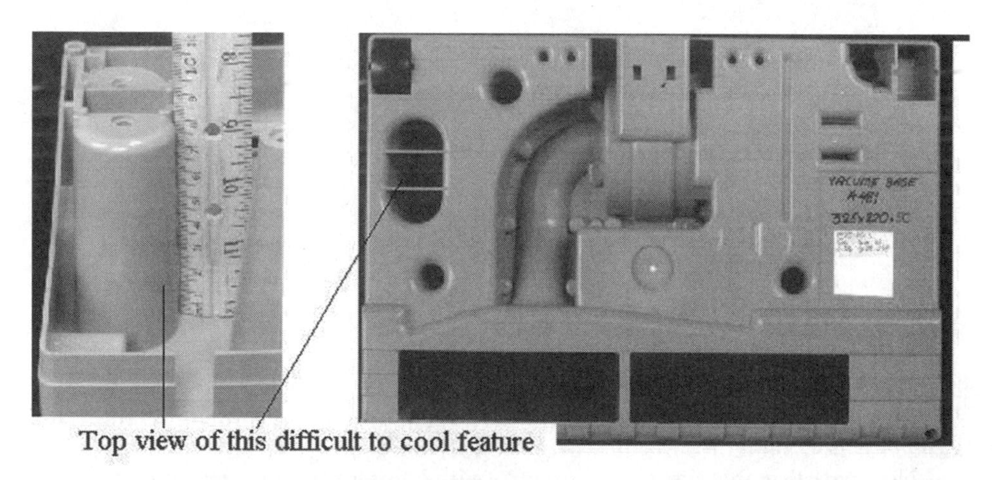

FIGURE 5.11 *Photograph of a part with a difficult to cool feature.*

5.4 PARTITIONING PARTITIONABLE PARTS

Part partitioning is a procedure for dividing or partitioning a part into a series of elemental plates similar to the ones shown in Figure 5.4. The procedure works best on parts with an uncomplicated geometrical shape, and features that can be cooled using cooling channels, bubblers, or baffles, as described below.

To partition or divide a part into a series of plates (see Figure 5.4) we proceed as follows:

1. Determine whether the part is slender or non-slender.
2. Divide the part into a series of plates with their corresponding add-on features (ribs, bosses, etc.).

Figure 5.12 shows a hollow rectangular prismatic part comprised of four side walls and a base. The four walls and base have equal and constant wall thicknesses. Also shown in Figure 5.12 are two alternative divisions of the part into elemental plates. To understand that the divisions are equivalent, it is necessary to understand how parts are cooled.

Figure 5.13 shows a schematic of a cooling system for such a part. Although the side walls (external plates) are efficiently cooled by cooling channels running parallel to the walls, the base (internal plate) needs to be cooled by more sophisticated but less efficient units such as baffles and bubblers. Thus, although plates (1), (2), (3), and (5) (Figure 5.12) are geometrically similar, plate (5) is a more difficult to cool internal plate, whereas the others are easier to cool external plates. Although plate (4) is also an external plate, it is a grilled plate.

Although the above explanation indicates that a difference in cooling time should exist between internal and external plates, in practice the increase in cycle time for injection-molded parts was not found to be statistically significant. This may be due in part to the fact that most engineering type partitionable parts have an L/H ratio small enough so that reasonable cooling occurs in any case.

Figure 5.14 shows the partitioning of some additional parts whose part envelope is rectangular. Figure 5.15 shows the partitioning of parts whose part

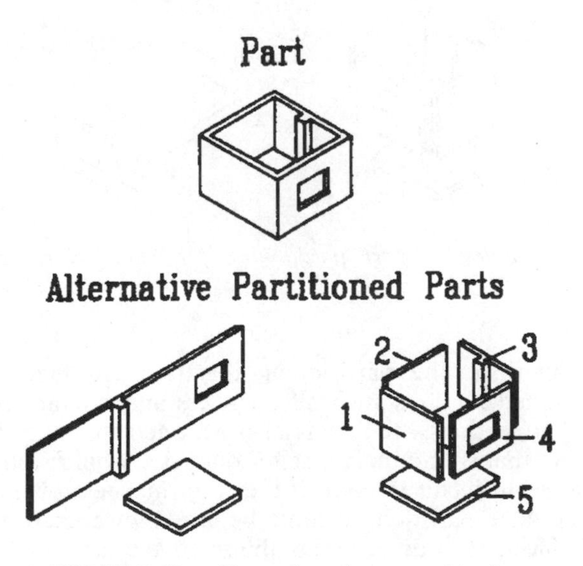

FIGURE 5.12 *Examples of part partitioning.*

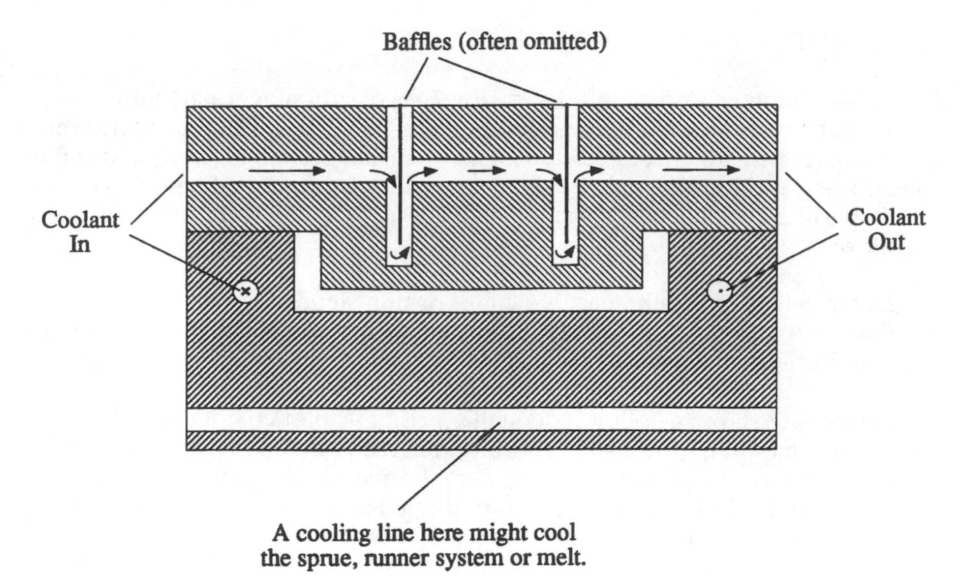

FIGURE 5.13 *Cooling system for part in Figure 5.12.*

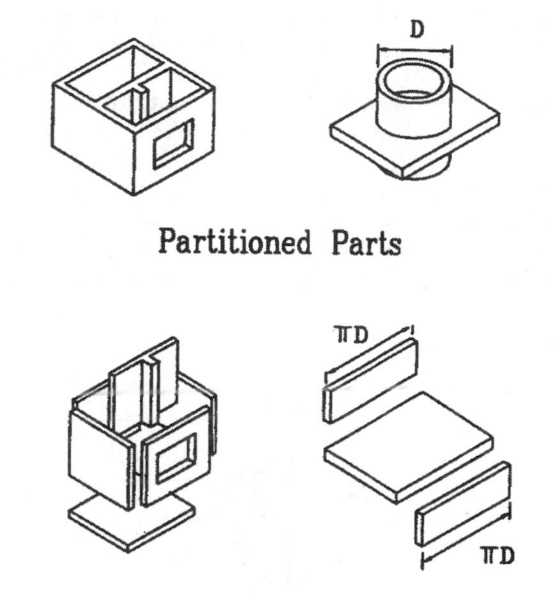

FIGURE 5.14 *Examples of part partitioning for parts whose part envelope is rectangular.*

envelope is cylindrical. The partitioning illustrated in Figure 5.15 assumes that the cylinders have a constant wall thickness and similar sets of subsidiary features (holes, projections, etc.). It is also assumed that the diameter of the cylinder is greater than 12.5 mm. A smaller diameter would result in a part that would be extremely difficult to cool. If the wall thickness of the cylinder were not constant, the same partitioning could be used; however, one would use the maximum wall thickness in determining the relative cycle time of the elemental plate.

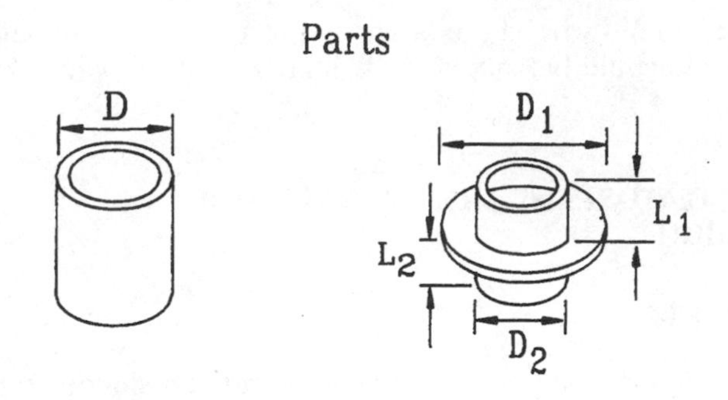

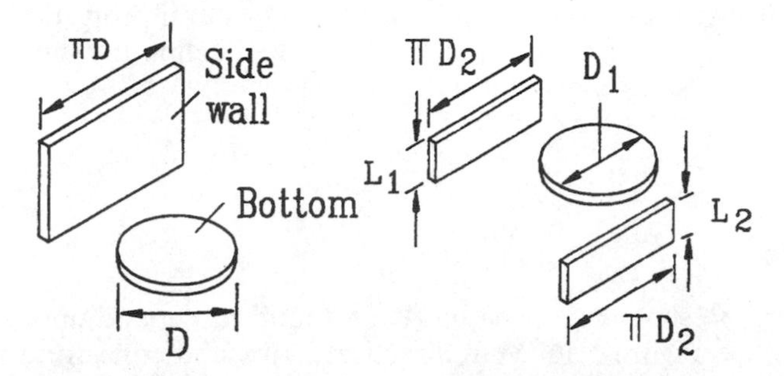

FIGURE 5.15 *Examples of part partitioning for parts whose part envelope is cylindrical.*

Before leaving this section it should be pointed out that slender parts (section (a) of Figure 5.2) tend to solidify faster than non-slender parts (section (b) of Figure 5.2). Slender parts cool faster because they are usually edge-gated, and non-slender parts are usually center-gated. Edge-gating permits cooling lines to be run along the top and bottom surfaces of the part. As indicated in Figure 5.13, center-gating precludes the use of cooling lines in the vicinity of the sprue and gate.

5.5 NON-PARTITIONABLE PARTS

Not all parts can be easily visualized as comprised of a series of elemental plates as depicted in Figure 5.4. That is to say, not all parts are partitionable. Some contain geometrically complex shapes (Figure 5.9) and others contain some extremely difficult to cool features (Figure 5.10) where even baffles and bubblers cannot be used. These parts are generally harder to cool and may produce difficulties in maintaining other requirements related to warping, surface finish, tolerances, and so on. These types of parts have relative cycle times greater than those predicted by use of the coding system applied to slender and non-slender partitionable parts. The coding system can, however, be used to determine the lower bound for the relative cycle time of non-partitionable parts. This is done by determining

1. The relative cycle time for that portion of the part that is "easy-to-cool," and
2. The relative cycle time for the most difficult to cool feature.

The lower bound will be the larger of the values determined in (a) and (b). The actual value could be 25% to 50% higher than the values obtained via the coding system.

5.6 OTHER FEATURES NEEDED TO DETERMINE THE FIRST DIGIT

5.6.1 Introduction

The ability to identify (and partition) slender and non-slender partitionable parts enables us to determine whether the basic relative cycle time, t_b, will be found in section (a), (b), or (c) of Figure 5.2. However, to completely identify the first digit and get the actual value of t_b, we must also be able to determine the presence or absence of such features as ribs (and their types), lateral projections, grilles and slots, and gussets. Therefore, in this section, we will define the meaning of these terms as used in the coding system.

5.6.2 Ribs

Types of Ribs

When ribs are present, cycle time has been found to depend upon the types of ribs present. See Figure 5.16. Multidirectional ribs and concentric ribs provide greater rigidity than unidirectional ribs and radial ribs. For this reason multidirectional ribbing is often preferred as a means of stiffening parts with thin walls. (A properly designed thin wall with ribs can be lighter than a non-ribbed thicker

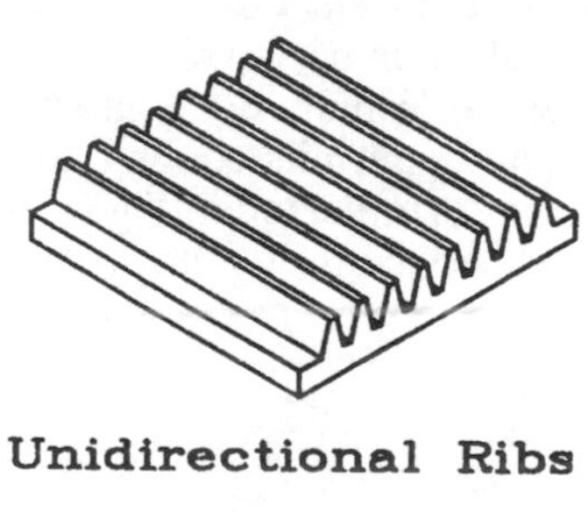

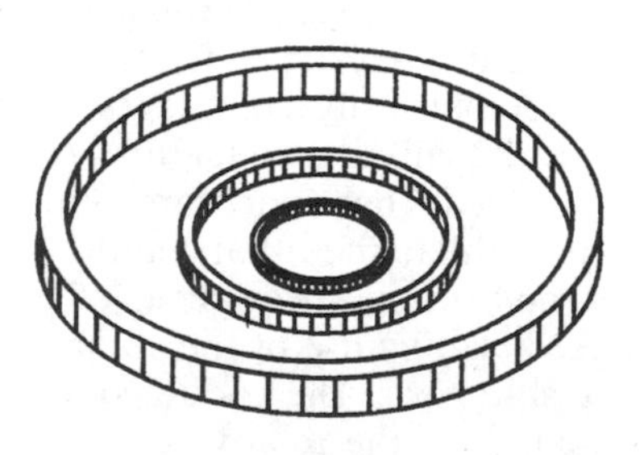

FIGURE 5.16 *Types of ribbing.*

wall with the same stiffness. In addition, plates with multidirectional ribbing have less of a tendency to warp than plates with unidirectional ribbing, thereby making shorter cycle times possible.) From the point of view of machine cycle time, peripheral ribs can be treated as walls because they do not increase the localized wall thickness of the part.

Significant Ribs

Significant ribs tend to increase the machine cycle time, because of increased local wall thickness at their base, and, in the case of unsupported unidirectional ribs, make tolerances difficult to hold. In addition, such ribs can be difficult to fill and result in shallow depressions (called sink marks) on the surface of the part; these are due to the collapsing of the surface following local internal shrinkage. Sink marks reduce part quality and should generally be avoided. To avoid sink marks ribs should be designed so that (1) the rib height, h, is less than or equal to three times the localized wall thickness, w, and (2) the rib width, b, is less than or equal to the localized wall thickness. Thus, we call ribs designed such that [3w < h < 6w] or [b > w] significant ribs (Figure 15.17).

5.6.3 Gussets

A rib of variable height, usually present at the junction of two elemental plates, is called a gusset plate (Figure 5.18). A gusset plate can also be present at the junction between a projection (boss, rib, etc.) and the wall to which it is attached (see Figures 4.5 and 4.7). Gusset plates facilitate mold filling, help hold tolerances, and provide stiffening that may permit a reduction in the thickness of bosses and walls.

5.6.4 Bosses

For bosses, cycle time is influenced by whether they are supported by gusset plates (see Figures 5.18 and 4.7). Plates with large (tall) projections are difficult to cool because these features significantly increase localized wall thickness, are difficult to fill, and have a tendency to warp. Significant projections should be

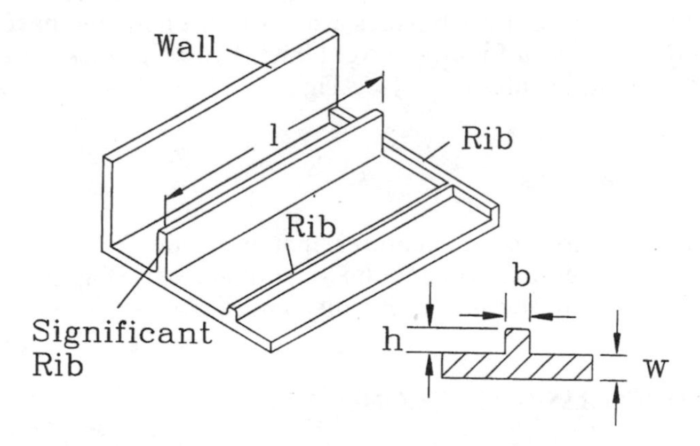

FIGURE 5.17 *Ribs. The length of the rib is denoted by l, the rib width by b, the rib height by h, and the wall thickness of the plate by w. A significant rib is one where 3w < h < 6w or b > w.*

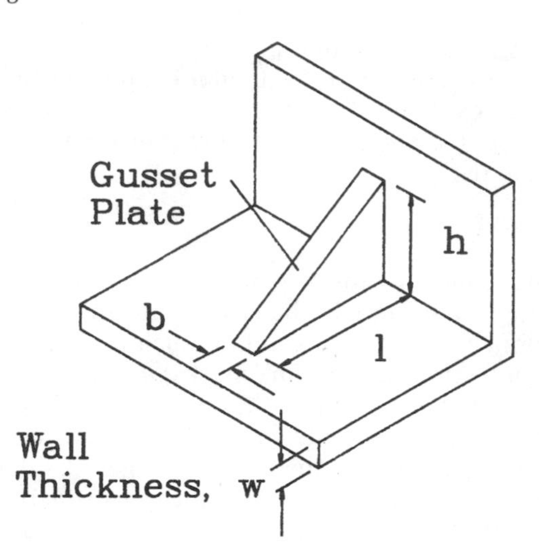

FIGURE 5.18 *Gusset Plate. The thickness of the gusset plate is denoted by b, the height by h, and the length by l. The wall thickness of the plate is w.*

supported by gusset plates to reduce their tendency to deflect due to residual stresses.

Significant Bosses

We consider bosses designed such that [$3w < h$] or [$b > w$] as significant bosses (Figure 5.19). Significant bosses tend to increase the machine cycle time, make tolerances difficult to hold, and can be difficult to fill. In addition, significant bosses often cause sink marks.

5.6.5 Grilles and Slots

Elemental plates that contain (1) multiple through holes, (2) no continuous solid section with a projected area greater than 20% of the projected area of the plate envelope, and (3) whose height is equal to its wall thickness are considered grilled/slotted (Figure 5.20). Such plates are low in strength and have low surface gloss due to the presence of multiple weld lines. The weld lines are caused when the flows from multiple gates meet (see Figure 5.21).

5.6.6 Lateral Projections

Lateral projections are add-on features that protrude from the surface of a slender plate in a direction normal to the longitudinal axis (Figure 5.22). For long slender parts, such projections are difficult to fill.

5.7 WALL THICKNESS—THE SECOND DIGIT

The definitions in Section 5.6 will enable designers to determine the basic part type, and to establish the first digit of the coding system in Figure 5.2. Determining the second digit is essentially based on the largest wall thickness (w) of the elemental plates.

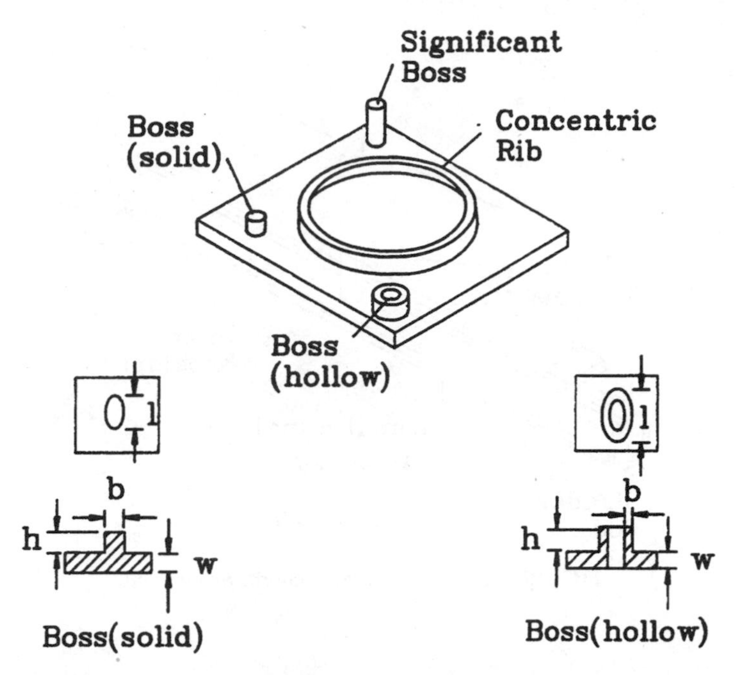

FIGURE 5.19 *Bosses. The boss length is denoted by l, the width by b_w, the wall thickness by b, the height by h, and the wall thickness of the plate by w.*

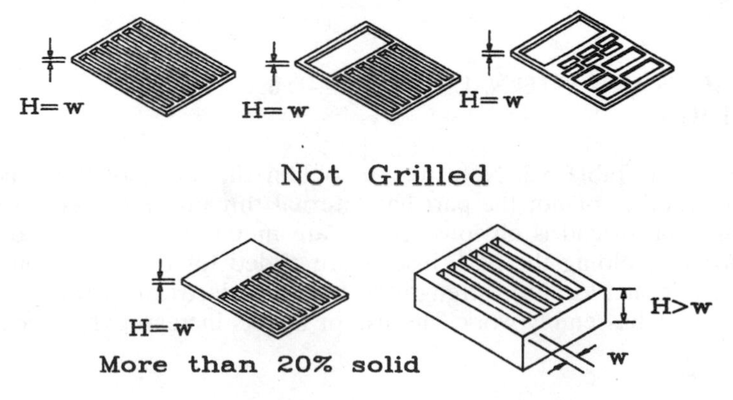

FIGURE 5.20 *Examples of grilled or slotted parts.*

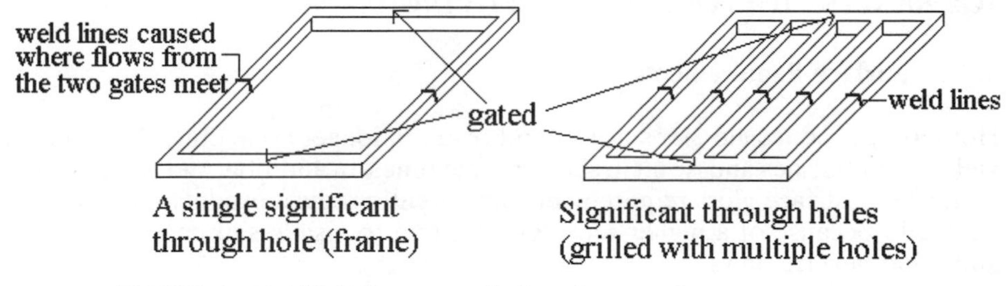

FIGURE 5.21 *Weld lines caused when the flows from two gates meet.*

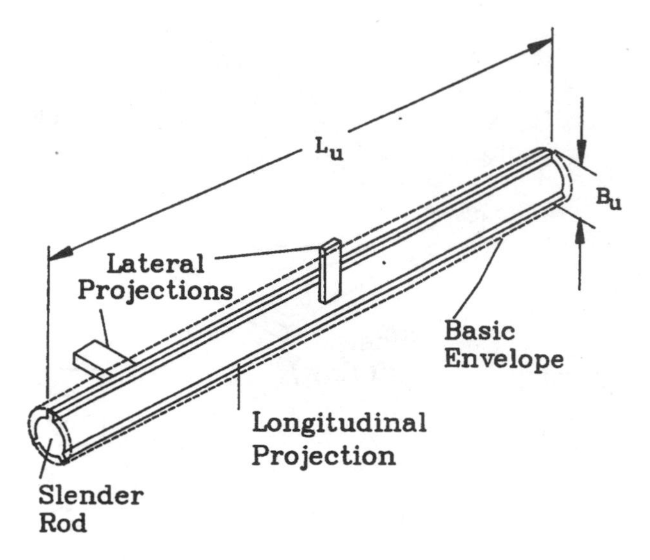

FIGURE 5.22 *Example of lateral projections.*

In the case of non-partitionable parts, two wall thicknesses need to be used. (Remember, we need to determine the cycle time based on the easy-to-cool features and the cycle time based on the difficult-to-cool features.) One wall thickness is the largest wall thickness of all the easy-to-cool portions of the part. The other wall thickness is the largest wall thickness of all the difficult-to-cool features.

5.8 INSERTS AND INTERNAL THREADS—THE THIRD DIGIT

Refer again to Table 5.3. Note that to obtain the value of t_e we need only ascertain whether or not the part has internal threads and inserts. The meaning of internal thread is obvious. Inserts are metal components added to the part prior to molding the part. Inserts are added for decorative purposes, to provide additional localized strength, to transmit electrical current, or to aid in assembly or subassembly work. The use of inserts increases the machine cycle time.

5.9 SURFACE REQUIREMENTS AND TOLERANCES—THE FOURTH AND FIFTH DIGITS

5.9.1 Surface Requirements

Hot molds produce glossier surfaces and result in longer cycle times. Cool molds yield duller finishes and result in shorter cycle times. In addition, and more importantly, high surface gloss requirements may greatly reduce production efficiency and yield because of a higher rejection rate due to visible sink marks, jet lines, and other surface flaws.

The preferred surface for a part is usually one that is produced from a mold having a Society of Plastics Industry (SPI) finish of 3 or 4. Such parts tend to be used on industrial products where high gloss is not required. Parts requiring a high gloss and good transparent clarity are produced from molds having an SPI of 1 or 2. For parts requiring a textured surface and low surface gloss, a mold having an SPI of 5 or 6 is used.

For the purposes of the coding system, the part surface requirement (see Table 5.4) is considered high when

1. Parts are produced from a mold having an SPI surface finish of 1 or 2.
2. Sink marks and weld lines are not allowed on an untextured surface.

In the coding system, parts without high surface requirements are considered to have low surface requirements.

A statistical analysis of piece parts collected from several molders shows that the adverse effect of a high-gloss requirement is more significant on thin parts than on thick parts. A possible explanation is as follows:

The colder the mold or the melt, the more viscous the flow. The more viscous the flow, the greater the tendency to leave visible "flow marks" on the surface of the part. These flow marks make it difficult to obtain a good surface gloss. Since thin parts cool faster than thick parts, poor surface gloss is more likely to occur on thin parts.

5.9.2 Tolerances

There are two types of tolerance requirements: (1) dimensional and (2) geometric. Dimensional tolerances refer to tolerances on the length, width, and height of a part as well as on the distance between features. Geometric tolerances refer to tolerances on flatness, straightness, perpendicularity, cylindricity, and other features.

For dimensional tolerances, an industry standard exists prepared by the Society of Plastic Industry, and each material supplier converts its data to suit a specific material. Thus, to determine whether the tolerances specified are tight or commercial, we must refer to either the data published by the Society of the Plastic Industry for the material in question, or to data supplied by the resin manufacturer. No industrywide standard exists as yet for geometrical tolerances.

Part yield is influenced by part tolerance requirements. Part yield decreases when part tolerances are relatively difficult to hold. Part tolerances are considered difficult to hold if:

(a) External undercuts are present.
(b) A tolerance is specified across the parting surface of the dies. The high pressures (10,000 psi) cause slides and molds to move slightly, making tolerances across the moving surfaces difficult to hold.
(c) The wall thickness is not uniform. Thick sections connected to thin sections tend to shrink more than the thin sections. This is because the thick sections continue to cool down and shrink after the thin sections have solidified. This variation in shrinkage can result in part warpage. When warping is a problem, the cycle time of a part is increased to allow the part to be more rigid when it is ejected.
(d) Unsupported projections (ribs, bosses) and walls are used. Unsupported projections can bend, making tolerances between them difficult to hold.

(e) More than three tight tolerances, or more than five commercial tolerances are required. Tolerances should be specified only where absolutely necessary. As the number of tolerances to be held increases, the proportion of defective parts produced increases.

5.10 USING THE CODING SYSTEM—OVERVIEW

To determine the relative cycle time for a part using the coding system shown in Figure 5.2 and Tables 5.3 and 5.4, we proceed as follows:

1. Determine whether or not the part is partitionable.
2. If the part is partitionable, partition it into elemental plates and assess the relative cycle time for each plate. The plate with the largest relative cycle time controls the cycle time for the part.
3. If the part is not partitionable, we first code the part using the maximum wall thickness of the part. If this is an easy to cool feature we then code the part a second time using the most difficult to cool feature. The feature with the largest relative cycle time controls the cycle time for the part.

Several Examples are presented in Sections 5.12 to 5.14 below.

5.11 EFFECT OF MATERIALS ON RELATIVE CYCLE TIME

In theory, the machine cycle time depends on the material used. In practice, however, there is little significant difference in cycle time between geometrically similar engineering parts made of different materials. Although it is true that some materials will solidify faster than others, there are other factors that tend to cause the actual cycle time to be very similar for the parts with the same geometry but made of different materials. Some of these factors include the following issues.

1. The most important objective for a precision molder is the production of parts with satisfactory dimensional stability and high gloss. Fast molding cycles cause greater variations in mold and melt temperatures than slow cycles. These large temperature variations often result in poor dimensional stability, rejected parts, and resulting lower yield. Furthermore, fast cycles require lower mold and melt temperatures that are likely to cause difficulties in producing parts with high gloss.

Therefore, in order to produce parts with high gloss and dimensional stability, precision molders are likely to manufacture parts at nearly the same rate for rapidly solidifying materials as they would for materials that solidify at a slower rate.

2. Many engineering parts are produced in annual production volumes of 5,000 to 10,000 and in batch sizes of 500 to 1,000. These batch sizes can be produced in two to four shifts (16–32 hours). For such small batches, it is not practical to "optimize" the cycle time for the particular part/material combination.

3. Each machine has its own peculiar characteristics and idiosyncrasies. Thus, a part made of the same material but molded on two different machines under the same set of operating conditions will require different cycle times in order to produce a part of the same high gloss and dimensional stability.

Consequently, optimal operating conditions will vary from machine to machine. The likelihood that a given part would, for each batch, be run on the same machine is very low.

5.12 EXAMPLE 5.1—DETERMINATION OF RELATIVE CYCLE TIME FOR A PARTITIONABLE PART

5.12.1 The Part

For the first example, we consider the part shown in Figure 5.23. In addition to the general overall size of the part, the wall thickness, and the size of the ribs and bosses are given. Although these latter dimensions are not needed in order to estimate the relative tooling cost for this part, they must be known in order to determine the relative cycle time.

5.12.2 The Basic Part Type

In this example, the length of the projections parallel to the surface of the part, c, are 10 mm. Since c/L is less than 1/3 then, as you may recall from Figure 4.3, these projections are considered isolated projections of small volume. Consequently, the dimensions of the basic envelope are

$$L = 160, \quad B = 130, \quad H = 10$$

Since the part is straight, the unbent dimensions L_u and B_u are equal to L and B, respectively. Thus, $L_u/B_u < 10$ and the part is non-slender. We will indicate the fact that the part is non-slender by using the letter N.

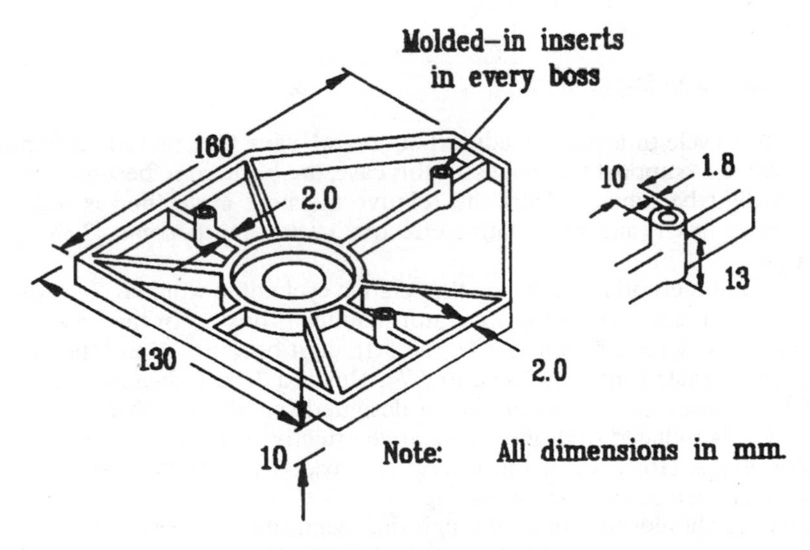

FIGURE 5.23 *Example 5.1.*

5.12.3 Part Partitioning

This part consists of a single elemental plate; thus, no partitioning is needed.

5.12.4 Relative Cycle Time

Basic Relative Cycle Time

The length, l, the width, b, and the height, h, of each projection (boss) is 10mm, 1.8mm, and 13mm, respectively. The plate thickness w is 2.0mm and uniform. Thus, $h/w > 3$, and the projections are considered significant bosses. Since the boss thickness (b) is less than the wall thickness (w) and gusset plates do not support them, the first digit is 6.

The maximum plate thickness is 2.0mm, hence, the second digit is 1, thus, the basic code is N61 and the basic relative machine cycle time, t_b, is 2.52. (Remember, relative cycle time is the cycle time relative to our reference part and is, thus, dimensionless.)

Internal Threads/Inserts

There are no internal threads, but there are three molded-in inserts (one in each boss). Hence, the third digit is 1 and the additional relative time, t_e, is 0.5 per insert or 1.5 for the part.

Surface Gloss/Tolerances

Let's assume that the part requires a low surface gloss (which corresponds to an SPI surface finish of 3 on the mold). Thus, the fourth digit is 0. Because the bosses are unsupported it is difficult to hold tolerances between them, hence, the fifth digit is 1 and the penalty factor, t_p, is 1.20.

Relative Effective Cycle Time

The relative effective cycle time is given by substituting into Equation 5.8, thus,

$$t_r = (t_b + t_e)t_p = (2.52 + 1.5)1.2 = 4.82$$

5.12.5 Redesign Suggestions

The relative cycle time can be reduced in several ways. One method is to provide gusset plates to support the bosses. In this case, the basic code becomes N51, and the fifth digit becomes 0. Thus, the relative machine cycle time is reduced to 2.38, t_p becomes 1, and the relative effective cycle time becomes 3.88—a 19% reduction.

A second method for reducing the relative cycle time would be to reduce the height of the bosses so they become nonsignificant bosses. In this case, the new basic complexity code becomes N41, the fifth digit becomes 0, and the new relative machine cycle time is reduced to 3.74. This is a 22% reduction.

If the bosses are left as originally designed but the molded-in inserts are removed, t_e is reduced to 0, and the relative effective cycle time becomes 3.02—a 37% savings. However, if the inserts are inserted after the part is molded, nothing has been gained by this saving.

Readers should note that although the magnitude of these savings, created by relatively small design changes, might be unimportant for small production volumes, they could become extremely valuable for high-volume parts.

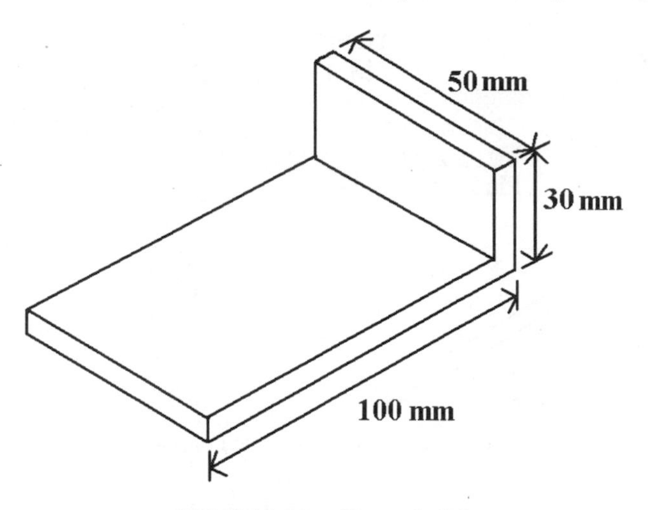

FIGURE 5.24 *Example 5.2.*

5.13 EXAMPLE 5.2—DETERMINATION OF RELATIVE CYCLE TIME FOR A PARTITIONABLE PART

5.13.1 The Part

As a second example we consider the part shown in Figure 5.24. The wall thickness of the part is a constant 3.1 mm. Since the height of the "peripheral projection" in this case is greater than 6w, it will be treated as a wall rather than a rib. (From the point of view of creating the tooling required to mold the part, peripheral ribs and peripheral walls are comparable. Hence, for the purposes of the coding system, all peripheral projections, including ribs, will be treated as peripheral walls.) The part geometry is rather simple and is, thus, partitionable.

5.13.2 Basic Part Type

Since $L_u = L + H = 130$ mm and $B_u = B = 50$ mm, then $L_u/B_u = 2.6$, and the part is non-slender, (N).

5.13.3 Part Partitioning

This part can be partitioned in two ways. It can be partitioned as shown in Figure 5.25 into two separate elemental plates. Alternatively, since the wall thickness of the part is constant and both plates are external plates, the part can be treated as a single plate, as shown in Figure 5.26.

5.13.4 Relative Cycle Time—Elemental Plate 1 (Figure 5.25)

Basic Relative Cycle Time

The plate is neither grilled nor slotted, and no projections are present. Thus, the first digit is 2. The wall thickness is 3.1 mm, hence, the second digit is 3. The basic code is N23. Thus, the basic relative machine cycle time, t_b, is 3.39.

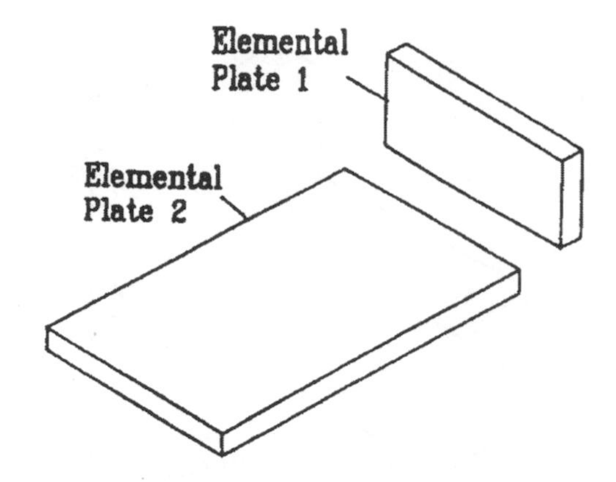

FIGURE 5.25 *Part partitioning.*

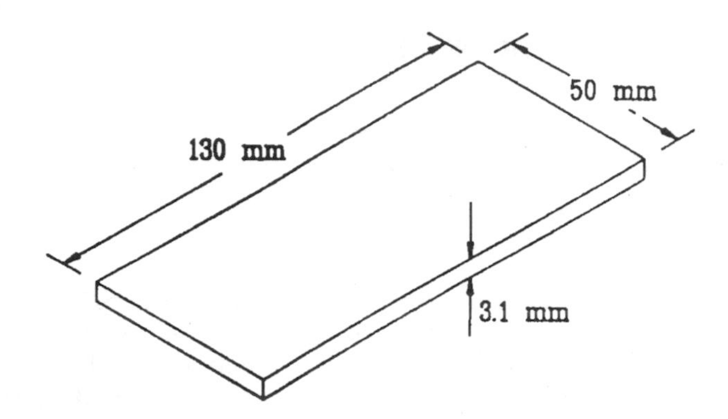

FIGURE 5.26 *Alternate partitioning.*

Internal Threads/Inserts

There are no internal threads or molded-in inserts. Hence, the third digit is 0 and the relative extra mold opening time, t_e, is 0.

Surface Gloss/Tolerances

The part is assumed to require a high surface gloss (SPI surface finish of 1 or 2 on the mold), thus the fourth digit is 3. The 90° angle between the side wall and the bottom is difficult to hold because of the sharp (unradiused) corner and the lack of supporting gusset plates. Thus, the part tolerance is difficult to hold, the fifth digit is 1, and the penalty, t_p, is 1.37.

Relative Effective Cycle Time

The relative effective cycle time for the original design is:

$$t_r = (t_b + t_e)t_p = (3.39)1.37 = 4.64$$

5.13.5 Relative Cycle Time—Elemental Plate 2 (Figure 5.25)

The complete code for this plate is identical to that for plate 1, thus the relative effective machine cycle time is the same.

5.13.6 Relative Cycle Time—Elemental Plate (Figure 5.26)

Basic Relative Cycle Time

Once again, the plate is neither grilled nor slotted and no projections are present. The code for this plate is also N23, and the relative machine cycle time is 3.39.

Internal Threads/Inserts

There are no internal threads or molded-in inserts. Hence, the third digit is 0 and the relative extra mold opening time, t_e, is 0.

Surface Gloss/Tolerances

The part is assumed to require high surface gloss, thus the fourth digit is 3. The 90° angle between the side wall and the bottom is difficult to hold because of the unradiused (sharp) corner and the lack of supporting gusset plates. Thus, the part tolerance is difficult to hold, the fifth digit is 1 and the penalty is 1.37.

Relative Effective Cycle Time

The relative effective cycle time for the design is:

$$t_r = (t_b + t_e)t_p = 4.64$$

5.13.7 Redesign Suggestions

The two cost drivers in this case are the high surface-gloss requirement and the difficult to maintain 90° angle between the side wall and the bottom surface. If the part surface can be replaced by a textured surface (SPI-3 on the mold) the fourth digit becomes 0. If in addition the corner between the two plates can be radiused, or if supporting gusset plates can be used, or if the maintenance of a precise 90° angle is not considered important, the fifth digit becomes 0 and the new penalty is lowered to 1.0. Thus the new relative effective cycle time is 3.39—a 27% reduction.

5.14 EXAMPLE 5.3—DETERMINATION OF RELATIVE CYCLE TIME FOR A NON-PARTITIONABLE PART

5.14.1 The Part

The part and all of the dimensions necessary for a determination of the relative cycle time are shown in Figure 5.27.

5.14.2 Part Partitioning

Because the spacing between fins is less than 12.5 mm, the closely spaced fins are difficult to cool. Thus, in spite of its rather simple geometry, the part is considered non-partitionable.

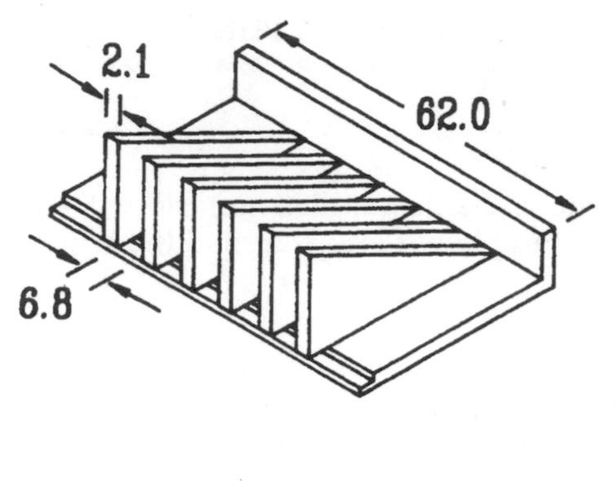

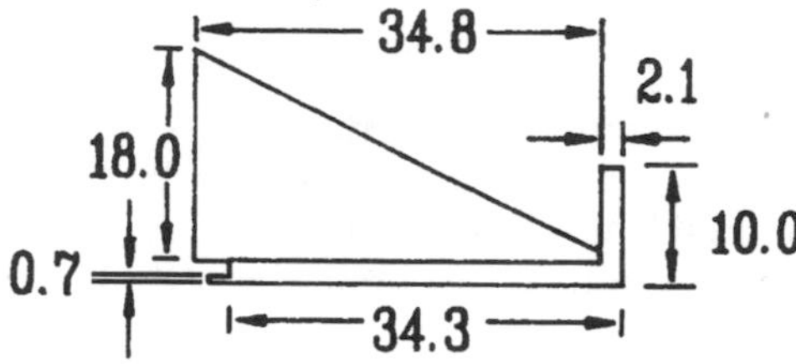

Note: All dimensions in mm.

FIGURE 5.27 *Example 5.3.*

5.14.3 Relative Cycle Time

Basic Machine Cycle Time

Since this part is non-partitionable, the basic code must be found for the thickest easy-to-cool feature as well as the thickest difficult-to-cool feature. However, since the thickness of the part is uniform, only the difficult-to-cool feature needs to be coded. Thus, since the wall thickness is 2.1 mm, the first and second digits are 1 and 2, respectively. Hence, the basic code for the part is NP12, and the basic relative time, t_b, is 4.50.

Internal Thread/Inserts

There are no internal threads or molded-in inserts, thus the third digit is 0 and $t_e = 0$.

Surface Gloss/Tolerances

The presence of the fins makes sink marks unavoidable on the plate to which the fins are attached. Thus, as pointed out in Section 5.9.1, the fourth digit is 2. Assuming that the spacing between the fins is critical, then the lack of gusset plates supporting the closely spaced fins makes the tolerance on the distance between fins difficult to hold—hence, the fifth digit is 1. Therefore, the penalty is 1.41.

Relative Effective Cycle Time

The relative effective cycle time is

$$t_r = (t_b + t_e)t_p = (4.50 + 0)(1.41) = 6.34$$

Table 5.5 Machine tonnage and relative hourly rate. (Data published in *Plastic Technology*, June, 1989.)

Machine Tonnage	*Machine Hourly Rate*
<100	1.00
100–299	1.19
300–499	1.44
500–699	1.83
700–999	2.87
>1000	2.93

5.14.4 Redesign Suggestions

For non-partitionable parts, design improvements are often difficult to make without major changes in part geometry. One possible improvement in the current case, in the sense of reducing processing costs, is to increase the spacing between fins so that the cooling difficulty can be reduced.

It is worth pointing out that the part was coded as though the spacing between fins is critical, and that sink marks are not acceptable. However, practical concerns may be such that stringent tolerance and surface requirements are not necessary. This would change both the fourth and fifth digits to 0, and the relative effective cycle time would also change accordingly.

5.15 RELATIVE PROCESSING COST

The preceding discussion and examples have shown how to compute the relative part cycle time, t_r. This is a critical requirement in estimating relative processing costs. As noted in the Section 5.1.2 of this chapter, the relative processing cost is given by:

$$C_e = t_r C_{hr} \qquad \text{(Equation 5.2)}$$

where C_{hr} represents the ratio C_h/C_{ho}, C_{ho} represents the machine hourly rate for the reference part, and C_h is the machine hourly rate ($/h) for a given part. The relative machine hourly rate, C_{hr}, can be determined from Table 5.5, but it is first necessary to determine the injection molding machine size (tonnage) required to mold the part.

The machine tonnage required to mold a part is approximately two to five tons per square inch, depending on the material to be molded and on the projected area of the part normal to the direction of mold closure. In general, it is assumed that a machine whose tonnage, F_p, exceeds a numerical value equal to three times the projected area of the part (expressed in in^2) will suffice. Thus, the required machine tonnage is approximately

$$F_p = 3A_p \qquad \text{(Equation 5.9)}$$

where the projected area, A_p, is in in^2 or

$$F_p = 0.005A_p \qquad \text{(Equation 5.10)}$$

where the projected area is in mm^2.

5.16 RELATIVE MATERIAL COST

As noted in Section 5.1.3 of this chapter, the material cost for a part, K_m, is given by

$$K_m = VK_p \qquad \text{(Equation 5.4)}$$

where V is the part volume and K_p is the material cost per unit volume, and the relative material cost is:

$$C_m = \frac{V}{V_o} C_{mr} \qquad \text{(Equation 5.5)}$$

Values for C_{mr} are given above in Table 5.2.

5.17 TOTAL RELATIVE PART COST

As stated earlier in this chapter (Section 5.1.4), the total production cost of a part, K_t, in units such as dollars or cents, is now computed as the sum of the material cost of the part, K_m, the tooling cost, K_d/N, and the processing cost, K_e:

$$K_t = K_m + \frac{K_d}{N} + K_e \qquad \text{(Equation 5.6)}$$

where K_d represents the total cost of the tool and N represents the number of parts, or production volume, to be produced using that tool.

If K_o denotes the manufacturing cost of some standard or reference part, then

$$K_o = K_{mo} + \frac{K_{do}}{N_o} + K_{eo}$$

where K_{mo}, K_{do}, and K_{eo} represent the material cost, tooling cost, and equipment operating cost for the reference part. Thus, the total cost of the part relative to the cost of the reference part, C_r, can be expressed as:

$$C_r = \frac{K_m + K_d/N + K_e}{K_o} \qquad \text{(Equation 5.7)}$$

where C_r is, of course, dimensionless.

The total relative cost C_r can be written as follows:

$$C_r = \frac{K_m}{K_{mo}}\frac{K_{mo}}{K_o} + \frac{(K_d/N)}{K_{do}}\frac{K_{do}}{K_o} + \frac{K_e}{K_{eo}}\frac{K_{eo}}{K_o}$$

If

$f_m = K_{mo}/K_o$,

$f_d = K_{do}/K_o$, and

$f_e = K_{eo}/K_o$,

which represent the ratio of the material cost, tooling cost, and processing cost of the reference part to the total manufacturing cost of the reference part, and if

$$C_m = K_m/K_{mo},$$
$$C_d = K_d/K_{do}, \text{ and}$$
$$C_e = K_e/K_{eo}$$

are, respectively, the material cost, tooling cost, and equipment operating cost of a part relative to the material cost, tooling cost, and equipment operating cost of the reference part, then the above equation becomes,

$$C_r = C_m f_m + (C_d/N) f_d + C_e f_e \qquad \text{(Equation 5.11)}$$

The value for C_d is obtained from Equation 4.4, as described in Chapter 4, "Injections Molding: Relative Tooling Cost," while the values for C_e and C_m are obtained from Equations 5.2 and 5.5, respectively.

If it is assumed that only single cavity molds will be used, then from the data given in Table 5.1, the cost of the reference part, in dollars, can be shown to be

$$K_o = 0.124 + 6980/N_o \text{ (dollars)}$$

where the first term is the sum of the material and processing costs for the reference part and the second term is the tooling cost for the reference part. It is seen here that at low production volumes most of the cost is due to the cost of the tooling. At very high production volumes, say when N_o approaches infinity, the cost is due primarily to material and processing costs and approaches \$0.124 or 12.4 cents. Thus, the cost of the reference part depends upon its production volume.

Since the main concern here is the comparison of alternative designs for a given part, thc actual cost of the reference part is not important. All that is really of interest is a comparison in relative costs between two competing designs. For this reason it becomes convenient to obtain the relative cost of a part with respect to the standard part when its production volume is 7,970 and its total production cost, K_o, is \$1. In this case the values of f_m, f_e, and f_d become

$$f_m = K_{mo}/K_o = 0.00182$$
$$f_e = K_{eo}/K_o = 0.1224$$
$$f_d = K_{do}/K_o = 6980$$

and

$$C_r = 0.00182 C_m + (6980/N) C_d + 0.1224 C_e \qquad \text{(Equation 5.12)}$$

5.18 EXAMPLE 5.4—DETERMINATION OF THE TOTAL RELATIVE PART COST

The part shown in Figure 5.28 is made of polycarbonate. The wall thickness is a uniform 3.5 mm and the ribs are assumed to be, according to the classification system, not significant.

As originally designed, the total relative die construction cost, C_d, is found to be 2.32. (See Problem 4.4 of Chapter 4.)

The relative cycle time, t_r, for this part can be shown to be

$$t_r = 3.67(1.20) = 4.40.$$

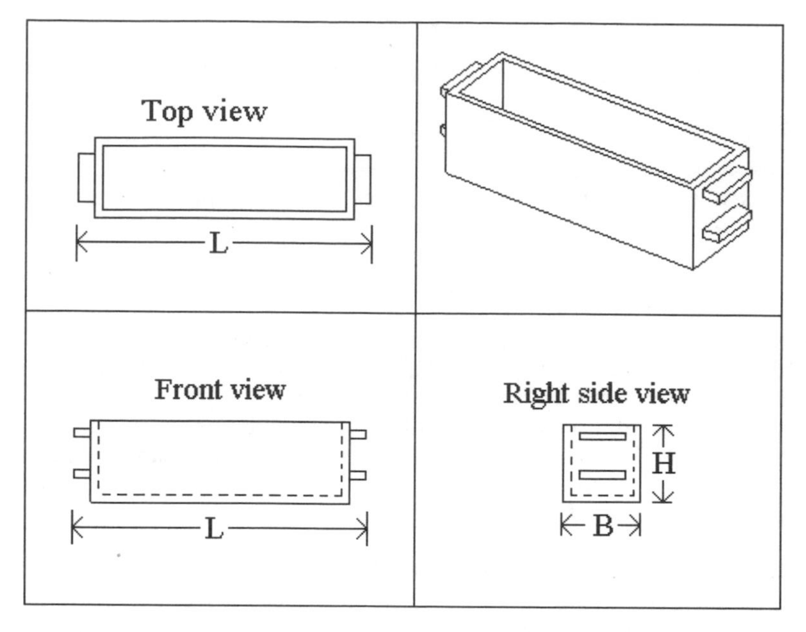

FIGURE 5.28 *Example 5.4. Original design. (L = 180 mm, B = H = 50 mm.)*

The projected area of the part is 9,000 mm^2, hence, from Equation 5.10, the required clamping force is

$$F_p = 0.005A_p = 45\,\text{tons}$$

Consequently, from Table 5.5, the relative machine hourly rate, C_{hr}, is 1.00. Thus,

$$C_e = t_r C_{hr} = (4.40)(1) = 4.40$$

The part volume is approximately 105,000 mm^3. Since the material is polycarbonate, then from Table 5.2 the relative material price, C_{mr}, is 2.96. From Equation 5.5, the relative material cost for the part is

$$C_m = (V/V_o)C_{mr} = (105{,}000/1244)(2.96) = 250$$

where $V_o = 1{,}244\,\text{mm}^3$ is the volume of the reference part.

Production Volume = 10,000

If the production volume of the part is 10,000 pieces, then from Equation 5.12 the total relative cost of the original part, C_r, is,

$$C_r = (0.00182)250 + (6980/10{,}000)(2.32) + (0.1224)(4.40) = 2.61$$

If the bottom rib is removed from each end of the part as shown in Figure 5.29, then C_d is reduced to 1.65 (See Problem 4.4 of Chapter 4). The values for t_r, C_{hr}, C_e, and C_m remain the same. Thus,

$$C_r = (0.00182)250 + (0.6980)(1.65) + (0.1224)(4.40) = 2.14$$

This is an overall savings of some 18% in piece part cost.

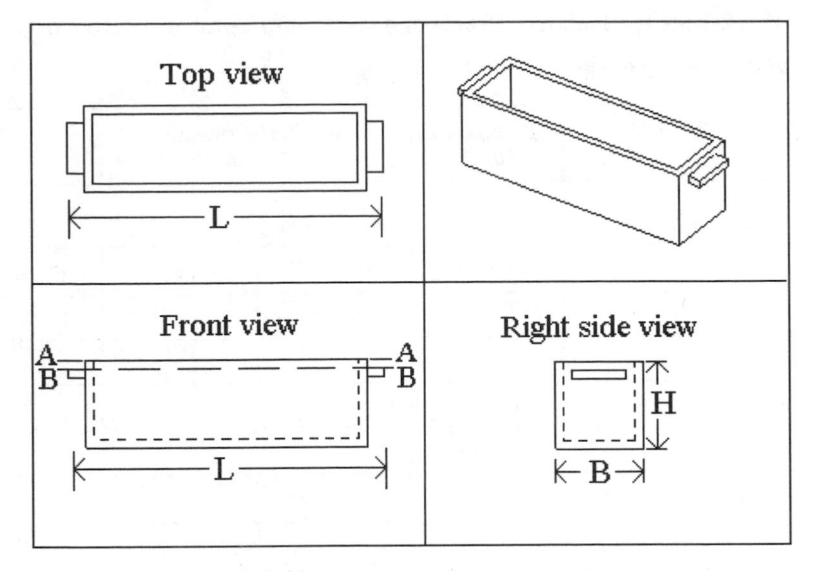

FIGURE 5.29 *Redesigned part.*

Production Volume = 100,000

If the production volume of the part is 100,000 pieces, then the total relative cost of the original design, C_r, is, from Equation 5.12,

$$C_r = (0.00182)250 + (0.06980)(2.32) + (0.1224)(4.40) = 1.15$$

For the redesigned part,

$$C_r = (0.00182)250 + (0.06980)(1.65) + (0.1224)(4.40) = 1.10$$

Thus, for a production volume of 100,000 parts, the savings are only 4.3%. This reduction in savings is due to the fact that, in this particular case all of the savings was due entirely to a reduction in tooling costs whose contribution to overall part cost diminishes with increasing production volume.

5.19 WORKSHEET FOR RELATIVE PROCESSING COST AND TOTAL RELATIVE PART COST

To facilitate the calculation of the relative processing cost and the overall relative part cost, a worksheet has been prepared. The copies shown on the next two pages have been completed for Example 5.1 in Section 5.12. A blank copy of the worksheet is available in Appendix 5.B and may be reproduced for use with this book.

Worksheet for Relative Processing Costs and Total Relative Cost

***Original Design*/Redesign** *Example 5.1*

$L_u = 160$	$B_u = 130$	$L_u/B_u = 1.23 < 10$	Slender/ Non-slender? *NS*	

Basic Relative Cycle Time

	Plate 1	Plate 2	Plate 3	Plate 4	Plate 5
Ext/Int					
1st Digit	*6*				
2nd Digit	*1*				
t_b	*2.52*				

Additional Time

3rd Digit	*1*				
t_e	*3(0.5) = 1.5*				

Time Penalty

4th Digit	*0*				
5th Digit	*1*				
t_p	*1.2*				

Relative Cycle Time for Plate

$t_r = (t_b + t_e)t_p$	$(2.52 + 1.5) \times (1.2) = 4.82$				
Relative Cycle Time for the part = ***4.82***					

Relative Processing Cost

$A_p =$	$F_p =$	$C_{hr} =$	$C_e = t_r C_{hr} =$

Relative Material Cost

$V =$	$V_o =$	$C_{mr} =$	$C_m = (V/V_o)C_{mr} =$

Total Relative Cost

$N =$
$C_r = 0.00182C_m + (6980/N)C_d + 0.1224C_e$

Redesign Suggestions

Support bosses to reduce the first and fifth digits OR reduce the boss height to make them nonsignificant. Could eliminate inserts, but these would need to be inserted after molding, thus little to be gained (i.e., pay me now or pay me later).

% Savings in processing costs:
% Savings in overall costs:

Worksheet for Relative Processing Costs and Total Relative Cost

Original Design/*Redesign Version 1*

$L_u =$	$B_u =$	$L_u/B_u =$	Slender/ Non-slender?	

Basic Relative Cycle Time

	Plate 1	Plate 2	Plate 3	Plate 4	Plate 5
Ext/Int					
1st Digit	*5*				
2nd Digit	*1*				
t_b	*2.38*				

Additional Time

3rd Digit	*1*				
t_e	*1.5*				

Time Penalty

4th Digit	*0*				
5th Digit	*0*				
t_p	*1*				

Relative Cycle Time for Plate

$t_r = (t_b + t_e)t_p$	*(2.38 + 1.5) = 3.88*				
Relative Cycle Time for the part = ***3.88***					

Relative Processing Cost

$A_p =$	$F_p =$	$C_{hr} =$	$C_e = t_rC_{hr} =$

Relative Material Cost

$V =$	$V_o =$	$C_{mr} =$	$C_m = (V/V_o)C_{mr} =$

Total Relative Cost

$N =$
$C_r = 0.00182C_m + (6980/N)C_d + 0.1224C_e$

Redesign Suggestions

% Savings in processing costs: (4.82 – 3.88)/4.82 = 19.5%
% Savings in overall costs:

5.20 SUMMARY

The purpose of this chapter is to present a systematic approach for identifying, at the parametric design stage, those features of the part that significantly affect the processing cost of injection-molded parts. The goal was to learn how to design so as to minimize difficult-to-process features.

A methodology for estimating the relative processing cost of proposed injection-molded parts based on parametric information was presented. In addition, a method for estimating the overall relative part cost for this same part was introduced.

REFERENCES

Dym, J. B. *Product Design with Plastics.* New York: Industrial Press, 1983.

Fredette, Lee. "A Design Aid for Increasing the Producibility of Die Cast Parts." M.S. Final Project Report, Mechanical Engineering Department, University of Massachusetts at Amherst, Amherst, MA, 1989.

Kuo, Sheng-Ming. "A Knowledge-Based System for Economical Injection Molding." Ph.D. Dissertation, University of Massachusetts at Amherst, Amherst, MA, February 1990.

Poli, C., Fredette, L., and Sunderland, J. E. "Trimming the Cost of Die Castings." *Machine Design* 62 (March 8, 1990): 99–102.

Poli, C., Kuo, S. M., and Sunderland, J. E. "Keeping a Lid on Mold Processing Costs." *Machine Design* 61 (Oct. 26, 1989): 119–122.

Shanmugasundaram, S. K. "An Integrated Economic Model for the Analysis of Die Cast and Injection-molded parts." M.S. Final Project Report, Mechanical Engineering Department, University of Massachusetts at Amherst, Amherst, MA, August 1990.

QUESTIONS AND PROBLEMS

5.1 As part of the same training program discussed in Problem 4.1 of Chapter 4, you are in the process of explaining to some new hires that the processing cost of injection-molded parts is a function of what you have called the *basic relative cycle time* of the part. Explain in detail exactly which features of a part affect the basic cycle time of a part.

5.2 In addition to the basic relative cycle time discussed in Problem 5.1, what other factors affect the overall relative cycle time of an injection-molded part?

5.3 Assuming that the dimensions for the parts shown in Figures P3.2 to P3.5 of Chapter 3 are such that the parts are easily cooled, which parts, if any, are partitionable? For those part that are partitionable, which plate controls the cycle time of the part? Assume that in all cases the wall thickness is constant.

As examples to illustrate both basic complexity and subsidiary complexity, explain which features of these parts, if any, affect basic complexity and which features, if any, affect subsidiary complexity.

5.4 Assume that you are still part of the integrated product and process design (IPPD) team formed as part of Problem 4.3 in Chapter 4. One of the components used in Widget A consists of the hollow cylindrical part shown in Figure P5.4. This component is presently made of nylon 6 in a mold having an SPI finish of 3. What redesign suggestions would you make to your IPPD team in order to reduce the relative cycle time of the part? What savings in processing costs would you achieve by your proposed redesign. Assume commercial tolerances.

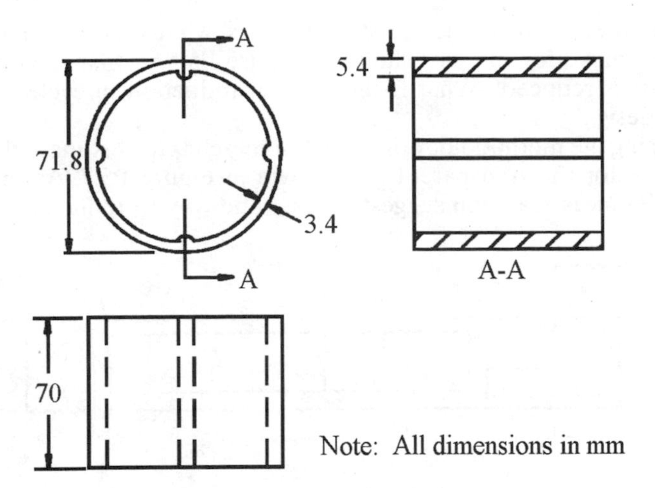

FIGURE P5.4

5.5 Imagine that you are still part of the IPPD team discussed above. You have been asked to roughly estimate the cycle time for the part shown in Figure P5.5 if it is produced with commercial tolerances. What would you estimate the cycle time to be if the part is made of nylon 6 in a mold with an SPI finish of 3. The maximum wall thickness of the part is 2.5 mm. The minimum wall thickness of the part is 1.5 mm.

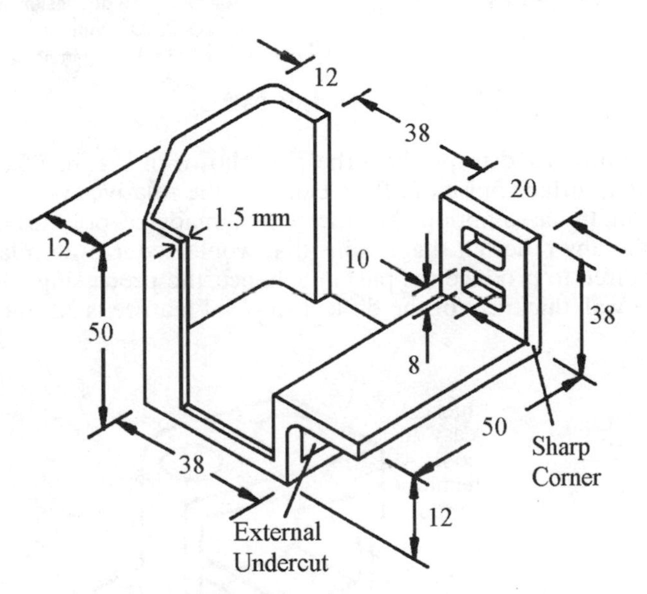

FIGURE P5.5

5.6 In an effort to reduce processing costs, what redesign suggestions would you make for the part shown in Figure P5.5 so that the cycle time of the part is reduced? What is the percent reduction in cycle time due to this redesign?

5.7 Using the methodology discussed in this chapter, estimate the relative cycle time for the transparent part shown in Figure P5.7. Assume commercial tolerances. Can you suggest at least one way to reduce the cycle time?

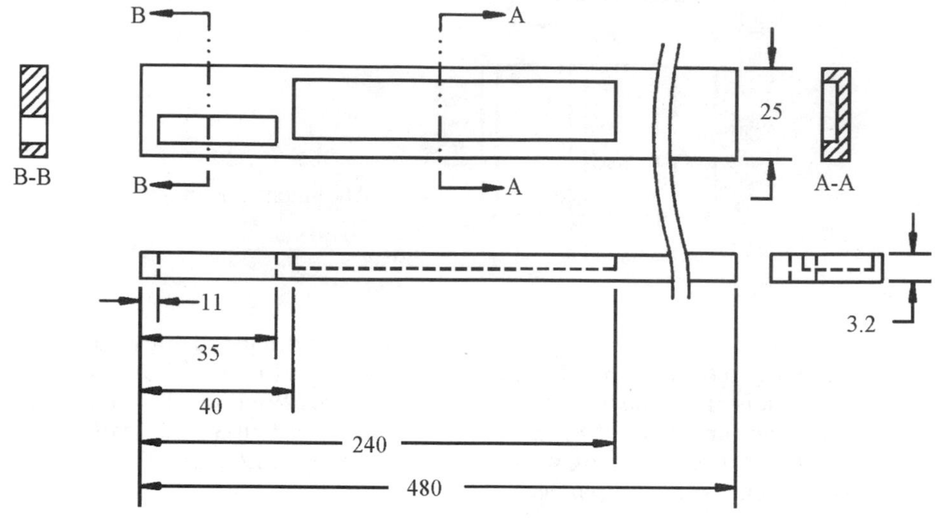

FIGURE P5.7

5.8 The mold used to produce the part shown in Figure P5.8 is assumed to have a surface finish of SPI-3. Estimate the relative cycle time for the part under the assumption that the part is made of polycarbonate. Can you make any redesign suggestions that would reduce the relative cycle time required to produce the part and, hence, the processing cost for the part? The wall thickness of the difficult-to-cool feature is 2.5 mm.

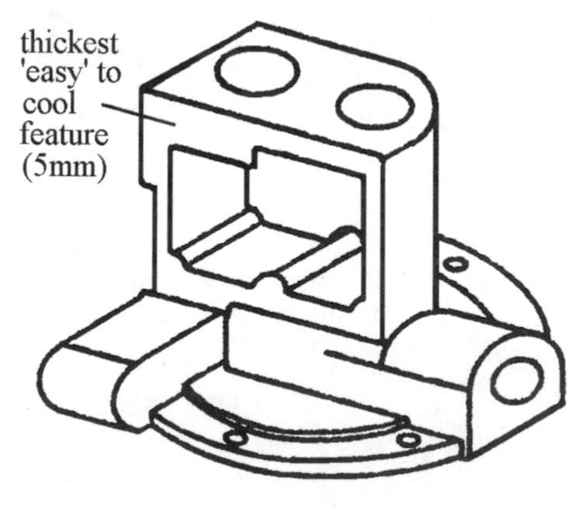

FIGURE P5.8

5.9 Some members of your design team seem unconvinced that overall costs for the part shown in Figure P5.5 can be significantly reduced by applying the redesign suggestions you proposed earlier (Problems 4.8 and 4.9) to reduce tooling costs along with your more recent suggestions for reducing processing costs (Exercise 5.6). Thus, for the part shown in Figure P5.5, determine the percent savings in cost achieved for production volumes of 25,000 and 100,000. Are these significant savings?

5.10 For Example 5.4 (in Section 5.18) it was assumed that the cycle time for the reference part, t_o, is 16 s (see Table 5.1). Assume that the cycle time t_o is 10 s and determine for the part analyzed in Example 5.4 the reduction in costs between the original design and the redesign for production volumes of 10,000 and 100,000.

APPENDIX 5.A

Notes for Figure 5.2 and Tables 5.3 and 5.4

(1) For parts or elemental plates with a bent or curved longitudinal axis, the unbent length, L_u, is the maximum length of the part with the axis straight. The width of this unbent part is referred to as B_u. See Figure 5.6.

(2) Frames are parts or elemental plates that have a through hole greater than 0.7 times the projected area of the part/plate envelope and whose height is equal to its wall thickness. See Figure 5.7.

(3) The thickness of the elemental plate is denoted by w. For parts/plates where the thickness is not constant, w is the maximum thickness of the plate or part.

(4) Lateral projections are shape features that protrude from the surface of a slender plate (rod) in a direction normal to the longitudinal axis. For long slender parts such projections are difficult to fill. See Figure 5.22.

(5) A rib is a narrow, elongated wall-like projection whose length, l, is greater than three times its width, b. Ribs may be located either at the periphery or on the interior of a part/plate. A rib may be continuous or discontinuous or part of a network of other ribs and projecting elements. To avoid sink marks ribs should be designed so that (a) the rib height, h, is less than or equal to three times the localized wall thickness, w, and (b) the rib width, b, is less than or equal to the localized wall thickness. Such ribs are considered as nonsignificant ribs. Ribs designed such that $[3w < h < 6w]$ or $b > w$ are called significant ribs. Significant ribs tend to increase the machine cycle time and make tolerances difficult to hold. For the purposes of this system, peripheral ribs are treated as walls. See Figure 5.17.

(6) A boss, like a rib, is a projecting element; however, its length, l, is less than three times its width, b. It takes a variety of forms such as a knob, hub, lug, button, pad, or "prolong." A boss should be designed such that (a) the boss height, h, is less than or equal to three times the localized wall thickness, and (b) the boss width, b, should be less than or equal to the localized wall thickness. These types of bosses are considered nonsignificant bosses. Bosses designed such that $3w < h$ or $b > w$ are called significant bosses. Significant bosses tend to increase the machine cycle time and make tolerances difficult to hold. See Figure 5.19.

(7) An elemental plate with
  a) multiple through holes,
  b) no continuous solid section with a projected area greater than 20% of the projected area of the plate envelope, and
  c) whose height is equal to its wall thickness

  is called grilled/slotted. See Figure 5.20.

(8) The wall thickness referred to here is the localized wall thickness. (See also Notes 6 and 7.)

(9) The use of foamed materials results in a surface gloss that is generally less acceptable than the one that results from the use of thermoplastic materials. To improve the surface gloss of parts made with foamed materials the parts are usually subjected to secondary finishing operations (painting, etc.). In addition, the minimum thickness achievable with foamed materials is greater than that obtainable with thermoplastics. For these reasons foamed materials are generally not used for parts whose wall thickness is less than or equal to 5 mm.

In addition, engineering thermoplastics are not generally used if the wall thickness is greater than 5 mm because the shrinkage of these plastics becomes difficult to control when $w > 5$ mm.

(10) Features such as holes or depressions that have an internal diameter smaller than 12.5 mm are considered difficult to cool.

(11) Holes or depressions with internal grooves, or undercuts such that a solid plug that conforms to the shape of the hole or depression cannot be inserted, are called internal undercuts. Such restrictions prevent molding from being extracted from the core in the line of draw. When these internal undercuts take the form of internal threads, a special unscrewing mechanism is used. When the number of threads becomes large, the time required to unscrew the mechanism can significantly increase the machine cycle.

(12) Inserts are metal components added to the part prior to molding the part. These metal components are added for decorative purposes, to provide additional localized strength, to transmit electrical current, and to aid in assembly or subassembly work. The use of inserts increases the machine cycle time.

(13) For the purposes of the present coding system, the part surface requirement is considered high when parts are produced from a mold having an SPI/SPE surface finish of 1 or 2, or sink marks and weld lines are not allowed on an untextured surface.

Parts without high surface requirements are considered to have low surface requirements.

(14) Part tolerances are considered difficult-to-hold if:
a) External undercuts are present.
b) The wall thickness is not uniform.
c) A tolerance is required across the parting surface of the dies.
d) Unsupported projections (ribs, bosses) and walls are used.
e) More than three tight tolerances or more than five commercial tolerances are required.

APPENDIX 5.B

Worksheet for Relative Processing Cost and Total Relative Cost

Original Design/Redesign

L_u =	B_u =	L_u/B_u =	Slender/ Non-slender?	

Basic Relative Cycle Time

	Plate 1	Plate 2	Plate 3	Plate 4	Plate 5
Ext/Int					
1st Digit					
2nd Digit					
t_b					

Additional Time

3rd Digit					
t_e					

Time Penalty

4th Digit					
5th Digit					
t_p					

Relative Cycle Time for Plate

$t_r = (t_b + t_e)t_p$					
Relative Cycle Time for the part =					

Relative Processing Cost

A_p =	F_p =	C_{hr} =	$C_e = t_r C_{hr}$ =

Relative Material Cost

V =	V_o =	C_{mr} =	$C_m = (V/V_o)C_{mr}$ =

Total Relative Cost

N =
$C_r = 0.00182C_m + (6980/N)C_d + 0.1224C_e$

Redesign Suggestions

% Savings in processing costs:
% Savings in overall costs:

Chapter 6

Metal Casting Processes

6.1 INTRODUCTION

As with polymer processing, there are also a number of metal casting processes. Although there are distinct differences between these many casting processes, there are also many common characteristics. For example, in all casting processes, a metal alloy is melted and then poured or forced into a mold where it takes the shape of the mold and is allowed to solidify. Once it has solidified, the casting is removed from the mold. Some castings require finishing due to the cast appearance, tolerance, or surface finish requirements.

During solidification, most metals shrink (gray cast iron is an exception) so molds must be made slightly oversize in order to accommodate the shrinkage and still achieve the desired final dimensions.

There are two basic types of molds used in castings, namely, expendable molds (sand casting and investment casting) that are destroyed to remove the part, and permanent molds (die casting). Expendable molds are created using either a permanent pattern (sand casting) or an expendable pattern (investment casting). Permanent molds, of course, do not require a pattern.

Two of the major advantages for selecting casting as the process of choice for creating a part are the wide selection of alloys available and the ability, as in injection molding, to create complex shapes. However, not all alloys can be cast by all processes.

The most common metal casting processes are sand casting, investment casting, and die casting. These are described in Sections 6.2 to 6.4 below. Other casting processes are more briefly described in Section 6.5. The nature of the molds used and the method for removing the part from the mold differs for the various processes. The tolerances and surface finishes achievable are also different.

For more information on casting processes and technology, consult *ASM Metals Handbook* in the reference list.

6.2 SAND CASTING

Sand casting is a process in which a sand mold is formed by packing a mixture of sand, a clay binder, and water around a wood or metal pattern that has the same external shape as the part to be cast. A pattern can come in two halves: a top half (called a cope) and a bottom half (called a drag) (Figure 6.1). Each half is placed in a molding box, and the sand mixture is then poured all around the pattern. After the sand is packed, holes, which are used to pour the molten metal into the mold (sprue) and to be used as a reservoir of molten metal (risers), are formed in the sand. Vents are also created in order to allow the escape of gases from the melt. Then the pattern is removed and a runner system or small

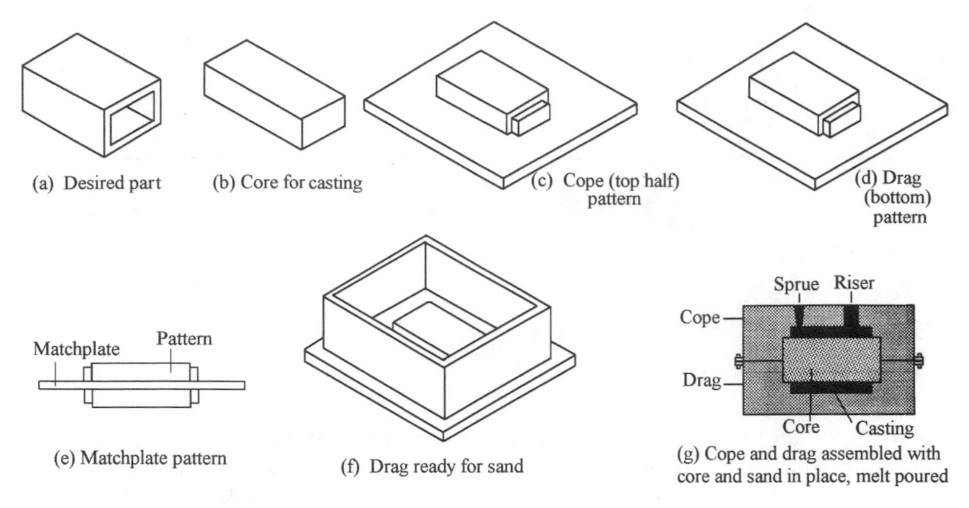

FIGURE 6.1 *Sand casting process.*

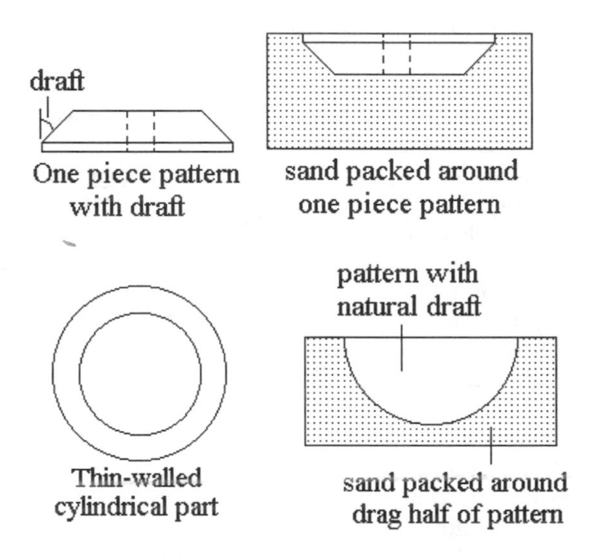

FIGURE 6.2 *An illustration of parts with draft and pattern ready for removal.*

passageway is created inside the die through which the melt can flow and be distributed. Gates are the sections where the melt enters the impression. Thus, sprues feed the runners, and the runners feed the gates.

To facilitate removal of the pattern from the sand mold, the pattern must be provided with an angle or taper called *draft*. If possible, parts should be designed so that natural draft is provided (Figure 6.2).

If the part to be cast has one completely flat surface, then the pattern can be made in one piece (Figure 6.2). If the production volume is sufficiently large, the two halves of the pattern are usually mounted on opposite sides of a single board or metal plate to form what is called a match-plate (Figure 6.1e). To avoid the necessity of forming the runner system by hand, the patterns that form the runners can also be mounted on the match-plate. For large castings a match-plate would become too large and heavy for convenient handling and the cope and drag half approach shown in Figure 6.1 is used.

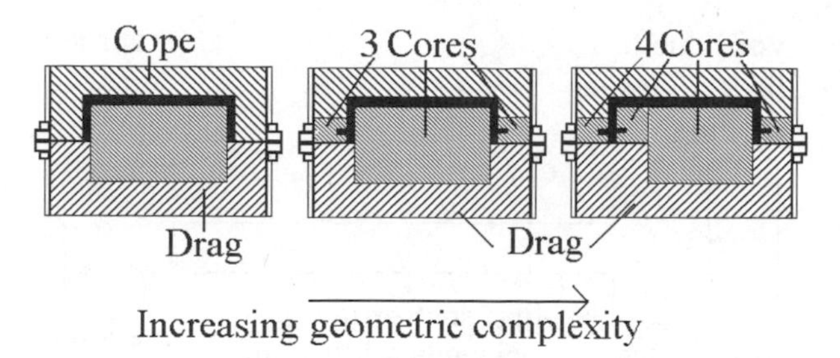

FIGURE 6.3 *Cores required with increasing geometric complexity and cost. Sprues and risers not shown.*

If the casting is to be a hollow shape, such as a thin-walled cylinder, then a separate sand core is placed in position so that the melt cannot fill what is to be the open portion of the casting (e.g., the inside of the cylinder). In sand casting, these sand cores provide a function similar to those provided by side cavities and side cores in injection molding (see Figures 3.12 and 4.15) and die casting (Section 6.4), that is, they provide those geometric features not easily obtainable using a conventional two-piece mold or pattern. As the part geometry becomes more complex, the number of cores required to provide the geometric shape increases (see Figure 6.3).

Once the core or cores are in place, the cope half of the mold box is then placed on top of the drag half. The melt is then poured, and the casting left to solidify. Once the casting has cooled, the sand mold is destroyed and the casting removed.

Sand castings are typically used to produce large parts such as machine-tool bases and components, structures, large housings, engine blocks, transmission cases, connecting rods, and other large components that, because of their size, cannot be cast by other processes. Although almost any metal that can be melted can be sand cast, sand castings have (as a result of the sand mold) a grainy surface with large dimensional variations. Thus, sand castings often require local finish machining operations (see Chapter 11, "Other Metal Shaping Processes") in order to obtain the necessary surface finishes and dimensional tolerances.

Sand casting also results in parts with internal porosity that causes leaking and reduces part strength. Porosity is the result of voids or pores caused by trapped air, liquids, or gases that come about during freezing of the melt. Trapped air and liquids are a result of the dendrites (a crystal that has a treelike branching pattern) that occur when the cooling rate is relatively slow, as in sand casting. Because the trapped liquids and gases continue to freeze and shrink, holes are created.

Because of the shrinkage that occurs before and during solidification, risers (see Figure 6.1), which contain a reservoir of molten metal, are connected to the casting. Thus, as the casting shrinks the riser supplies additional melt. To be effective, the risers must freeze last, otherwise the supply of melt to the mold is shut off and shrink holes are created (see Figure 6.4). The location of risers is also critical, and if they are properly located they can reduce the number and size of shrink holes.

In order to produce a sound casting the number and location of risers is important. However, since

1. Time is devoted in providing risers, and
2. Time is required to remove the risers, and
3. The yield of the metal poured is reduced

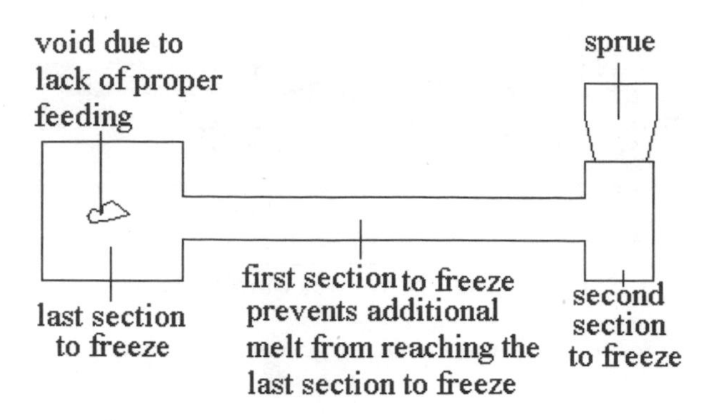

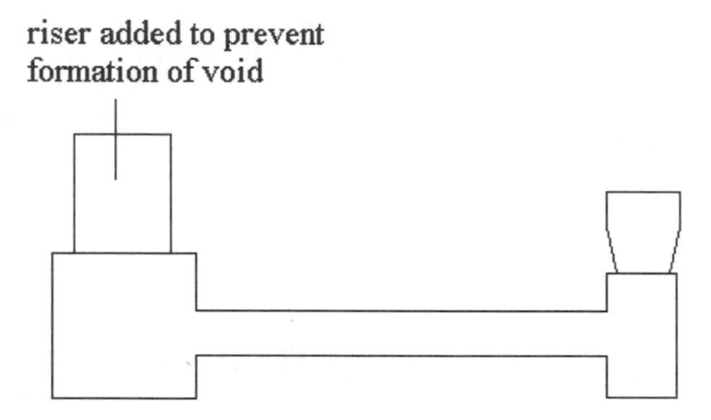

FIGURE 6.4 *Creation of shrink holes due to improper location of a riser.*

risers add to the cost of a casting and, thus, the number of risers should be kept to a minimum. Often redesign of a casting by the addition of webs and ribs permit the reduction in the number and/or placement of risers (see *American Society of Metals Handbook*, Vol. 15).

Sand castings are not generally used for the production of parts that require high production volumes. They are sometimes used to produce prototype parts and at times they are the only method available for the creation of large parts that require a large crane capacity to remove from the mold.

6.3 INVESTMENT CASTING

Investment casting, as well as die casting (which is discussed in the next section), can produce parts of similar geometric shapes and size. Since, as you will learn below, the disposable pattern is made by injecting wax into a mold, features that are difficult or costly to injection mold or die cast (e.g., undercuts) are also costly to investment cast.

Investment casting is typically used when low production volumes are expected (e.g., less than 10,000 pieces), whereas die casting tends to be used when high production volumes are expected.

Investment cast parts can be made of a wide range of metal alloys including aluminum and copper alloys, carbon and low alloy steels, stainless steels, tool steels, and nickel and cobalt alloys. Die castings, as you will learn in the next section, must be restricted to metals with relatively low melting temperatures—primarily zinc and aluminum.

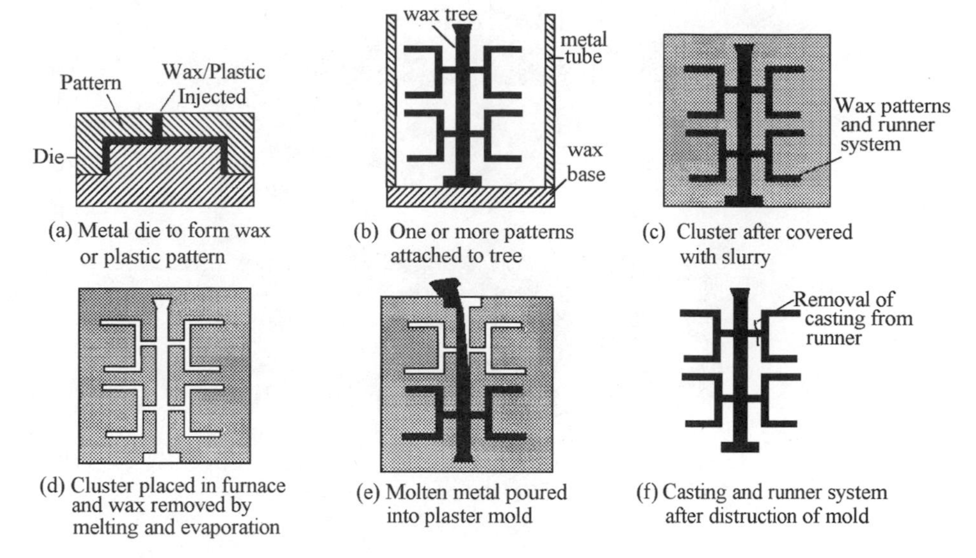

(a) Metal die to form wax or plastic pattern
(b) One or more patterns attached to tree
(c) Cluster after covered with slurry
(d) Cluster placed in furnace and wax removed by melting and evaporation
(e) Molten metal poured into plaster mold
(f) Casting and runner system after distruction of mold

FIGURE 6.5 *Investment casting process.*

In investment casting (Figure 6.5), a metal die or mold is made by either machining or casting. The more complicated the shape (because of undercuts, for example), the more costly the metal dies.

After the mold is formed, wax is injected to form a pattern. The external shape of the wax pattern resembles the internal shape of the mold. The wax pattern is removed from the mold and attached to a wax base that contains a gate. If the production volume is large enough several wax patterns are attached to a tree that contains the runners, gates, and other features that will feed and distribute the molten metal. A metal hollow tube is now placed over the wax patterns and a slurry—such as plaster of Paris—is poured to entirely cover the patterns. The completed mold is placed in an oven and the wax removed by melting and evaporation. Following this the mold is usually placed in a second oven to cure for 12 to 24 hours.

To make parts, the mold cavity is filled with molten metal that is allowed to solidify. To facilitate filling of the mold the melt is poured while the mold is still hot. When the part has cooled, the mold is destroyed and the part removed. The tolerances and surface finishes achievable by investment casting are such that machining is not generally required.

6.4 DIE CASTING

Like injection molding, die casting is a process in which a melt is injected under pressure into a metal mold. The melt then cools and solidifies, conforming to the internal shape of the mold.

As in injection molding, as the part geometry becomes more complex, the cost of the mold increases. Also, as the wall thickness increases, the cycle time required to produce the part also increases. While the thin film, called flashing (Figure 6.6), that extrudes out through the spaces between parts of a mold is easily removed by hand in the case of injection-molded parts, the same cannot be said for die-cast parts. Hence, because of the difficulty of

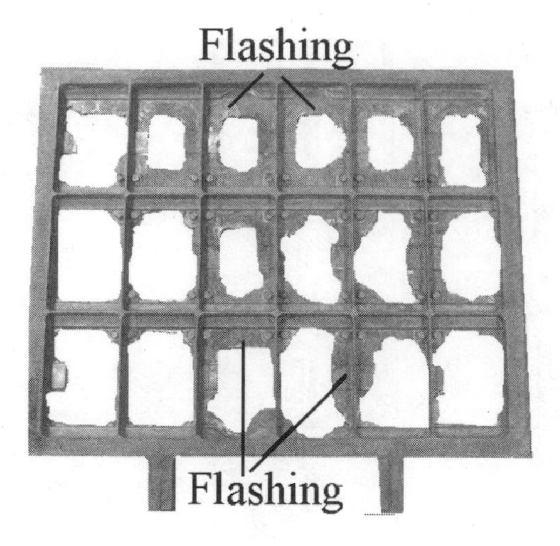

FIGURE 6.6 *Die cast part with flashing.*

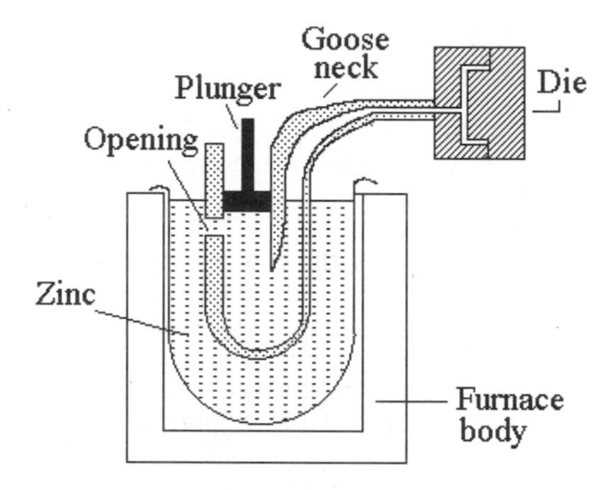

FIGURE 6.7 *Hot chamber die casting machine.*

flash removal, internal undercuts are not generally die cast. Nevertheless, both injection molding and die casting can economically produce parts of great complexity.

There are two types of die casting machines: a hot chamber machine (Figure 6.7) and a cold chamber machine (Figure 6.8). Both have four main elements: (1) a source of molten metal, (2) an injection mechanism, (3) a mold, and (4) a clamping system.

In a hot chamber machine, the injection mechanism is submerged in the molten metal (Figure 6.7). Because the plunger is submerged in the molten metal, only alloys such as zinc, tin, and lead (which do not chemically attack or erode the submerged injection system) can be used. Aluminum and copper alloys are not suitable for hot chamber machines.

When the die is opened and the plunger retracted the molten metal flows into the pressure chamber (gooseneck). After the mold (die) is closed, the hydraulic cylinder is actuated and the plunger forces the melt into the die at pressures between 14 and 28 MPa (2,000–4,000 psi). After the melt solidifies, the die is opened, the part ejected, and the cycle repeated.

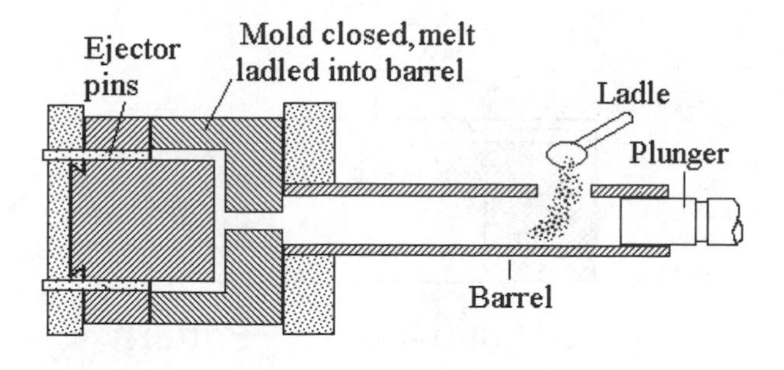

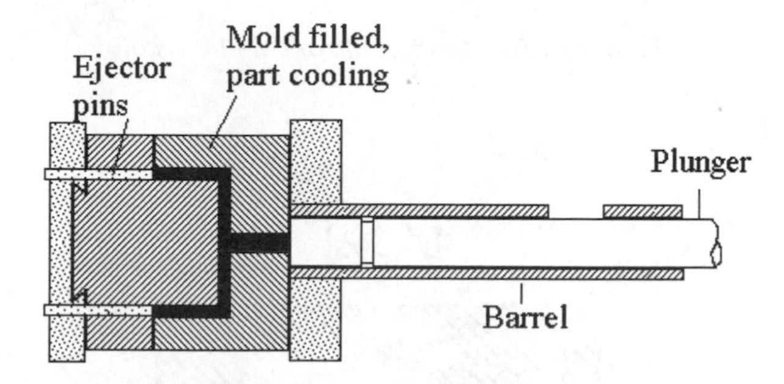

FIGURE 6.8 *Cold chamber die casting machine.*

Because the higher temperatures used in casting aluminum and copper alloys significantly shorten the life of hot chamber machines, cold chamber machines are often used (Figure 6.8). In a cold chamber machine, molten metal from a separate holding furnace is ladled into the cold chamber sleeve after the mold is closed. The melt is then forced into the mold, and after solidification the mold is opened and the part ejected. Injection pressures in this type of machine usually range from 17 to 41 MPa (2,500–6,000 psi). Pressures as high as 138 MPa (20,000 psi) are possible.

Since the molds used in die casting are made of steel, only metals with relatively low melting points can be die cast. The vast majority of castings are made of either zinc alloys or aluminum alloys. Zinc alloys are used for most ornamental or decorative objects; aluminum alloys are used for most nondecorative parts.

6.5 OTHER CASTING PROCESSES

Die casting, investment casting, and sand casting are the most commonly used casting processes. However, other casting processes, such as centrifugal casting, carbon dioxide mold casting, permanent mold casting, plaster mold casting, shell mold casting, and ceramic casting are also used.

In centrifugal casting, molten metal is poured into a mold that is revolving about a horizontal or vertical axis. Horizontal centrifugal casting is used to produce rotationally symmetric parts, such as pipes, tubes, bushings, and other parts. Vertical centrifugal casting can be used to produce both symmetrical as well as nonsymmetrical parts. However, since only a reasonable amount of imbalance can be tolerated for a nonsymmetrical part, the most common shapes produced are cylinders and rotationally symmetric flanged parts. Centrifugal casting of metal produces a finer grain structure and thinner ribs and webs than can be achieved in ordinary static mold casting.

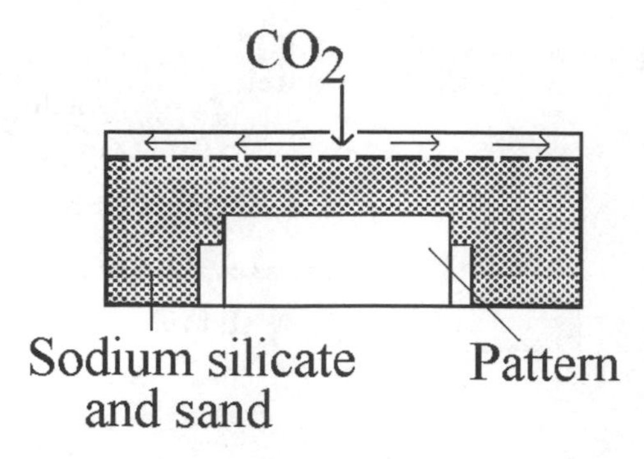

FIGURE 6.9 *Carbon dioxide mold casting.*

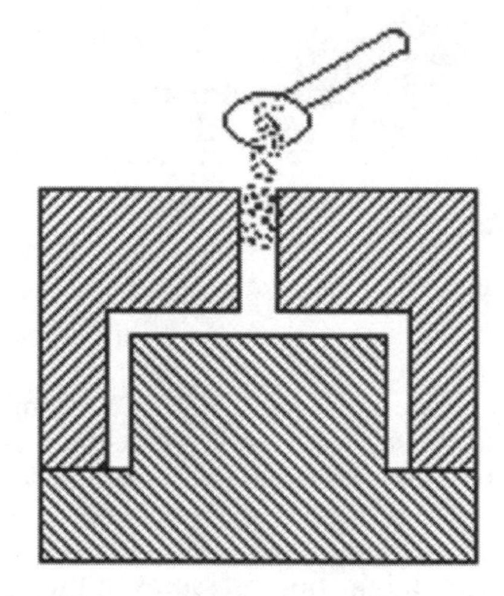

FIGURE 6.10 *Permanent mold casting.*

In carbon dioxide mold casting (Figure 6.9), sodium silicate (water glass) is used as a binder in place of the clay binders used in conventional sand molds and cores. In this process a low strength mold is made with a mixture of sodium silicate and sand. When CO_2 gas is sent through the mixture the mold is hardened. The pattern can then be removed and the melt added. CO_2 molds are used when closer tolerances than those attainable through sand casting are needed.

In permanent mold casting, also referred to as gravity die casting, molten metal is poured by gravity into a reusable permanent mold made of two or more parts (Figure 6.10). This process is closely related to die casting; however, the tolerances and surface finishes achievable by this process are not as good as those obtainable by "pressure" die casting. Because of the high pressures used during filling of the mold during die casting, die casting can produce more complex shapes than achievable via permanent mold casting. Gravity die casting accounts for less than 5% of all die castings produced.

In plaster mold casting, molds are made by coating a pattern with plaster and allowing it to harden. The pattern is then removed and the plaster mold baked.

Ceramic molding is similar to plaster molding. In ceramic molding a fine-grain slurry is poured over the pattern and allowed to set chemically.

Shell mold casting is a process in which an expendable mold is formed by pouring a resin-coated sand onto a heated pattern. The sand bonds together to

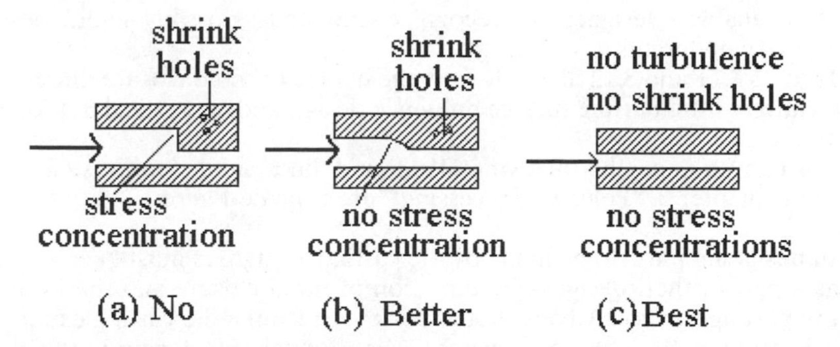

FIGURE 6.11 *Alternative designs to eliminate shrink holes and stress concentrations.*

form a hardened shell that corresponds to the outer shape of the pattern. Two shell halves are put together to form the single-use mold. This process is used for the production of small parts that require a finer tolerance than is obtainable via sand casting. If better tolerances and surface finishes are required, then investment casting and pressure die casting are necessary.

Details concerning each of these processes can be found in *ASM Metals Handbook*.

6.6 QUALITATIVE DFM GUIDELINES FOR CASTING

All casting processes are internal flow processes in which molten metal flows into and fills a die cavity. Then the liquid is cooled to form a solid, and finally the part is removed from the mold by either destroying the mold or, as in the case of die casting, ejecting the part from the mold. The physical nature of these processes—flow, cooling to solidify, and, in the case of die casting, ejection—provides the basis for a number of the qualitative DFM guidelines or rules of thumb that have been established. Many of these rules are similar to the ones discussed earlier for injection molding, compression molding, and transfer molding (Chapter 3, "Polymer Processing"). For example, parts should ideally be designed so that:

1. The flow can be smooth and fill the cavity evenly;
2. Cooling, and hence solidification, can be rapid to shorten cycle time and uniform to reduce warpage; and
3. If ejection is needed, it can be accomplished with as little tooling complexity as possible.

Figure 6.11 shows some examples of alternative designs used to reduce the development of stress concentrations and shrink holes. As shown in Figure 6.11, major differences in section thickness cause turbulent flow and uneven cooling, which results in shrink cavities and voids. Abrupt changes of sections also create shrink stresses. *ASM Metals Handbook* contains additional qualitative design guidelines for castings.

As in injection molding, to design parts properly for die casting, designers must consider the effect of mold closure direction and parting surface location on tooling costs. The location of the parting surface, the direction of closure, and the design of the part must be considered simultaneously in order to provide for ejection of the part from the mold after solidification. Knowing the mold closure direction enables designers to recognize and, thus, possibly avoid designing unnecessary undercuts.

Figures 3.11 and 3.12 illustrate how the choice of mold closure direction and the location of the parting surface influence design and, in particular, tool design and tool cost.

For convenience, the following DFM guidelines similar to those first introduced in Chapter 3, "Polymer Processing," are repeated here:

1. In designing parts to be made by die casting, designers must keep in mind—as a part of their design—the direction of mold closure and the location of the parting surface. Advice should be sought from a die-casting expert, since it is really impossible to do much design for manufacturing in this process without considering the mold closure direction and parting surface location.
2. An easy to manufacture part must be easily ejected from the die, and dies will be less expensive if they do not require special moving parts (such as side cores) that must be activated in order to allow parts to be ejected. Since undercuts require side cores, parts without undercuts are less costly to cast. Some examples of undercuts are shown in Figures 3.2 and 3.3. With knowledge of the mold closure direction and parting surface, designers can make tentative decisions about location(s) of features (holes, projections, etc.) in order to avoid undercuts wherever possible.
3. Because of the need for metal to flow through the die cavity, parts that provide relatively smooth and easy internal flow paths with low flow resistance are desirable. For example, sharp corners and sudden changes or large differences in wall thickness should be avoided because they both create flow problems. Such features also make uniform cooling difficult and result in the development of shrink cavities.
4. Thick walls or heavy sections will slow the cooling process. Thus, parts with no thick walls or other thick sections are less costly to produce. Although reducing wall thickness does generally reduce strength, decreasing section thickness in die casting does not proportionately reduce casting strength. Reducing wall thickness will reduce cycle time. This rapid cooling rate for thin sections yields castings with better mechanical properties. Thick sections, on the other hand, suffer from a coarse crystalline structure that results in internal voids and porosity that reduces strength.
5. In addition, every effort should be made to design parts of uniform, or nearly uniform, wall thickness. If there are both thick and thin sections in a part, solidification may proceed unevenly causing difficult to control internal stresses and warping. Remember, too, that the thickest section largely determines solidification time, and hence total cycle time.
6. We do not discuss gate location in this book except in this paragraph and in a similar paragraph in Chapter 3. However, in large or complex parts, two or more gates may be required through which metal will flow in two or more streams into the mold. There will therefore be fusion lines in the part where the streams meet inside the mold. The line of fusion may be a weak region, and it may also be visible. Therefore, designers who suspect that multiple gates may be needed for a part should discuss these issues with casting experts as early as possible in the design process. With proper design and planning, the location of the fusion lines can usually be controlled as needed for appearance and functionality.

As noted earlier in Chapter 3, "Polymer Processing," these DFM "rules" are not absolute, rigorous laws. If there are designs that have great advantages for function or marketing, then those designs can be given special consideration. Manufacturing engineers can sometimes solve the problems that may be associated with highly desirable functional but difficult to manufacture designs at a cost low enough to justify the benefit.

However, relatively easy to manufacture designs should always be sought. More often than not, a design can be found that will be both efficient from a functional viewpoint and relatively easy to manufacture.

6.7 SUMMARY

This chapter has described some of the most common metal casting methods and materials used for the economical production of both complex parts (with significant levels of geometric detail) and simple parts (with little geometric detail). Included in this chapter was a discussion of design for manufacturing issues as they apply to the production of metal castings. The chapter concluded with a set of qualitative DFM guidelines for die casting.

REFERENCES

Ashby, M. F. "Materials Selection in Mechanical Design," Pergamon Press, New York, 1992.

American Society of Metals International. *American Society of Metals Handbook*, Vol. 15. "Casting." Metals Park, OH: ASM International, 1988.

Bralla, J. G., editor. *Handbook of Product Design for Manufacturing.* New York: McGraw-Hill, 1986.

Kalpakjian, S. *Manufacturing Engineering and Technology.* Reading, MA: Addison-Wesley, 1989.

Poli, C., Dastidar, P., and Graves, R. A. "Design Knowledge Acquisition for DFM Methodologies." *Research in Engineering Design*, Vol. 4, no. 3, 1992, pp. 131–145.

Wick, C., editor. *Tool and Manufacturing Engineers Handbook*, Vol. 2. "Forming, 9th edition." Dearborn, MI: Society of Manufacturing Engineers, 1984.

QUESTIONS AND PROBLEMS

6.1 As we saw in the previous chapter in the case of injection molding, as the production volume increased the unit cost of the part decreased. Would you expect the same to be true in die casting? Why? Would you also expect the same to be true in the case of sand casting? Explain.

6.2 Explain the difference between an expendable mold and a permanent mold. What are the advantages and disadvantages of each?

6.3 Patterns are often made oversize to provide for so called "allowances." List three reasons why a pattern should be made oversize if a part is to be made as a sand casting.

6.4 Quite often in the past consumer products made of plastic were considered to be low-quality products but those made of metal were considered to be high-quality products. Assume that you work for a company considering a switch from plastic to metal for some of the parts used in their top of the line products. As a result of this decision you have been asked to reexamine those parts discussed in Problems 3.1 to 3.6 under the assumption that the parts are to be die cast instead of injection molded. Would you change any of the recommendations you made earlier when you assumed the parts were to be injection molded? Explain!

6.5 Shown in Figure P6.5 are pictorial drawings of some of the thin-walled parts considered in Problem 6.4. Would you anticipate any difficulty in sand casting these parts? Explain.

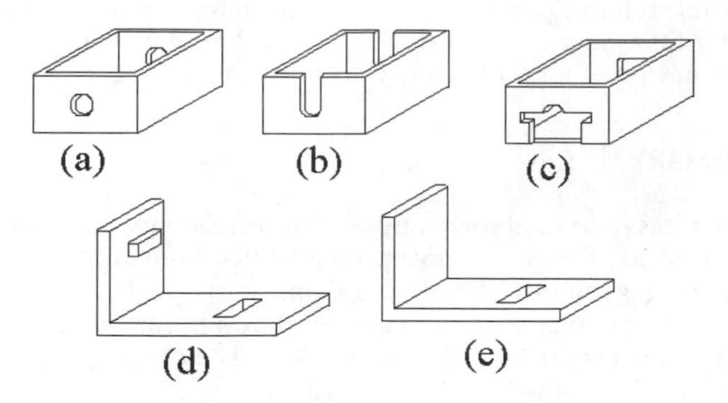

FIGURE P6.5.

6.6 For the same parts shown in Figure P6.5 would you anticipate any difficulty in creating these parts via investment casting?

Chapter 7

Die Casting: Total Relative Part Cost

7.1 DIE CAST PART COSTS—OVERVIEW

The processes of injection molding and die casting are very similar (see Figure 7.1). In both cases a melt is injected under pressure into a metal mold. The melt then cools, shrinks, and solidifies, taking on the shape of the mold. The mold then opens and the part is ejected. Because the processes are so similar, much of what has been said concerning the influence of part geometry and tolerance specifications on injection molding, tooling costs, and processing costs applies equally well to die casting. For example, molds should contain as few moving parts as possible, hence, the mold closure direction and the parting surface location are important and the part should be easy to eject (Figure 7.2). For ease of flow of the melt, be it an alloy or a resin, smooth paths and low flow resistance are important (Figure 7.3). Finally, thin uniformly thick walls are desirable to shorten cycle times and solidification.

In die casting, as in injection molding, the three major cost components of a part are material cost, tooling cost, and processing cost. These three cost components are influenced, to varying degrees, by the geometry, size, and material of the part as well as by subsidiary factors such as part quality requirements. The same coding system for tooling costs applies, with only minor modifications, to die casting.

The purpose of the sections that follow are to discuss the application of the previous coding systems and cost models to die casting. The definition of terms introduced earlier will not be repeated here.

7.2 RELATIVE TOOLING COST

7.2.1 Relative Tooling Construction Cost

As in injection molding, tooling cost is a function of die construction cost and die material cost. Die construction cost in turn is a function of basic complexity, subsidiary complexity, and tolerance. Basic complexity itself depends on part size (small, medium, large), the number and type of undercuts present, and the parting surface location. Thus, as in the case of injection molding, the total relative mold construction cost is given by

$$C_{dc} = C_b C_s C_t \qquad \text{(Equation 7.1)}$$

Figures 7.4 and 7.5, as well as Tables 7.1 and 7.2, apply to die casting and are used in precisely the same way that Figures 4.1 and 4.19, and Tables 4.1 and 4.2, are used for injection molding.

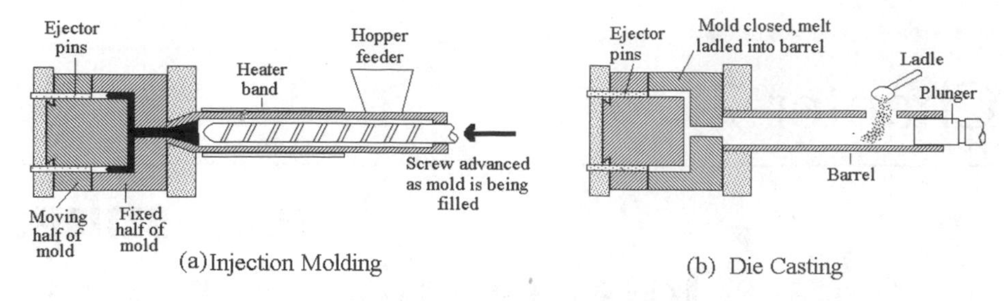

FIGURE 7.1 *A box-shaped part being injection molded and die cast.*

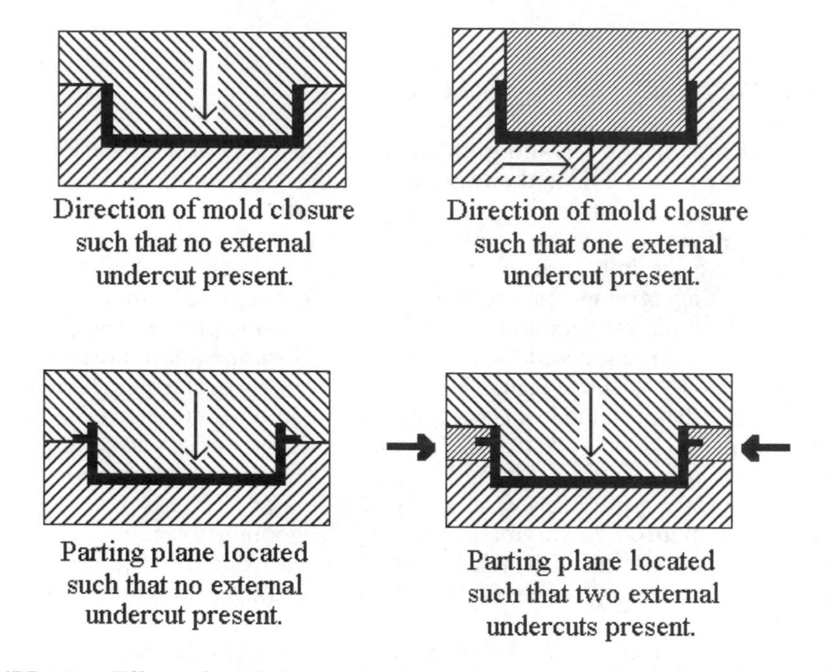

FIGURE 7.2 *Effects of mold closure direction and parting line location on die complexity.*

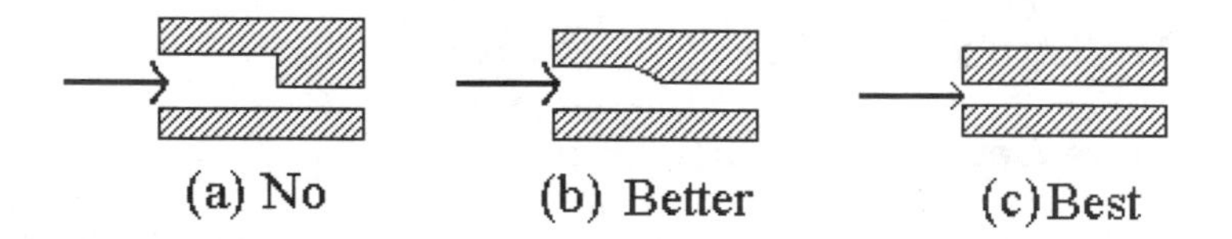

FIGURE 7.3 *Effect of part design on melt flow.*

As indicated in Figure 7.4, internal undercuts and internal threads generally cannot be die cast. This is due primarily to the combination of high temperatures, high pressure, and high injection velocities required for die casting. This combination of temperatures, pressures, and velocities result in rapid wear of the form pins used to create internal undercuts that, in turn, creates excessive flash that is difficult and costly to remove. In addition to the problem of flash removal, in the case of internal threads, high shrinkage makes extraction of the thread-forming unit very difficult.

1 in = 25.4 mm; 100 mm/25.4 mm = 3.94 in

Key: 6 — Flat Parts 5.37 / Box-Shaped Parts 6.28

BASIC COMPLEXITY				SECOND DIGIT										
				L ≤ 250 mm (4)				250mm < L ≤ 480mm				L > 480 mm		
				Number of External Undercuts (5)				Number of External Undercuts (5)				Number of External Undercuts (5)		
				zero	one	two	More than two	zero	one	two	More than two	zero	one	More than one
FIRST DIGIT				0	1	2	3	4	5	6	7	8	9	10
Parts Without Internal Undercuts (1)	Parts whose peripheral height from a planar dividing surface is constant (2)	Part in one half(3)	0	1.00 / 1.64	1.23 / 1.87	1.38 / 2.02	1.52 / 2.16	1.42 / 2.89	1.65 / 3.12	1.79 / 3.27	1.94 / 3.41	1.83 / 4.28	2.07 / 4.51	2.33 / 4.77
		Part not in one half(3)	1	1.14 / 1.86	1.37 / 2.09	1.52 / 2.24	1.66 / 2.38	1.61 / 2.99	1.84 / 3.22	1.99 / 3.37	2.13 / 3.51	2.09 / 4.42	2.32 / 4.66	2.58 / 4.92
	Parts whose peripheral height from a planar Dividing Surface is not constant – or – parts with a non-planar Dividing Surface(2)		2	1.28 / 1.92	1.51 / 2.15	1.66 / 2.29	1.80 / 2.44	1.81 / 3.38	2.04 / 3.61	2.19 / 3.76	2.33 / 3.90	2.34 / 5.01	2.58 / 5.24	2.84 / 5.50
Parts With Internal Undercuts (1)	Internal undercuts can not be generally die cast.													

FIGURE 7.4 *Classification system for basic complexity, C_b, die casting. (The numbers in parentheses refer to notes found in Chapter 4, Appendix 4.A.)*

Feature		Number of Features (n)	Penalty per Features	Penalty
Holes or Depressions	Circular		2n	
	Rectangular		4n	
	Irregular		7n	
Bosses	Solid (8)		n	
	Hollow (8)		3n	
Non-peripheral ribs and/or walls and/or rib clusters (8)			3n	
Side Shutoffs	Simple (9)		2.5n	
	Complex (9)		4.5n	
Lettering (10)			n	
			Total Penalty	

SMALL PARTS (L ≤ 250 mm)

Total Penalty ≤ 10 => Low cavity detail
10 < Total Penalty ≤ 20 => Moderate cavity detail
20 < Total Penalty ≤ 40 => High cavity detail
Total Penalty > 40 => Very high cavity detail

MEDIUM PARTS (250 < L ≤ 480 mm)

Total Penalty ≤ 15 => Low cavity detail
15 < Total Penalty ≤ 30 => Moderate cavity detail
30 < Total Penalty ≤ 60 => High cavity detail
Total Penalty > 60 => Very high cavity detail

LARGE PARTS (L > 480 mm)

Total Penalty ≤ 20 => Low cavity detail
20 < Total Penalty ≤ 40 => Moderate cavity detail
40 < Total Penalty ≤ 80 => High cavity detail
Total Penalty > 80 => Very high cavity detail

1 in = 25.4 mm; 100 mm/25.4mm = 3.94 in

FIGURE 7.5 *Cavity detail—die casting. (The numbers in parentheses refer to notes found in Chapter 4, Appendix 4.A.)*

Table 7.1 Subsidiary complexity rating, C_s, for die casting. (The numbers in parentheses refer to notes found in Chapter 4, Appendix 4.A.)

				Fourth Digit	
				Without Extensive (7) External Undercuts (5)	With Extensive (7) External Undercuts (5)
				0	1
Third Digit	Cavity Detail (6)	Low	0	1.00	1.25
		Moderate	1	1.25	1.45
		High	2	1.60	1.75
		Very High	3	2.05	2.15

All of the tooling used in die castings is produced with essentially the same surface finish. The effect of molten metal on the surface of the dies is such that deterioration of the surface occurs so quickly that various grades of surface finish are unwarranted. Thus, while the part itself may have different levels of surface quality (as will be explained later in Sections 7.6 and 7.7), all of the tooling is produced with the same surface finish. This fact is reflected in Table 7.2.

7.2.2 Relative Mold Material Cost

As in the case of injection molding, most of the tooling used to produce die castings is composed of pre-engineered, standardized mold base assemblies and components. Thus, the equations and curves developed earlier (Section 4.8) for the determination of the relative mold material cost, C_{dm}, are equally applicable here. For convenience, most of Section 4.8 is repeated below.

To compute the total relative tooling costs, we must be able to estimate the mold material cost as well as its construction costs. This is relatively easy to do from a knowledge of the approximate size of a part—which in turn dictates the required size of the mold.

Table 7.2 Tolerance rating, C_t, for die casting.

			Sixth Digit	
			Commercial Tolerance, T_a	Tight Tolerance, T_a
			0	1
Fifth Digit	Surface Finish, R_a	0	—	—
		1	1.00	1.05
		2		
		3		

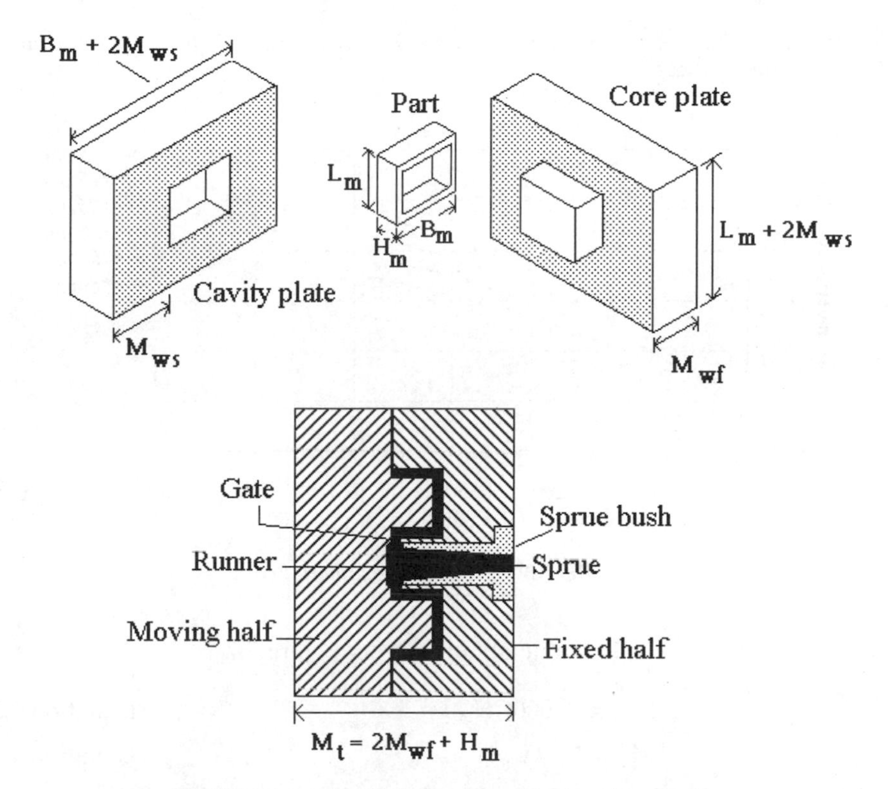

FIGURE 7.6 *Mold dimensions for two-plate mold.*

Referring to Figure 7.6, we define the following mold dimensions:

M_{ws} = Thickness of the mold's side walls (mm)
M_{wf} = Thickness of the core plate (mm)
L_m and B_m = The length and width of the part in a direction normal to the mold closure direction (mm)
H_m = The height of the part in the direction of mold closure (mm) (H_m not necessarily equal to H)
M_t = The required thickness of the mold base (mm)

With these definitions, the following equations can be used sequentially to determine the projected area of the mold base, M_a, and the required thickness of the mold base, M_t, which in turn are used to obtain the relative mold material cost (C_{dm}) from Figure 7.8.

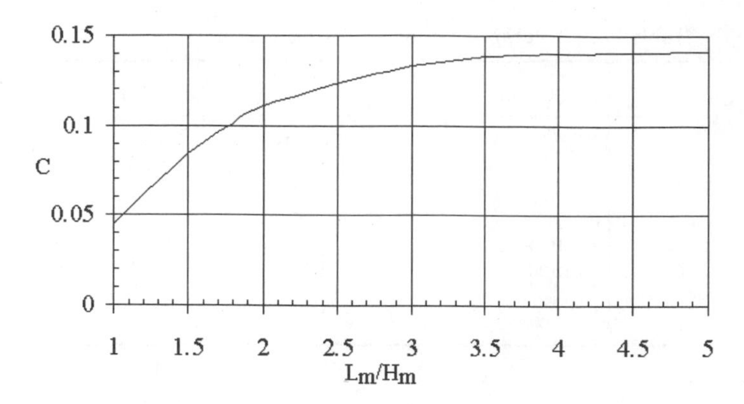

FIGURE 7.7 *Value of C for use in Equation 7.2. (If $L_m/H_m < 1$, then use the value of H_m/L_m to determine C.)*

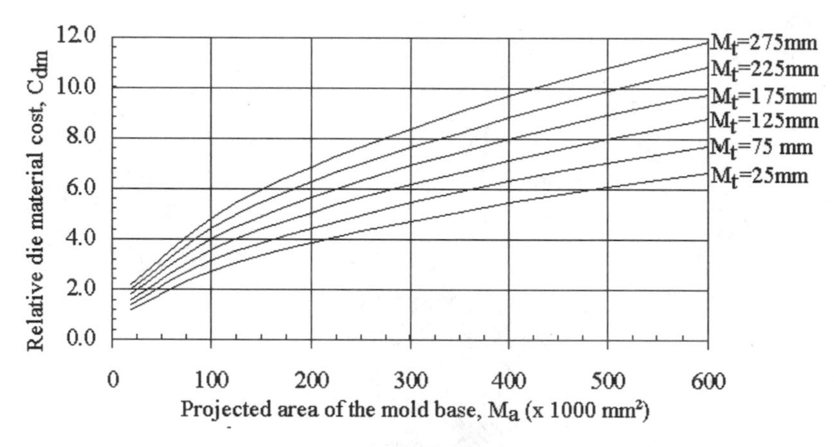

FIGURE 7.8 *Relative die material cost.*

C = value obtained from Figure 7.7.

$$M_{ws} = [0.006CH_m^{\ 4}]^{1/3} \quad \text{(Equation 7.2)}$$

$$M_{wf} = 0.04\,L_m^{\ 4/3} \quad \text{(Equation 7.3)}$$

$$M_a = (2M_{ws} + L_m)(2M_{ws} + B_m) \quad \text{(Equation 7.4)}$$

$$M_t = (H_m + 2M_{wf}) \quad \text{(Equation 7.5)}$$

C_{dm} = value obtained from Figure 7.8

7.2.3 Total Relative Mold Cost

The total relative mold cost of a part, C_d, is given by the expression

$$C_d = 0.8C_{dc} + 0.2C_{dm} \quad \text{(Equation 7.6)}$$

where C_d is the total mold cost of a part relative to the mold cost of the standard part, C_{dc} is the mold construction cost relative to the standard, and C_{dm} is the mold material cost relative to the mold material cost of the standard part.

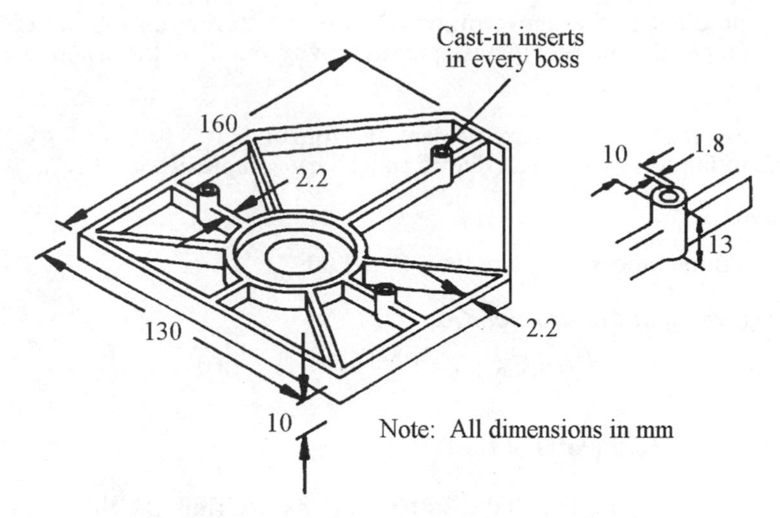

FIGURE 7.9 *Original design for Example 7.1.*

7.3 EXAMPLE 7.1—RELATIVE TOOLING COST FOR A DIE-CAST PART

7.3.1 The Part

As an example, we consider the part shown in Figure 7.9. Although the wall thickness of the part is specified along with the boss height and boss thickness, for an evaluation of the relative tooling cost of the part these dimensions are not necessary.

7.3.2 Relative Tool Construction Cost

Basic Complexity

In this example, the length of the projections parallel to the surface of the part, c, is 10 mm. The ratio of c to the envelope dimension parallel to c is less than 1/3 and so the projections are considered isolated projections of small volume. Consequently, the dimensions of the basic envelope are

$$L = 160\,\text{mm}, \quad B = 130\,\text{mm}, \quad H = 10\,\text{mm}$$

Since L/H is greater than 4, the part is flat.

The direction of mold closure is in the direction normal to the LB plane; thus a planar dividing surface exists for the principal shape of the part.

No internal undercuts are present, the part has a planar dividing surface, the peripheral height from that dividing surface is constant, and the part is in one-half. Thus, the first digit is 0.

There are no external undercuts and L is less than 250 mm. Hence, the second digit is also 0.

Thus, from Figure 7.1, the basic mold manufacturability cost, C_b, is 1.00.

Subsidiary Complexity

There are 8 radial ribs (24 penalty points), 1 concentric rib (3 penalty points), 3 hollow bosses (9 penalty point), and 1 circular hole (2 penalty point). Although it is not indicated in the above drawing, there is lettering on the underside of the part (1 penalty points). Consequently, cavity detail is high, and the third digit is 2.

Since no external undercuts are present, the fourth digit is 0.
Thus the multiplying factor, C_s, due to subsidiary complexity, is 1.60.

Tolerance

Commercial tolerances are used; thus, C_t is 1.00.

Total Relative Mold Construction Cost, C_{dc}

$$C_{dc} = C_b C_s C_t = 1.00(1.60)(1) = 1.60$$

7.3.3 Relative Mold Material Cost

From Figure 7.7, for L_m/H_m (160/13) of 12.3, C is 0.14. Thus, the thickness of the mold wall is

$$M_{ws} = [0.006CH_m^{\ 4}]^{1/3} = [0.006(0.14)(13)^4]^{1/3} = 2.9\,mm$$

and the thickness of the base is

$$M_{wf} = 0.04\,L_m^{\ 4/3} = 0.04(160)^{4/3} = 34.7\,mm$$

Consequently, the projected area of the mold base is

$$M_a = [2(2.9) + 160][2(2.9) + 132] = 22{,}847\,mm^2$$

and the required plate height is

$$M_t = [13 + 2(34.7)] = 82.4\,mm$$

Hence, from Figure 7.8, the relative die material cost, C_{dm}, for this part is approximately 1.2.

7.3.4 Total Relative Mold Cost

$$C_d = 0.8C_{dc} + 0.2C_{dm} = 0.8(1.60) + 0.2(1.2) = 1.52$$

7.3.5 Redesign Suggestions

Die manufacturability costs can be reduced if the cavity detail can be reduced. For example, if the radial ribs are removed the cavity detail would be reduced from high to moderate, and C_s would become 1.25. The total relative mold construction cost, C_{dc}, then becomes 1.25. This is a 22% reduction in die construction costs.

7.4 WORKSHEET FOR RELATIVE TOOLING COST

The determination of the relative die construction costs, the relative die material costs, and the overall relative die costs is a straightforward, though sometimes cumbersome, procedure. The Worksheet introduced in Chapter 4 for injection molding can also be used here for die casting. A blank version of the worksheet is shown in Appendix 4.B.

7.5 PROCESSING COSTS—OVERVIEW

As has already been stated, the processes of injection molding and die casting are somewhat similar. In fact, they are sufficiently similar that the classification system for the determination of relative cycle time developed for injection molding, with only minor modifications, is applicable to die casting.

As in the case of injection molding, cycle time, hence processing costs, is more a function of localized features and details than of overall part geometry. Once again parts are classified as partitionable or non-partitionable (see Figure 7.10). In addition, as in the case of injection molding, for those parts that can be partitioned into elemental plates, machine cycle time is affected by:

1. Whether or not the part is slender,
2. Maximum plate or wall thickness,
3. The type of elemental plate (framed, grilled, etc),
4. The type of subsidiary features (significant ribs, significant bosses, multidirectional ribs, presence of gusset plates on projections), and, unlike the case of injection molding,
5. Whether or not the elemental plate is an internal (difficult to cool) plate or external (easy to cool) plate.

In addition, unlike injection molding, wall thicknesses up to 14 mm are commonly die cast.

Figure 7.11 shows the part-coding system and database used to account for the above factors on machine cycle time for die-cast parts.

7.6 PRODUCTION YIELD AND EFFECTIVE CYCLE TIME

As in all manufacturing situations, production yield is defined as the percentage of acceptable parts produced in a given period of time. As in injection molding,

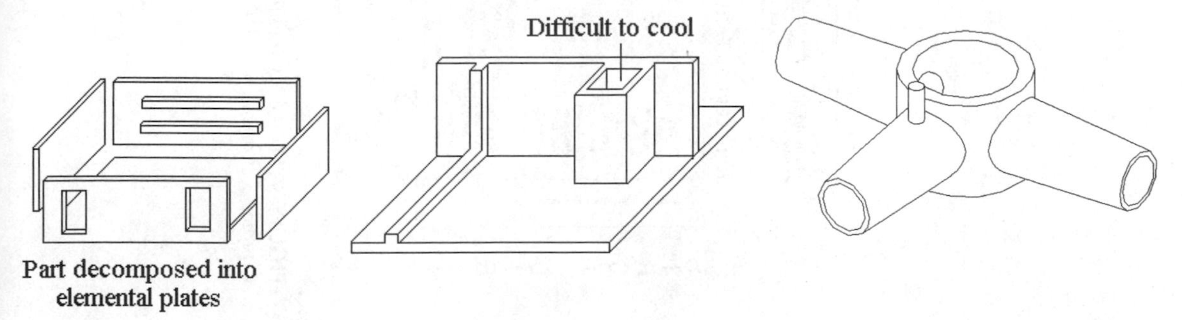

FIGURE 7.10 *Partitionable (part on the left) and non-partitionable parts due to difficult to cool feature (part in the center) or complex geometry (part on the right).*

External Plate / Internal Plate (each cell: 1.77 = External Plate, 2.01 = Internal Plate, written as External / Internal)

SECOND DIGIT

NOT RECOMMENDED (columns 6–10)

SLENDER PARTITIONABLE PARTS

First digit				w < 1 mm (0)	1mm ≤ w ≤ 2mm (1)	2mm < w ≤ 3mm (2)	3mm < w ≤ 4mm (3)	4mm < w ≤ 5mm (4)	5mm < w ≤ 6mm (5)	6mm < w ≤ 8mm (6)	8mm < w ≤ 10mm (7)	10mm < w ≤ 12mm (8)	12mm < w ≤ 14mm (9)	w > 14mm (10)
Parts with $L_u/B_u > 10$ (1) or Frames (2)	Plates with $L_u/2w < 100$ (3) without lateral projections (4)	Without ribs	0	Difficult to fill or eject	1.00 / 1.25	1.24 / 1.61	1.48 / 1.97	1.72 / 2.33	1.96 / 2.68	2.33 / 3.22	2.81 / 3.94	3.29 / 4.65	3.77 / 5.37	Too Thick
		With ribs	1		1.24 / 1.61	1.48 / 1.97	1.72 / 2.33	1.96 / 2.68	2.21 / 3.04	2.57 / 3.58	3.05 / 4.29	3.53 / 5.01	4.01 / 5.72	
	Plates with $L_u/2w \geq 100$ and/or plates with lateral projections (4)	Without ribs	2		1.48 / 1.97	1.72 / 2.33	1.96 / 2.68	2.21 / 3.04	2.45 / 3.40	2.81 / 3.94	3.29 / 4.65	3.77 / 5.37	4.25 / 6.06	
		With ribs	3		1.72 / 2.33	1.91 / 2.68	2.21 / 3.04	2.45 / 3.40	2.69 / 3.76	3.05 / 4.29	3.53 / 5.01	4.01 / 5.72	4.49 / 6.44	

NON-SLENDER PARTITIONABLE PARTS

First digit					w < 1 mm (0)	1mm ≤ w ≤ 2mm (1)	2mm < w ≤ 3mm (2)	3mm < w ≤ 4mm (3)	4mm < w ≤ 5mm (4)	5mm < w ≤ 6mm (5)	6mm < w ≤ 8mm (6)	8mm < w ≤ 10mm (7)	10mm < w ≤ 12mm (8)	12mm < w ≤ 14mm (9)	w > 14mm (10)
Parts with $L_u/B_u \leq 10$ (1)	Plates without significant rib (5) or significant bosses (6) with or without non-peripheral ribs or bosses	Plates which are grilled or slotted (7)	Without non-peripheral ribs	0	Difficult to fill or eject	1.77 / 2.01	1.84 / 2.12	1.91 / 2.26	1.98 / 2.33	2.05 / 2.42	2.14 / 2.57	2.28 / 2.76	2.42 / 2.99	2.56 / 3.20	Too Thick
			With non-peripheral ribs	1		1.84 / 2.12	1.91 / 2.26	1.98 / 2.33	2.05 / 2.42	2.12 / 2.53	2.21 / 2.69	2.35 / 2.90	2.49 / 3.09	2.63 / 3.30	
		Plates which are not grilled or slotted (7)	Without non-peripheral ribs	2		1.91 / 2.26	1.98 / 2.33	2.05 / 2.42	2.12 / 2.53	2.19 / 2.63	2.28 / 2.76	2.42 / 2.99	2.56 / 3.20	2.70 / 3.41	
			With concentric or cross ribbing	3		1.98 / 2.33	2.05 / 2.42	2.12 / 2.53	2.19 / 2.63	2.24 / 2.74	2.35 / 2.90	2.49 / 3.09	2.63 / 3.30	2.77 / 3.52	
			With radial or unidirectional ribbing	4		2.05 / 2.42	2.12 / 2.53	2.19 / 2.63	2.24 / 2.74	2.31 / 2.84	2.42 / 2.99	2.56 / 3.20	2.70 / 3.41	2.84 / 3.60	
	Plates with significant ribs (5) and/or significant bosses (6)	Plates with rib and/or boss thickness less than the wall thickness (8)	Ribs/bosses supported by gusset plates	5		2.12 / 2.53	2.19 / 2.63	2.24 / 2.74	2.31 / 2.84	2.38 / 2.95	2.49 / 3.09	2.63 / 3.30	2.77 / 3.52	2.92 / 3.71	
			Ribs/bosses not supported by gusset plates	6		2.19 / 2.63	2.24 / 2.74	2.31 / 2.84	2.38 / 2.95	2.46 / 3.04	2.56 / 3.20	2.70 / 3.41	2.84 / 3.60	2.99 / 3.82	
		Plates with rib and/or boss thicknes (8) greater than or equal to the wall thickness		7		2.24 / 2.74	2.31 / 2.84	2.38 / 2.95	2.46 / 3.04	2.53 / 3.14	2.63 / 3.30	2.77 / 3.52	2.92 / 3.71	3.06 / 3.92	

NON-PARTITIONABLE PARTS

First digit			w < 1 mm (0)	1mm ≤ w ≤ 2mm (1)	2mm < w ≤ 3mm (2)	3mm < w ≤ 4mm (3)	4mm < w ≤ 5mm (4)	5mm < w ≤ 6mm (5)	6mm < w ≤ 8mm (6)	8mm < w ≤ 10mm (7)	10mm < w ≤ 12mm (8)	12mm < w ≤ 14mm (9)	w > 14mm (10)
Parts which are not partitionable	Easy to cool (10)	0	Difficult to Fill	2.24	2.31	2.38	2.46	2.53	2.63	2.77	2.92	3.06	Too Thick
	Difficult to cool (10)	1		2.74	2.84	2.95	3.04	3.14	3.30	3.52	3.71	3.92	

FIGURE 7.11 *Classification system for basic relative cycle time (aluminum die castings). (The numbers in parentheses refer to notes found in Chapter 5, Appendix 5.A.)*

surface "quality" requirements of a part are the main cause for variations in part yield. A low production yield reduces the number of acceptable parts that are produced in a given time, and thus increases the "effective cycle time" to some value that is higher than the actual machine cycle time.

In die casting, as in injection molding, mold temperatures affect the surface quality of a part. Cool molds result in parts with more visible flaws such as blisters (Figure 7.12), flow lines (Figure 7.13), and others, and result in a shorter cycle time. Hot molds yield a smoother more flaw-free surface (Figure 7.14) and result

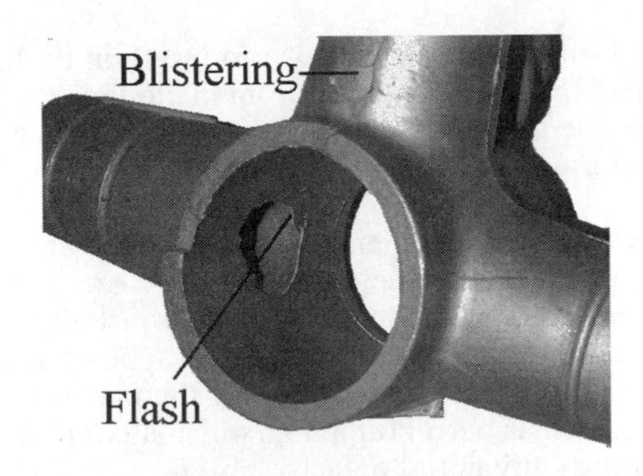

FIGURE 7.12 *Die-cast part with blistering.*

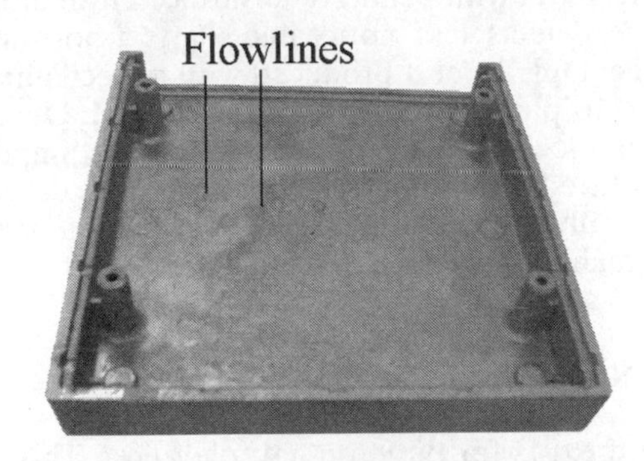

FIGURE 7.13 *Die-cast part with flow lines.*

FIGURE 7.14 *Die-cast part with no blisters or flow lines.*

in a longer cycle time. In addition, and more importantly, high surface quality requirements may greatly reduce production yield because of a higher rejection rate due to visible sink marks, flow (jet) lines, and other surface flaws.

Part yield is also influenced by part tolerance requirements. Part yield decreases—and effective cycle time increases—when part tolerances are difficult to hold. Tolerances are discussed in more detail below.

7.7 SURFACE FINISH

Although a standard surface smoothness is produced on tooling when it is constructed, the surface of the parts produced from the tooling can vary, as explained above in Section 7.6, due to diverse operating conditions. In addition, die age and die temperature can have serious effects on the as-cast surface finish of the resulting part.

As-cast surface finish requirements can be divided into three categories: mechanical grade, paint grade, and high grade.

A *mechanical grade* surface is one where all surface defects are acceptable. Such parts are generally used in a location where appearance is not important. A mechanical grade surface is also acceptable if the part is to be subjected to a secondary operation, such as barrel tumbling, which modifies the original surface and completely obscures any initial defects.

A *paint grade* surface requirement refers to a part on which minor surface defects are allowable. Parts with paint grade surface finish are generally painted to make the surface defects less noticeable. Parts whose strength would be reduced to an unacceptable level if produced with a mechanical grade finish are sometimes upgraded to paint grade and left unpainted. Once again such parts are generally used in a location where appearance is not important.

A *high grade* surface finish is one in which no surface defects are acceptable. These parts are generally subjected to a secondary plating operation that would make surface flaws highly noticeable.

7.8 PART TOLERANCES

An industry standard exists for dimensional tolerances. This standard was prepared by the American Die Casting Institute. To determine whether the tolerances specified are tight or commercial, one must refer to the data published by this Institute. No industrywide standard exists for geometrical tolerances.

Part yield is influenced by part tolerance requirements. Part yield decreases when part tolerances are difficult to hold. Part tolerances that are difficult to hold in injection moldings are also difficult to hold in die casting. Thus, the presence of external undercuts, nonuniform wall thickness, unsupported projections, and walls make part tolerances difficult to hold. In addition, the specification of tolerances across the parting surface of the dies, as well as the specification of more than three tight or five commercial tolerances, makes tolerances difficult to hold.

As is always the case, tolerances should be specified only where absolutely necessary. As the number of tolerances to be held increases, molders and die casters have difficulty in maintaining these tolerances and the proportion of defective parts produced increases. In general, molders and die casters can routinely hold two or three tolerances and have difficulty in maintaining five or six tolerances.

Tables 7.3 and 7.4 provide the part coding system and database used to account for the effect of part surface finish and tolerance on the relative cycle time of aluminum die castings.

Table 7.3 Additional relative time, t_e, due to inserts. Note 12 can be found in Chapter 5, Appendix 5.A.

Third Digit	Parts without cast-in inserts (12)	0	0.0
	Parts with cast-in inserts (12)	1	0.5*

*Per insert

Table 7.4 Time penalty, t_p, due to surface requirements and tolerances. Note 14 can be found in Chapter 5, Appendix 5.A.

				Fifth Digit	
				Tolerances not difficult to hold (14)	Tolerances difficult to hold (14)
				0	1
Fourth Digit	Cast surface finish	Mechanical Grade	0	1.00	1.05
		Paint Grade	1	1.11	1.16
		High Quality	2	1.16	1.22

7.9 EXAMPLE 7.2—DETERMINATION OF RELATIVE CYCLE TIME FOR A PARTITIONABLE PART

7.9.1 The Part

The part is shown in Figure 7.9. All of the necessary part dimensions, wall thickness, rib and boss sizes, and other specifications are specified. In addition, the part requires a high-grade surface finish since it is to be subjected to a secondary plating operation by the customer. The part also contains three cast-in inserts.

7.9.2 Part Partitioning

The part shown in Figure 7.9 consists of a single elemental plate and thus no partitioning is needed.

7.9.3 Relative Cycle Time

Basic Machine Cycle Time

As in Example 7.1 in Section 7.3, neglecting the three isolated projections of small volume, the dimensions of the basic envelope of this part are:

$$L = 160, \quad B = 130, \quad H = 10$$

Since the part is straight, then the unbent dimensions L_u and B_u are equal to L and B, respectively. Thus, $L_u/B_u < 10$, and the plate is non-slender, N.

The length, l, the width, b, and the height, h, of each projection (boss) are 10 mm, 1.8 mm, and 13 mm, respectively. The plate thickness w is 2.2 mm. Thus, $h/w > 3$, and the projections are considered significant bosses. Since the boss thickness is less than the wall thickness and the bosses are not supported by gusset plates, the first digit is 6.

The nominal plate thickness is 2.2 mm, hence the second digit is 2. The plate, as discussed in Section 5.4, is an external plate, thus the basic code is N62E and the basic relative machine cycle time, t_b, is 2.24.

Inserts

There are three cast-in inserts. Hence, the third digit is 1 and the relative extra mold opening time, t_e, is 0.5 per insert or 1.5 for the part.

Surface Finish/Tolerances

Since a high-grade surface finish is required, the fourth digit is 2. Because the bosses are unsupported it is difficult to hold tolerances between them, hence, the fifth digit is 1 and the production penalty, t_p, is 1.22.

Relative Effective Cycle Time

The relative effective cycle time for the original design is,

$$t_r = (t_b + t_e)t_p = (2.24 + 1.5)(1.22) = 4.56$$

7.9.4 Redesign Suggestions

The relative cycle time can be reduced in several ways. One method is to provide gusset plates to support the bosses. In this case the basic code becomes N52E and the fifth digit becomes 0. Better still, the bosses could be redesigned so that they are no longer significant. The basic code then becomes N42E and t_b becomes 2.12. In this case the fifth digit is still 0 and t_r becomes 4.20—about an 8% reduction.

A more significant method for reducing the relative cycle time would be to remove the cast-in inserts. If this is done, t_e is reduced to 0 and the relative effective cycle time becomes 2.46, about a 46% reduction.

7.10 TOTAL RELATIVE PART COST

As in the case of injection molding, the total production cost of a die casting is

$$K_t = K_m + \frac{K_d}{N} + K_e \qquad \text{(Equation 7.7)}$$

where K_m represents the material cost of the part, K_d represents the total cost of the tool, N represents the number of parts, or production volume, to be produced using that tool, and K_e represents the processing cost.

The total cost of the part relative to the cost of the reference part, C_r, is given by Equation 5.11, or

$$C_r = C_m f_m + (C_d/N) f_d + C_e f_e \qquad \text{(Equation 7.8)}$$

where

$$C_m = K_m/K_{mo},$$

$$C_d = K_d/K_{do},$$

$$C_e = K_e/K_{eo}$$

and C_m, C_d, and C_e represent the material cost, tooling cost, and processing cost of a part relative to the reference part. The reference part in this case is an aluminum die casting, a washer, whose OD = 65 mm and ID = 55 mm, and whose thickness is 2 mm.

The values of f_m, f_d, and f_e represent the ratio of the material cost, tooling cost, and processing cost of the reference part to the total manufacturing cost of the reference part, that is

$$f_m = K_{mo}/K_o,$$

$$f_d = K_{do}/K_o, \text{ and}$$

$$f_e = K_{eo}/K_o$$

As in the case of injection molding, the values for C_d, C_e, and C_m are obtained from the following expressions, namely,

$$C_d = 0.8C_{dc} + 0.2C_{dm} \quad \text{(Equation 7.9)}$$

$$C_e = t_r C_{hr} \quad \text{(Equation 7.10)}$$

$$C_m = (V/V_o)C_{mr} \quad \text{(Equation 7.11)}$$

The values for C_{dc} and C_{dm} are obtained by the methods discussed in Section 7.2, while the value of t_r is obtained by use of the method illustrated in Example 7.2 in Section 7.9. The values for C_{hr} for die-casting machines is obtained from Table 7.5.

Since, in general, the vast majority of die castings used for nondecorative purposes are made of aluminum, the only material considered here and the only material for which the database contained in Figure 7.4 is applicable to is aluminum. Thus, $C_{mr} = 1$ and

$$C_m = V/V_o \quad \text{(Equation 7.12)}$$

Table 7.6 contains some relevant data for the reference part. Using that data, the cost of the reference die-cast part, in dollars, can be shown to be

$$K_o = 0.311 + 6980/N_o \text{ (dollars)}$$

where the first term is the sum of the material and processing costs for the reference part and the second term is the tooling cost for the reference part. Once again it is seen that at low-production volumes most of the cost is due to the cost of the tooling and that at very high production volumes, the cost is due primar-

Table 7.5 Machine tonnage and relative hourly rate for die-casting machines. (Data obtained from the die-casting industry, June, 1989.)

Machine Tonnage	*Relative Hourly Rate, C_{hr}*
<100	1.00
100–199	1.05
200–299	1.08
300–399	1.12
400–499	1.17
500–599	1.21
600–699	1.26
700–799	1.29
800–899	1.33
900–999	1.38
1000–1199	1.45
1200–1499	1.59
>1500	1.73

Table 7.6 Relevant data for the reference part.

Part Material	*Aluminum*
Material Cost	0.0006 cents/mm^3
Part Vol., V_o	1885 mm^3
Die Mat. Cost	$980
Die Const. Time	200 hours
Labor Rate, Die Const	$30/hr
Cycle Time	17.23 s
Die Cast Machine, Hourly Rate	$62.57/hr

ily to material and processing costs. When N approaches infinity the cost of the reference part approaches $0.31. Thus, as expected, the cost of the reference part depends upon its production volume.

As in the case of injection molding, the main concern here is the comparison of alternative designs for a given part. Hence, the actual cost of the reference part is not important. For this reason it becomes convenient to obtain the relative cost of a part with respect to the standard part when its total production cost, K_o, is $1. For this particular reference part K_o is $1 when the production volume is 10,127. At this production volume the values of f_m, f_e, and f_d become

$$f_m = K_{mo}/K_o = 0.0113$$

$$f_e = K_{eo}/K_o = 0.2995$$

$$f_d = K_{do}/K_o = 6980$$

and

$$C_r = 0.0113C_m + (6980/N)C_d + 0.2995C_e \qquad \text{(Equation 7.13)}$$

7.11 EXAMPLE 7.3—DETERMINATION OF THE TOTAL RELATIVE PART COST

For the part considered earlier in Figure 7.9 it was found, in Section 7.3, that the relative tool cost for the part is $C_d = 1.52$. In addition, it was also found that the relative cycle time for the part is $t_r = 4.56$.

The volume of the part is about 52,000 mm^3 hence,

$$C_m = V/V_o = 52{,}000/1885 = 27.6$$

As in injection molding, the press tonnage required is given by the expression

$$F_p = 0.005A_p$$

where it is assumed that A_p is in mm^2. In this case $A_p = 20{,}000$ mm^2, thus,

$$F_p = 100 \text{ tons}$$

and, from Table 7.6, $C_{hr} = 1.05$. Thus,

$$C_e = t_r C_{hr} = 4.79$$

Production Volume = 10,000

From Equation 7.13, for a production volume of 10,000

$$C_r = 0.0113(27.6) + (6980/10{,}000)(1.52) + 0.2995(4.56) = 2.74$$

If the part is redesigned as discussed in Section 7.3 and 7.9, then $C_d = 1.24$, $t_r = 2.46$, and $C_e = 2.58$. Thus,

$$C_r = 0.0113(27.6) + (6980/10{,}000)(1.24) + 0.2995(2.58) = 1.95$$

This is an overall savings of some 30% in piece part cost.

Production Volume = 100,000

If the production volume of the original part is 100,000 pieces, then

$$C_r = 0.0113(27.6) + (6980/10{,}000)(1.52) + 0.2995(4.56) = 1.78$$

For the redesigned part,

$$C_r = 0.0113(27.6) + (6980/100{,}000)(1.24) + 0.2995(2.58) = 1.17$$

Thus, for a production volume of 100,000 parts, the overall savings in part cost is almost 34%. This increase in savings is due to the fact that the redesign of the part reduced the relative cycle time. Thus, as the production volume increased, the overall savings also increased.

7.12 WORKSHEET FOR RELATIVE PROCESSING COST AND TOTAL RELATIVE PART COST

To facilitate the calculation of the relative processing cost and the overall relative part cost, the worksheet introduced in Chapter 5, "Injection Molding: Total Relative Part Cost," for injection molding can also be used here, with two small changes, for die casting. The two changes required to the worksheet are changes in the coefficients of C_m and C_e. The correct coefficients for die casting can be found in Equation 7.13. A blank copy of the worksheet is available in Appendix 5.B, and may be reproduced for use with this book.

7.13 SUMMARY

This chapter has described a systematic approach for calling designers' attention to those features of a die casting which tend to increase the cost to manufacture parts—and for estimating the relative tooling costs, processing costs, material costs, and the overall part cost. The system is quite similar to the one for injection molding and highlights those features that significantly increase cost so that designers can minimize difficult to produce features.

REFERENCES

Fredette, Lee. "A Design Aid for Increasing the Producibility of Mold Cast Parts." M.S. Final Project Report, Mechanical Engineering Department, University of Massachusetts at Amherst, Amherst, MA, 1989.

Kuo, Sheng-Ming. "A Knowledge-Based System for Economical Injection Molding." Ph.D. Dissertation, University of Massachusetts at Amherst, Amherst, MA, Feb. 1990.

Poli, C, Fredette, L., and Sunderland, J. E. "Trimming the Cost of Die Castings." *Machine Design* 62, (March 8, 1990): 99–102.

Poli, C., Kuo, Sheng-Ming, and Sunderland, J. E. "Keeping a Lid on Mold Processing Costs." *Machine Design* 61, (Oct. 26, 1989): 119–122.

Shanmugasundaram, S. K. "An Integrated Economic Model for the Analysis of Mold Cast and Injection Molded Parts." M.S. Final Project Report, Mechanical Engineering Department, University of Massachusetts at Amherst, Amherst, MA, August 1990.

QUESTIONS AND PROBLEMS

7.1 As part of the same training program discussed in Problem 4.1 of Chapter 4, you are in the process of explaining to some new hires at DFM.com that the tool construction cost of die-cast parts is a function of what you have called the *basic complexity* of the part. Explain in detail exactly which features of a part affect the basic complexity of a die casting.

7.2 As a continuation of Problem 7.1, imagine that you are in the process of comparing the basic complexity of both die-cast parts and injection-molded parts. During this particular session, a trainee asks if there are any features that in theory can be die cast or injection molded but for practical reasons should be avoided. How would you respond to this question?

7.3 In addition to the basic complexity of a part discussed in Problem 7.1, what other factors affect the tool construction costs of a die-cast part?

7.4 Describe the differences, if any, between those factors that affect the tool construction costs of die-cast parts and those factors that affect the tool construction costs of injection-molded parts.

7.5 As a continuation of Problem 5.4 in Chapter 5, assume that you are considering changing the component, which is presently made of nylon 6 in a mold having an SPI finish of 3, to an aluminum die casting. Estimate and compare the relative cycle times and the total cycle times for the two alternative designs. Assume that the die casting is to be of paint quality.

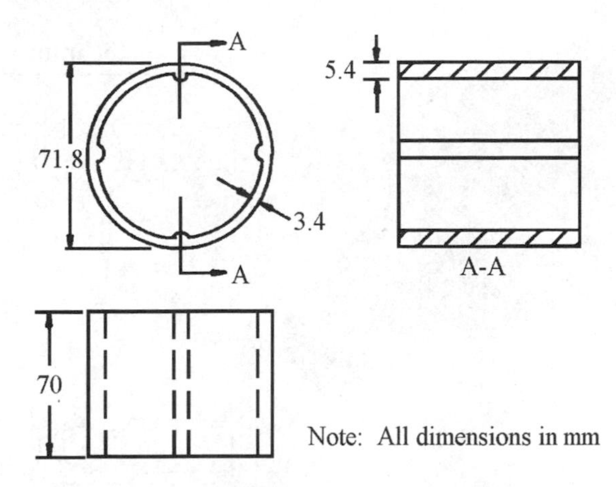

FIGURE P7.5

7.6 Assume that you are considering two possible designs for the part shown in Figure P7.5. In one case the part is to be made of nylon 6 in a mold having an SPI finish of 3 (see Problem 5.4), and in the other case the part is to be made as a paint-quality die-cast aluminum part (see Exercise 7.5). Assuming that the total cost of the reference part in each case is $1, which part is less costly to produce at a production volume of 10,000? Which part is less costly to produce at a production volume of 100,000? What are the limitations to the calculations made to determine the total part cost in this exercise?

7.7 Assume that the part shown in Figure P7.7 is to be an aluminum die casting. Determine the relative cycle time for the part under the assumption that it is to be produced with a paint-grade finish.

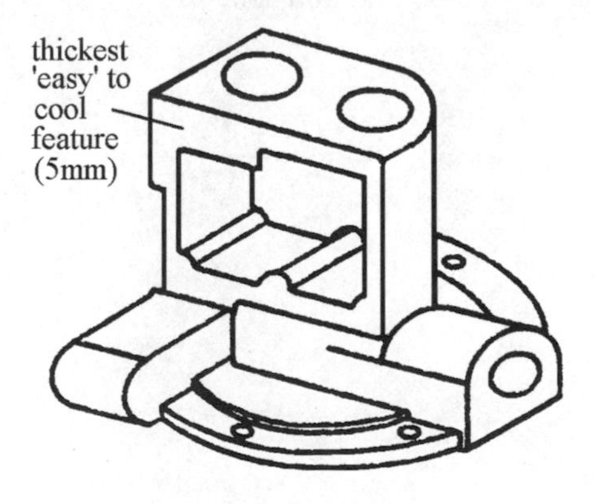

FIGURE P7.7

7.8 The aluminum die-cast part shown in Figure P7.8 was originally produced with a mechanical grade finish. What would you estimate to be the percentage increase in cost to produce the part if it is to be produced with a high-quality finish at a production volume of 25,000? Assume that the maximum wall thickness of the part is 5 mm.

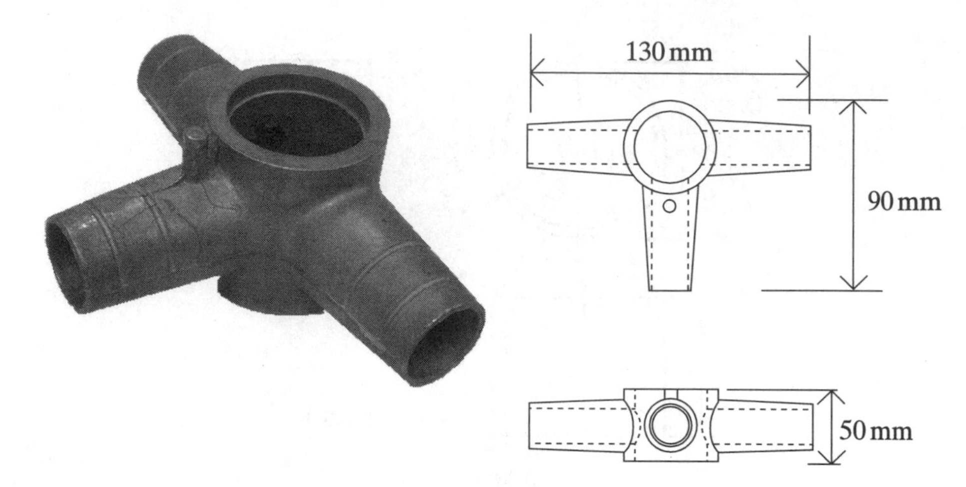

FIGURE P7.8

7.9 A study of Figure 5.2 and Figure 7.11 shows that the group technology-based coding system for injection-molded parts and die castings is essentially the same. These figures also show that the relative cycle time database for injection moldings is higher than those for a geometrically equivalent die casting. This difference is particularly true as the parts become more difficult to cool. Does this seem reasonable? Explain!

7.10 For Example 7.3 it was assumed that the cycle time for the reference part, t_o, is 17.23 s (see Table 7.6). Suppose that the cycle time t_o is in fact 10 s. If this is the case, what would be the reduction in costs between the original design and the redesign for the part analyzed in Example 7.3 at production volumes of 10,000 and 100,000?

Chapter 8

Sheet-Metal Forming

8.1 INTRODUCTION

From your previous study of materials you are probably aware that properties of metal alloys are accompanied by such terms as cast and wrought. Cast metals are those metals that have been formed by casting. Wrought metals are those that have been formed by such bulk deformation processes as rolling, drawing, extruding, or forging (see Chapter 11, "Other Metal Shaping Processes").

In Chapter 6, "Metal Casting Processes," we discussed various casting processes. In this chapter we will be discussing sheet-metal forming in which wrought metals, formed by rolling, are used to produce thin-walled parts.

Sheet-metal forming consists of two broad categories, namely, stamping and fabricating. Stamping is a process in which thin-walled metal parts (less than about 6.25 mm or 0.25 inches) are shaped by means of punches and dies driven by mechanical or hydraulic presses. Examples of parts made by stamping are can openers, fan blades, pulleys, ash trays, razor blades, buckles, kitchen utensils, cans, bottle caps, range tops, and other items. Stamping is also used to produce a wide variety of parts for machines, power tools, appliances, automobiles, hardware, office equipment, electrical equipment, and clothing.

Fabricating is a process of forming thick-walled parts (greater than about 6.25 mm or 0.25 inches) or large thin-walled parts using single or compound dies with or without non-press operations.

8.2 THE STAMPING PROCESS

Stamping can be divided into two broad categories of press operations:

1. Shearing or cutting operations (Figure 8.1) in which holes, slots, grooves, and other features are created by exceeding the shear strength of the material; and
2. Non-shearing operations such as forming (Figure 8.2), drawing (Figure 8.3), bending (Figure 8.4), and other operations in which various shapes are created by exceeding the tensile or compressive strength of the material.

Shearing is carried out by cutting a sheet that has been placed between a sharp punch and sharp die (Figure 8.5) where the die imposes a rapid, high shearing stress to the sheet. Shearing can be used for several purposes:

1. To produce blanks, which are then subjected to further shearing, bending, and/or forming;
2. To produce features such as holes, slots, notches, perforations, and lances;

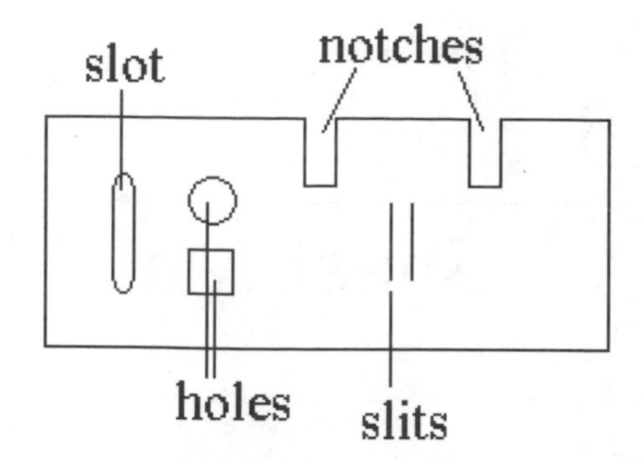

FIGURE 8.1 *Various features created using a shearing operation.*

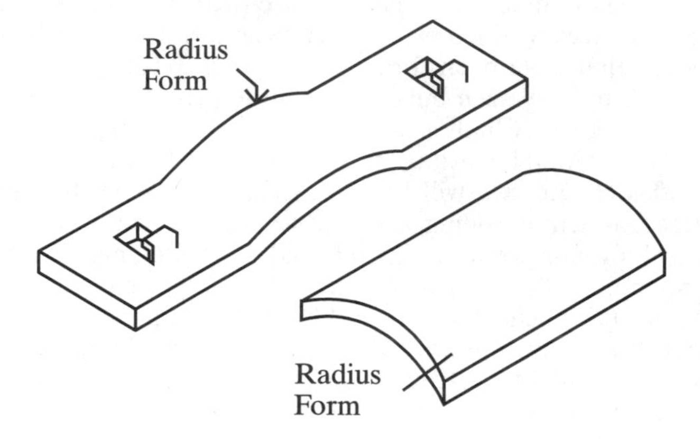

FIGURE 8.2 *Radius form features created using one or more large radius punch and die sets.*

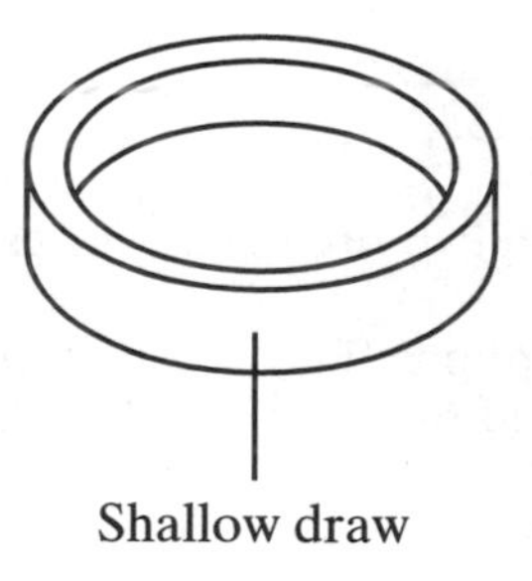

FIGURE 8.3 *A shallow drawn part created using one or more drawing operations.*

3. To separate and remove a part from a strip of sheet metal; and
4. To trim and slit a part for size control.

To be certain that shearing, and not drawing or bending, occurs, the clearance between the die and punch must be less than the sheet thickness.

Bending is done using a matched punch and die set (Figure 8.6) or, as is more commonly the case, by using a descending punch to fold or "wipe" the workpiece over the edge of the die (Figure 8.7).

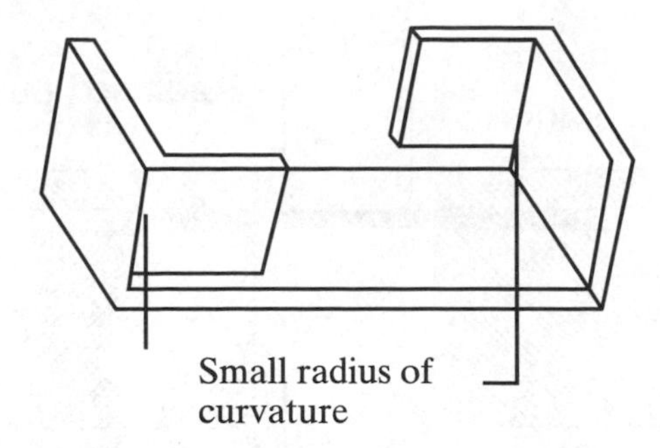

FIGURE 8.4 *A bent part created using a wipe forming operation (see Figure 8.7).*

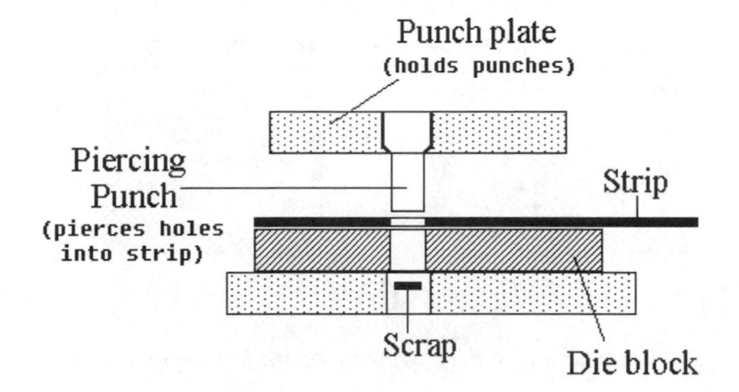

FIGURE 8.5 *Shearing.*

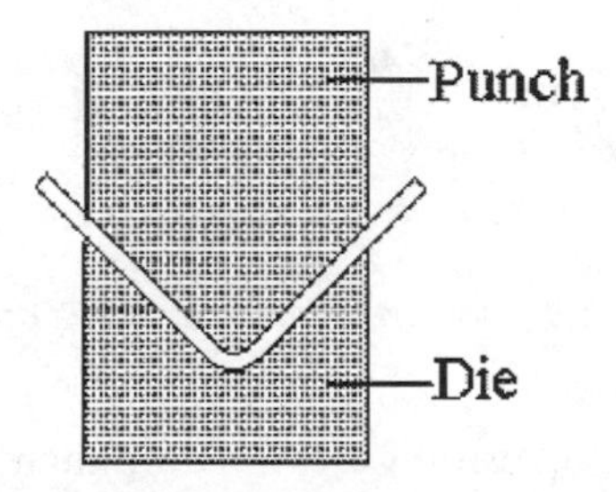

FIGURE 8.6 *Bending about a straight line using a matched punch and die.*

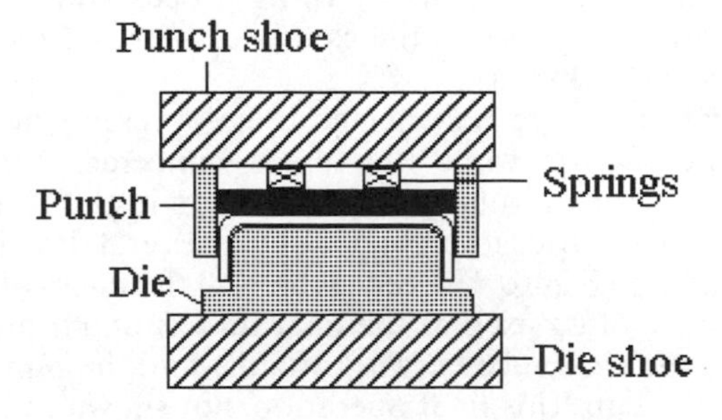

FIGURE 8.7 *Bending about a straight line using wipe down forming. Wipe up is also sometimes used.*

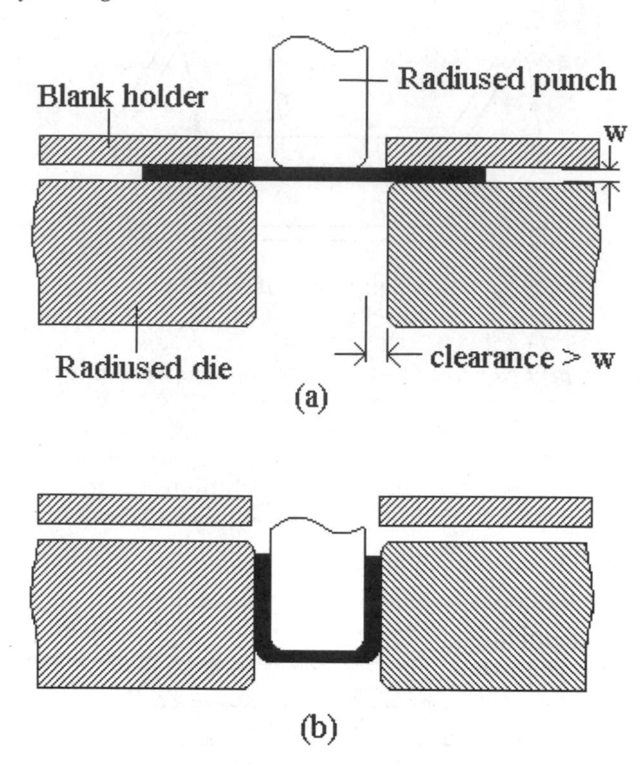

FIGURE 8.8 *An illustration of a deep drawing operation using a circular blank to create a cup-shaped part.*

FIGURE 8.9 *A three-station drawing operation.*

Drawing is a forming operation in which a punch draws a thin flat circular blank of diameter D_b into a die to form a recessed cup whose inner diameter D_p is equal to that of the punch (Figure 8.8). Drawing is used to form such products as pots and pans, beverage containers, bottle caps, fire extinguishers, and other items. To prevent shearing during the drawing process, the punch and dies are provided with rounded corners and the clearance between the punch and die is greater than the sheet thickness.

During drawing the flange is subjected to compressive hoop stresses as the blank is drawn into the die while the cup is subjected to tensile stresses. If the tensile stresses become too large, failure occurs in the form of tearing near the bottom of the cup. In order to prevent failure, several drawing operations are used to produce the cup (Figure 8.9). Since the wall thickness of the cup is usually not uniform, *ironing* is often used as the final station of a drawing operation. In the case of ironing, the clearance between the die and the punch is less than the sheet thickness of the cup. This final operation, not shown in Figure 8.9, is used to produce a cup with a thinner more uniform wall thickness.

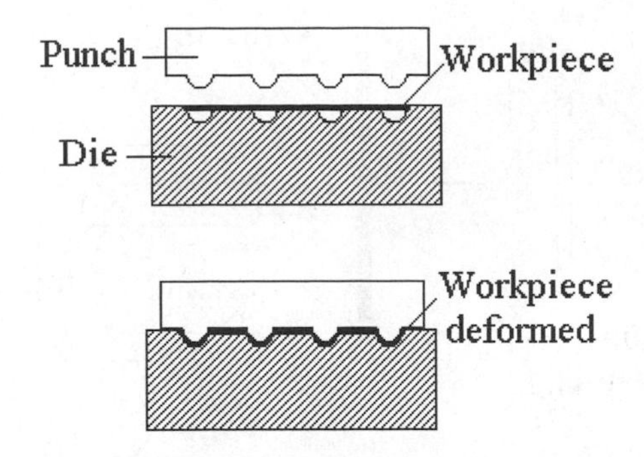

FIGURE 8.10 *An example of embossing.*

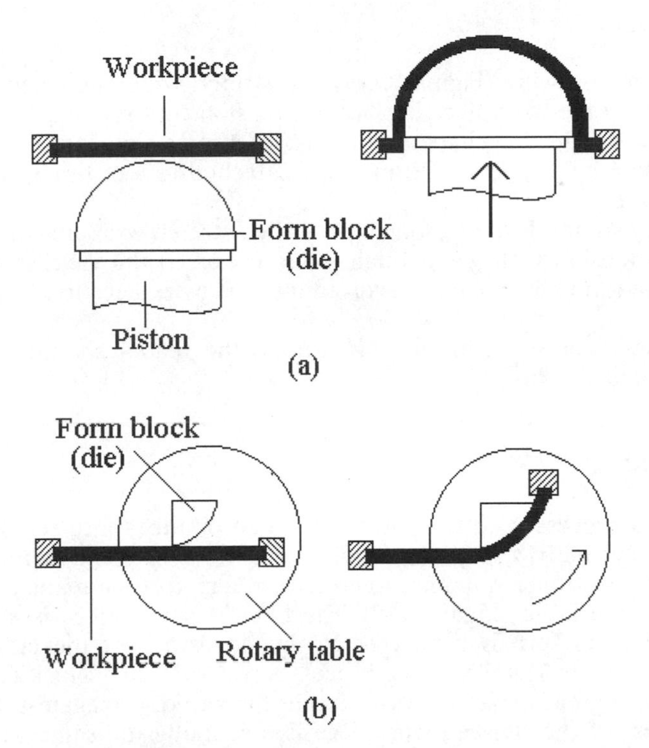

FIGURE 8.11 *Two examples of stretch forming. Stretch draw forming is shown in (a) while rotary stretch wrapping is shown in (b).*

Embossing is a forming operation used to form beads, ribs, and lettering by use of shallow indentations (draws) (Figure 8.10).

8.3 STRETCH FORMING

Stretch forming is a process used to shape a sheet or bar of uniform cross-section, by stretching the sheet or bar over a block or die of the desired shape while holding it in tension. Stretch forming is used extensively in the aerospace industry to produce parts with a large radius of curvature and in the auto industry to sometimes shape body panels. Figure 8.11 shows two forms of stretch forming,

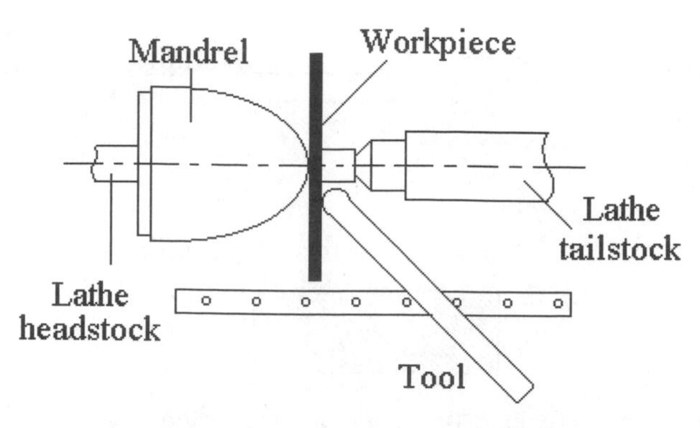

FIGURE 8.12 *Conventional spinning.*

namely, stretch drawing (Figure 8.11a) and stretch wrapping (Figure 8.11b). In stretch forming the workpiece is placed in tension by stretching it over a block that contains the desired shape. In the case of stretch drawing the workpiece is held in tension by grippers. A form block, attached to a piston, gives the workpiece its shape.

In rotary stretch forming, sometimes called stretch wrapping, the workpiece is placed in tension with just enough force to exceed the yield strength of the material. The form block is then revolved into the workpiece to form the desired shape.

For more details concerning this process the reader should consult *ASM Metals Handbook*, Vol. 14.

8.4 SPINNING

Spinning is a process used to form sheet metal or tubes into seamless, hollow, rotationally symmetric shapes such as cylinders, cones, and hemispheres, for example. Conventional spinning, sometimes referred to as manual spinning, is carried out on a lathe (Figure 8.12). The blank or workpiece is attached to a mandrel which in turn is connected to the headstock of the lathe (see also Chapter 11, "Other Metal Shaping Processes"). As the headstock rotates a tool lubricated with grease, soap, or waxes forces the workpiece against the mandrel. The thickness of the finished workpiece is essentially the same as that of the blank.

Manual spinning was originally used primarily for the production of household utensils of tin, silver, and aluminum. Today almost all metals can be spun but some may require several stages in order to produce the final shape and some alloys may require mechanical assistance. Musical instruments, missile bodies, truck wheel discs, and jet engine compressor cases are examples of spun parts.

Figures 8.13 and 8.14 show two examples of power or shear spinning. Power spinning is used primarily for the production of cones and tubes. In cone spinning (Figure 8.13) a roller shears the workpiece so that each element remains at the same distance from the spin axis.

Compared with other sheet-metal forming processes, such as stamping, the tooling costs for spinning are less, setup time is shorter, and design changes can generally be made at lower cost. However, manual spinning requires highly skilled operators and the rates of production are low.

Although sheet-metal forming operations such as stretch forming, spinning, and others are extensively used, from the point of view of consumer products,

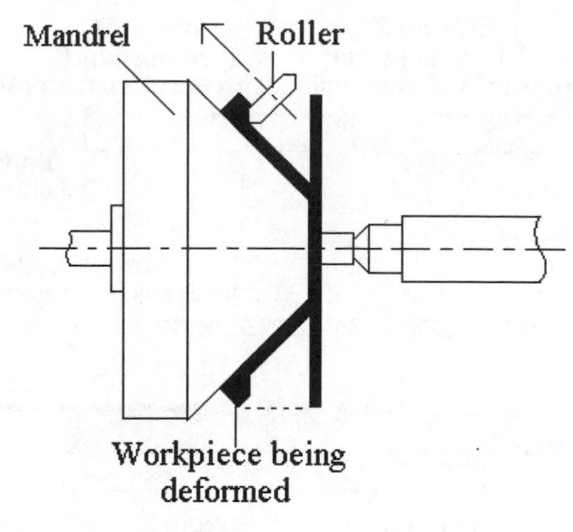

FIGURE 8.13 *Cone spinning.*

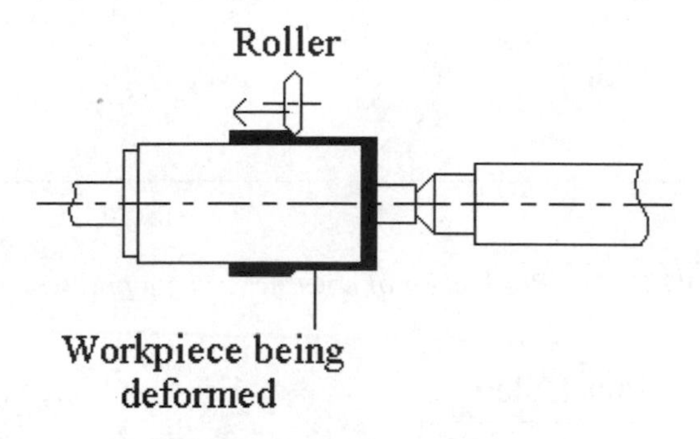

FIGURE 8.14 *Tube spinning.*

the stamping operations are the ones most commonly encountered. For this reason the emphasis on the remainder of this chapter is on stamping. For more details on spinning as well as other sheet-metal forming processes the reader is encouraged to consult *ASM Metals Handbook*, Vol. 14.

8.5 STAMPING DIES

A stamping die is a collection of parts mounted on a press used to produce a part. Although each die is custom-made to produce a given part, many die components are identical or similar. In this section we will concentrate on dies; in the next section we will discuss more about presses.

Most stampings require a series of operations. These operations can be carried out using a single die, called a *progressive die*, or on *compound* or combination dies. *Progressive dies* are divided into sections (called stations) that perform multiple operations with each stroke of the press (Figure 8.15). The part is carried from station to station by the stock strip that is left attached to the part throughout the operation. As the part moves forward it is positively positioned in each workstation. The part is cut out or separated from the strip in the final station.

A *compound* or combination *die* is essentially a single station die that performs two or more operations (one inside of the other) with each stroke of the press.

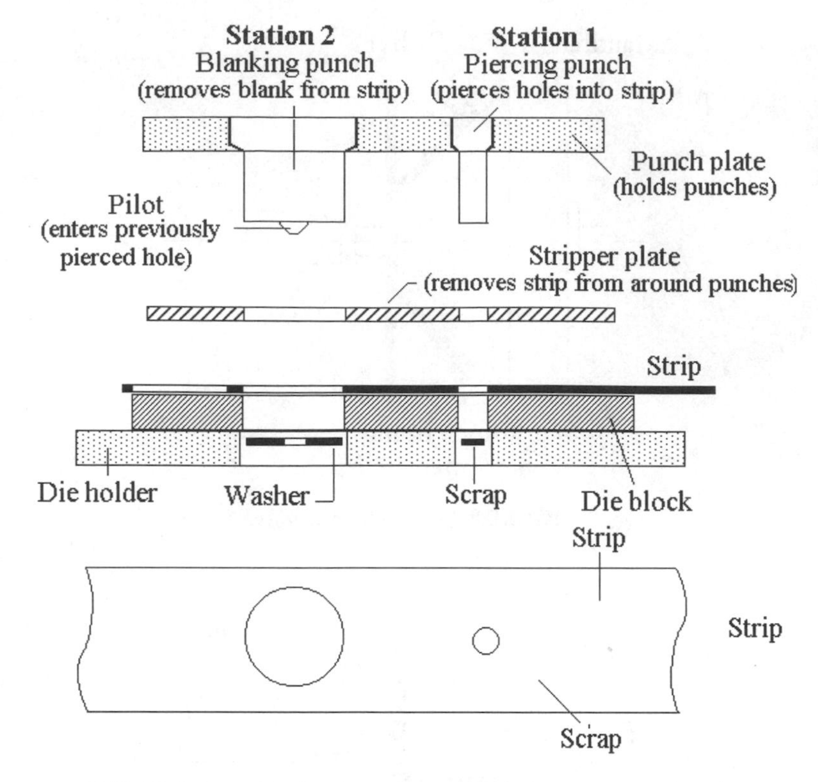

FIGURE 8.15 *Production of a washer using a progressive die.*

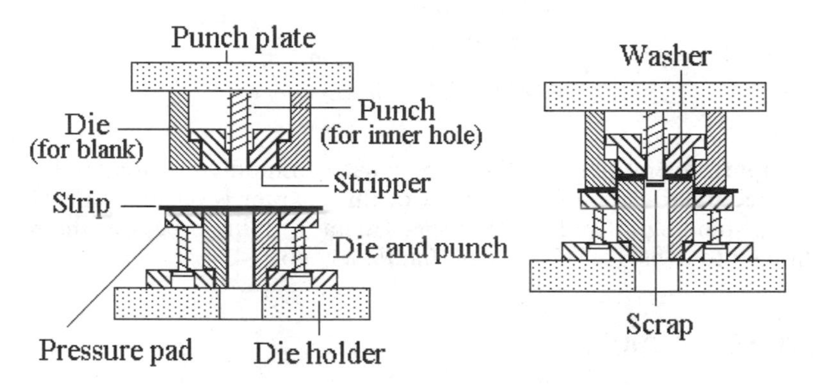

FIGURE 8.16 *Production of a washer using a compound die.*

Figure 8.15 shows a two-station progressive die used for producing a flat washer. At station 1 the inner hole of the washer is punched, and at station 2 the washer is separated from the strip. A top view of the strip as it moves from right to left is shown below the die holder. Figure 8.16 shows a compound die used to produce the same washer.

The geometric configuration of most stampings is such that the part cannot be made in a single compound die. Thus, in addition to the compound die, additional dies, referred to as secondary dies, are needed to perform such operations as bending, trimming, or other such operations on a part that has already been blanked and pierced. Hence, the use of compound dies usually requires the use

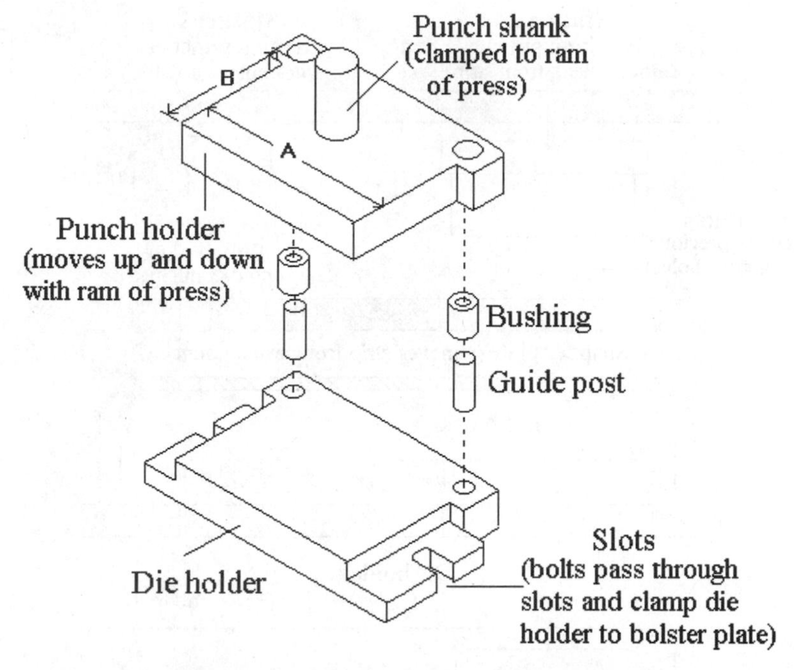

FIGURE 8.17 *Typical die set used for stamping.*

of secondary dies along with "secondary" presses. A progressive die needs only one press. Thus, when annual production volume is high (greater than 50,000 parts) and part geometry permits, progressive dies tend to be used. When annual production volume is low (20,000 parts) or part geometry is such that bends, forms, or hemming occurs around the entire periphery necessitating the use of precut blanks, then compound dies tend to be used. For medium-production volumes both progressive and compound dies are used.

Most stamping operations are carried out on parts less than 200 mm in size using progressive dies. Large parts are not usually produced on progressive dies because the size of the die becomes exorbitantly large.

Compound dies are used primarily for flat blanks. The process "simultaneously" performs piercing of the inside holes and features together with blanking of the outside periphery. Figure 8.16 shows a stamped washer produced using a compound die and one station.

Figure 8.17 shows a typical die set consisting of a stationary lower plate (called the die holder) to which the dies are attached, a movable upper plate to which the punches are attached (called the punch holder), and two sets of precisely fitted guide posts and guide bushings. Such units can be commercially purchased in various sizes and shapes.

Figure 8.18 shows a simplified version of a progressive die assembly including the die set, punches, punch plate (which holds the punches and is attached to the punch holder), dies (in this case in the form of a die block), and a stripper plate for producing a link. A top view of the strip as it moves from right to left is also shown in Figure 8.18.

When the part geometry is such that the workpiece cannot remain attached to the strip, transfer dies are used. With transfer dies, a blank is transferred from station to station by equipment built into or mounted on the press itself. As an example, imagine the link shown in Figure 8.18 with all its edges curled. In this

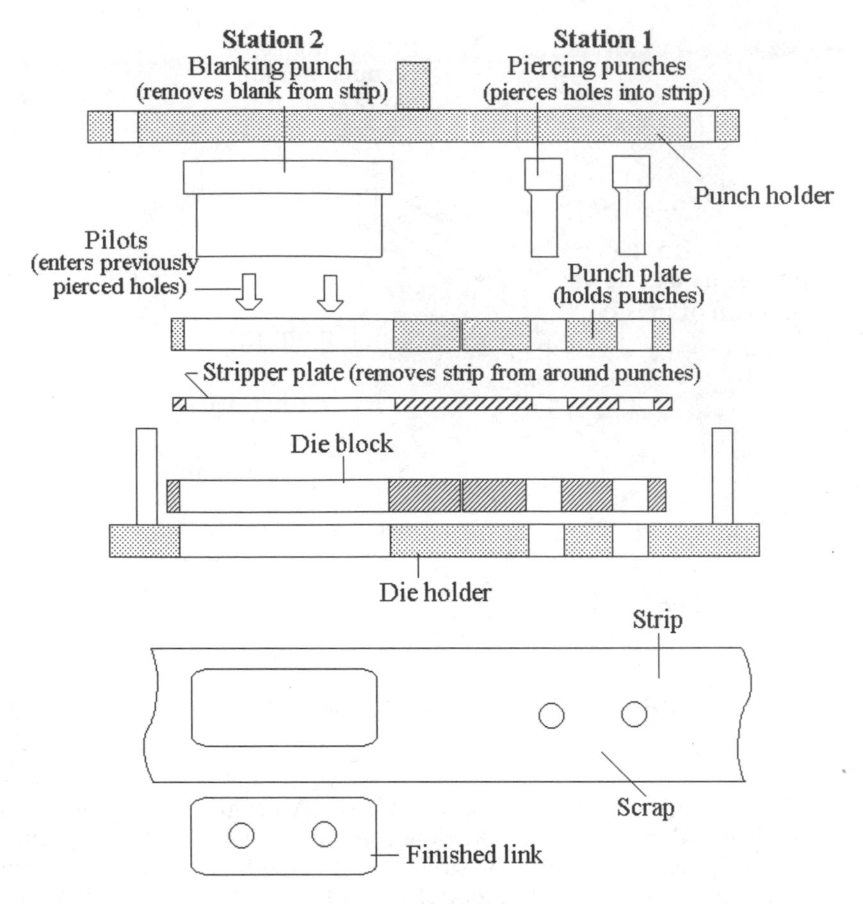

FIGURE 8.18 *Die assembly for producing a link using a progressive die.*

case the curls cannot be formed with the part attached to a strip. Hence, precut flat blanks that have been formed in another press, and which conform to the peripheral geometry of the part, are hand-loaded into a magazine. A set of transfer arms removes the part from the magazine and places it into the first station of the die. The part is then transferred from station to station as successive operations are performed.

8.6 STAMPING PRESSES

In order to appreciate the relative importance of tooling and processing costs for a stamped part, it is necessary for designers to understand the basic operations of the presses used.

All stamping presses contain a frame and bed, a ram or slide, a drive for the ram, and a power source and transmission. Power is supplied either mechanically or hydraulically (see *ASM Metals Handbook*).

Figure 8.19 shows a single-action (one ram) mechanical, open (C-frame) press. The bed, which forms the lower part of the frame, supports a bolster plate that in turn is used to support the die. T-slots in the top surface of the bolster plate are used when clamping the die to the plate. The ram holds the punch and, for a mechanical press, has a fixed stroke.

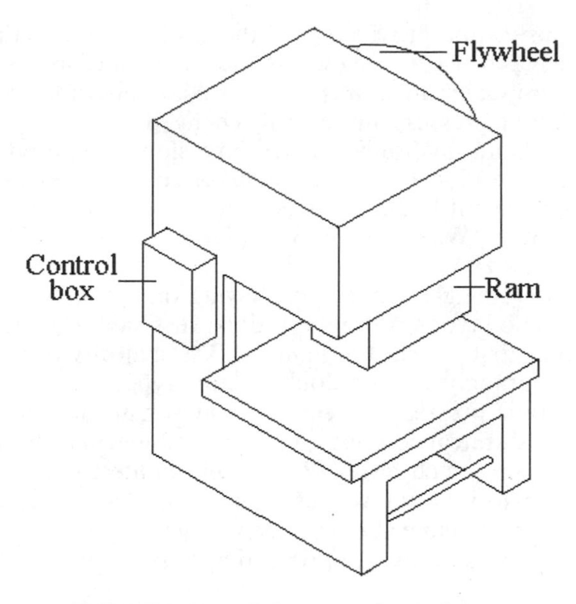

FIGURE 8.19 *C-frame mechanical press.*

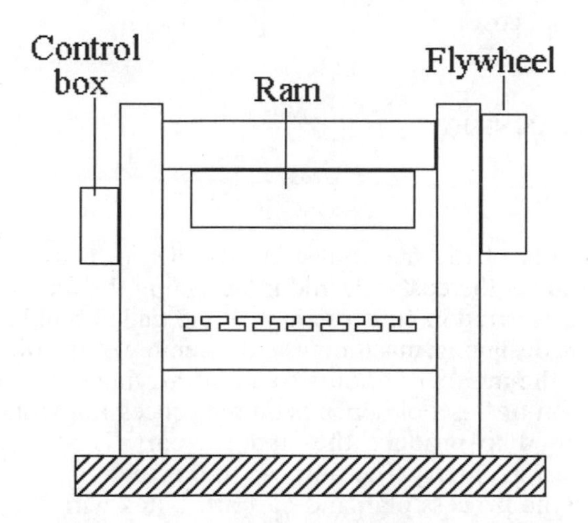

FIGURE 8.20 *Straight-sided mechanical press.*

Two basic frame types are available for mechanical presses: straight-sided presses with closed frames and four posts (Figure 8.20), and gap or C-frame (Figure 8.19). Although straight-sided presses are stiffer and deflect less than gap-frame presses, the dies are accessible only from the front or rear of the press.

In most mechanical presses a flywheel is used to store the energy required for the operation. The flywheel runs continuously, engaged by a clutch only when a press stroke is needed. Slider crank mechanisms, gears, eccentric shafts, and toggle mechanisms are used to convert the rotary motion of the flywheel to the straight-line motion of the ram.

A pressurized fluid, usually oil, provides the source of power for hydraulic presses by acting against one or more pistons. Unlike mechanical presses, whose driving force varies with the length of the stroke, the full force of the hydraulically driven pistons can be provided over the entire length of the stroke. Thus, for operations such as deep drawing where the maximum force is required at the beginning of the draw, hydraulic presses are used. In addition, the stroke length and force of hydraulic presses can be easily changed.

On the other hand, hydraulic presses are slow compared to mechanical presses. For example, a 150-ton mechanical press can easily operate at 40 strokes per minute. An equivalent hydraulic press rated at 100 tons can operate at only 12 strokes per minute. (When steady operation is established, each stroke will produce a finished part.)

Some presses are single-action type, having only one ram, and others are double- or triple-action types. A double-action press has a ram within a ram. The outer ram descends first to seat the material. The majority of hydraulic presses used in drawing operations are the double-action type.

Though hydraulic presses are less expensive than mechanical types, most stamping operations are carried out on high-speed mechanical presses because mechanical presses are much faster (24 strokes/minute to 700 strokes/minute). The cycle time per part is very low—less than 2.5 seconds—a value that is much less than the cycle time achievable in injection molding or die casting, which is 10 to 60 seconds. Consequently, the proportion of part cost due to processing is essentially negligible.

At low and medium production volumes (10,000 to 100,000), most of the cost of a stamped part is due primarily to tooling. At very high-production volumes (2,000,000), material cost is the dominating cost factor.

8.7 PROCESS PLANNING

8.7.1 Introduction

Tooling cost consists of die material cost and die construction cost. Die construction cost includes the cost incurred in designing the die, called engineering cost, and the cost incurred in constructing the die, called build cost. The amount of money spent on designing, machining, and assembly of the die parts and tryout is proportional to the number of hours spent in carrying out these activities. The actual configuration of the tool depends on the processing sequence, or processing plan, to be used to produce the stamped part. The process plan largely depends on the part geometry. The purpose of this section is to explain the relationship between the process plan and part attributes, with the assumption that a progressive die is being used.

8.7.2 Strip Development

The first step in designing a progressive die is what is referred to as the development of a process plan or strip layout. A strip layout is a visual representation of the sequence of operations used to form a part in a progressive die. The strips shown in Figures 8.15 and 8.18 can be considered the strip layout for the flat washer and link, respectively, being created with the tooling depicted in these two cases. Thus, as shown in Figures 8.15 and 8.18, a strip layout is simply the top view of the strip as it moves through a progressive die. It provides a picture of the position and relationship of each workstation in the die and indicates the distance between each station. This visual representation acts as a guide to determining whether or not

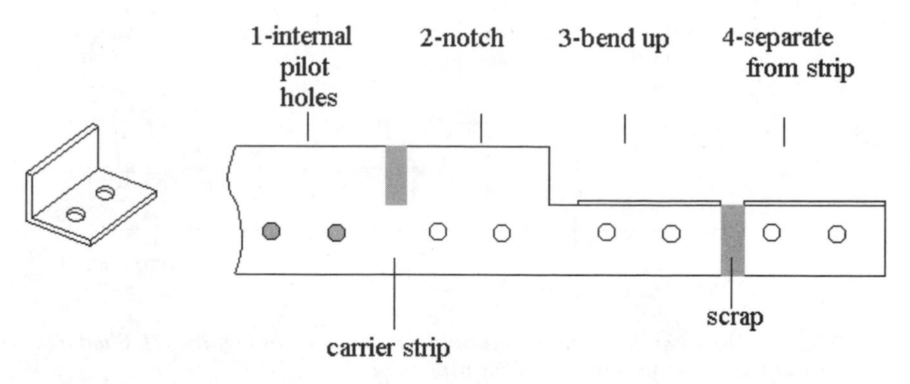

FIGURE 8.21 *A four-station strip layout for an L-bracket with holes. Shading indicates material being removed by use of shearing punches. The bend in this case is created parallel to the direction of strip feed.*

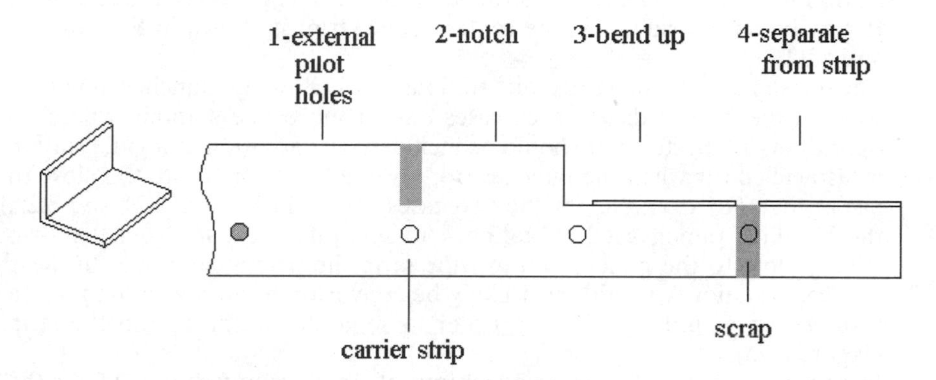

FIGURE 8.22 *A four-station strip layout for an L-bracket without holes. Shading indicates material being removed by use of shearing punches. The bend in this case is created parallel to the direction of strip feed.*

1. The die may be too complex to manufacture.
2. The operations involved may be too difficult for a progressive die, hence the use of compound and secondary dies may be preferred.
3. The operations required may be too long or too wide or require too much tonnage for the presses available at a particular site.

While there are no rigid rules to be followed for producing a good strip layout there are some rules of thumb that should be followed. For example:

1. All holes that can be pierced without danger of dislocation or deformation should be pierced prior to bending or forming.
2. If the part contains holes, some of these holes should be used as piloting holes for accurately locating the partially completed part with respect to the punches used in subsequent operations. These holes are referred to as *internal pilots* (Figure 8.21) and should be punched in the first station of the die so that they can be used throughout the remaining operations.
3. When no suitable holes are available for use as internal pilots, it will be necessary to pierce specific holes just for the purpose of piloting. These holes are outside the geometry of the part and are called *external pilots* (Figure 8.22). These external pilots are placed in the carrier strip, which is that portion of

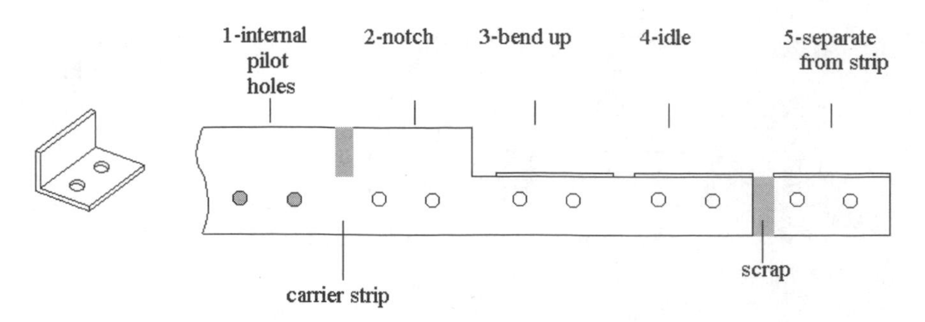

FIGURE 8.23 *Use of an idle station to form the part shown in Figure 8.21. Shading indicates material being removed by use of shearing punches.*

the sheet metal that helps to carry a partially completed part from one workstation to the next. The carrier strip does not form a portion of the final part. It may be on the side or in the center of the strip, as shown in Figures 8.21 and 8.22.

4. The existence of numerous die stations having many punches tends to weaken the overall die. In such cases one or more idle stations, where no operations are performed, should be included. The addition of an idle station is also called for when the punches from two active stations are too close to each other. For example, for the two cases shown in Figures 8.21 and 8.22, the blanking punch used at station 4 to separate the part from the strip is very close to the punch used to wipe form the part at station 3. In these two cases station 4 would most likely be converted to an idle station and a fifth station would be added in order to separate the part from the strip (Figure 8.23).
5. In order to generate the external contour of a part or to free metal from the strip to allow room to wipe form (bend) the sheet metal, a notching operation is required (Figures 8.21 and 8.22). This notching operation must precede wipe forming or bending.
6. Although idle stations are often included to strengthen a die, at times they are also included in order to ease die maintenance or to provide flexibility in the event that additional operations need to be added later. For example, after production has already begun a decision is made to add ribs to a part. This is easily accomplished if an already existing idle station can be converted to an active station. Idle stations are also included when a "programmable" tool is used to produce a family of parts. In this way active stations are added and removed as needed to produce a particular part geometry.

8.7.3 Creating a Strip Layout

In general, multiple strip layouts can be used to produce a part. For example, Figures 8.24 and 8.25 illustrate alternative process plans or strip layouts for the L-bracket shown in Figure 8.21. For the plan illustrated in Figure 8.21 the bend is created parallel to the direction of strip feed and a center carrier strip is used. For the process plans shown in Figures 8.24 and 8.25 the bends are created normal to the direction of strip feed and the carrier strips are on the side.

In all three cases two holes are punched and used as internal pilot holes at station 1. For the process plans depicted in Figures 8.21 and 8.24 notching occurs at station 2 so that the part can be wipe formed at station 3. In the latter case,

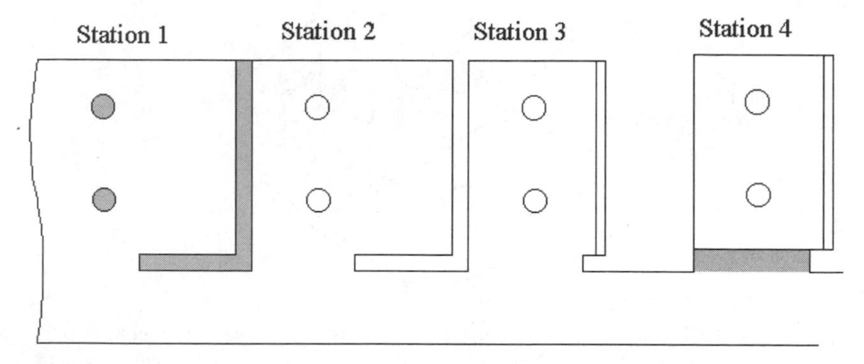

FIGURE 8.24 *An alternate four-station strip layout for the L-bracket shown in Figure 8.21. Shading indicates material being removed by use of shearing punches.*

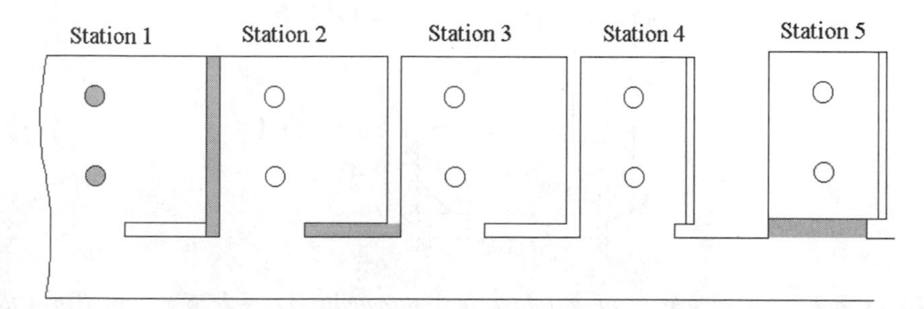

FIGURE 8.25 *An alternative five-station strip layout for producing the part shown in Figure 8.24.*

however, the notching punch is a more complex L-shaped punch. To avoid the necessity of a custom-made L-shaped notching punch, the process plan depicted in Figure 8.25 is proposed as another alternative. In this case two simple standard notching punches are used in place of a single L-shaped notching punch. This, however, requires an additional workstation.

In all cases, by wiping the bend up instead of down, the need to embed mechanical lifters in the tool in order to clear the strip of the die as it is fed to the next station is avoided. The layouts shown in Figures 8.24 and 8.25, however, result in more scrap.

The development of strip layouts for the parts shown in Figures 8.21 and 8.22 is straightforward. However, as the part geometry becomes more complex (Figure 8.26), development of a strip layout becomes more involved. The purpose of this section is to illustrate one possible procedure that can be used to develop a process plan for creating a stamped part for a progressive die. The part shown in Figure 8.26 will be used to illustrate the procedure.

Step 1: Make a sketch of the unfolded part and enclose the part in a flat rectangular envelope as shown in Figure 8.27.

Step 2: Make three or four copies of the sketch created in Step 1 and lay the sketches side-by-side to create two alternative patterns (Figure 8.28). Lay one pattern out so that the longest dimension of the flat envelope of the part, L_{ul}, is parallel to the direction of strip feed, while in the other case L_{ul} is perpendicular to the direction of strip feed. This should help to determine which is the best way to feed the part, and the number and

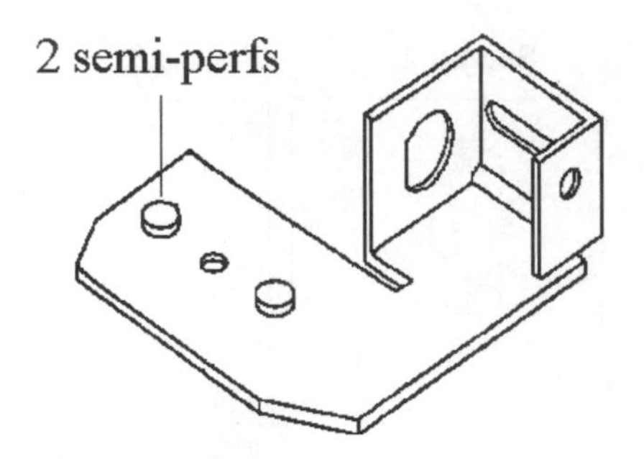

FIGURE 8.26 *A stamped part with a more complex geometry.*

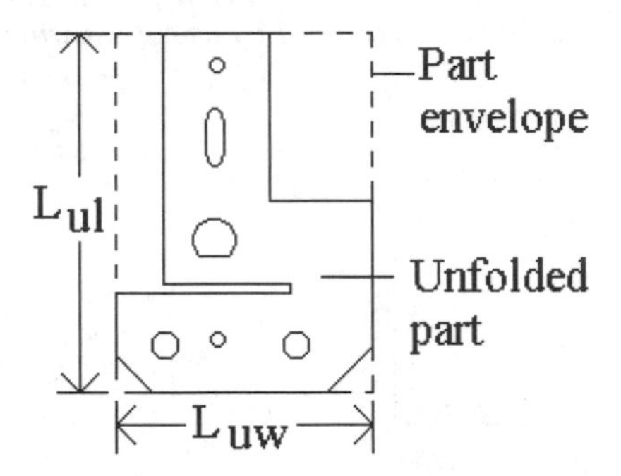

FIGURE 8.27 *A sketch of the unfolded part shown in Figure 8.26 enclosed in a flat rectangular envelope.*

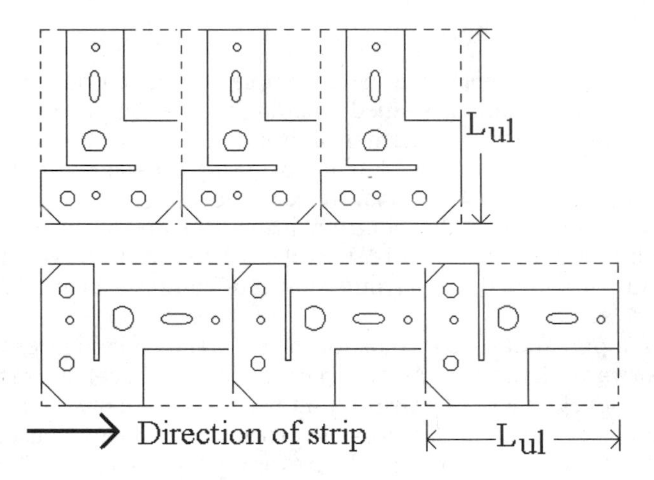

FIGURE 8.28 *Two alternative patterns for beginning the development of a process plan for the part shown in Figure 8.26.*

shape of the notching punches needed to create the peripheral shape of the unfolded part. For this particular case, feeding with L_{ul} perpendicular to the feed seems to make both notching and bending a bit easier.

Step 3: Attach a carrier strip to the pattern and begin to label the various workstations.

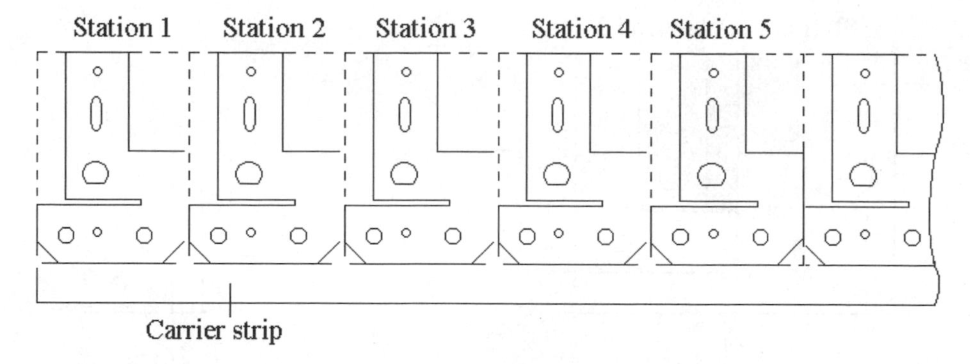

FIGURE 8.29 *Carrier strip added to the pattern selected from Figure 8.28 and workstations labeled.*

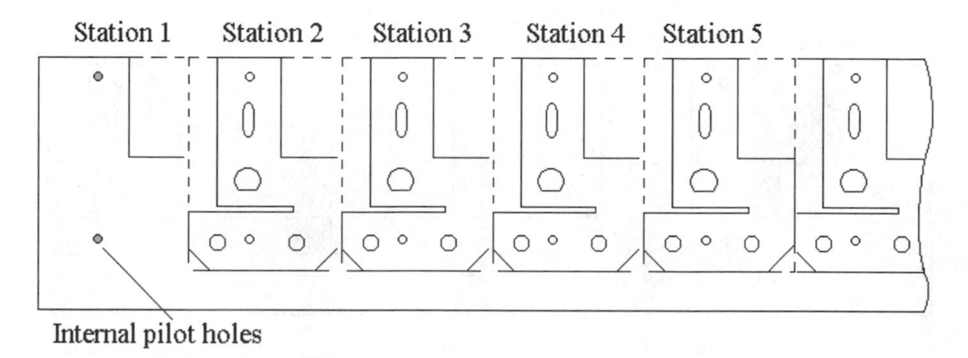

FIGURE 8.30 *Creation of workstation 1.*

Step 4: At this point we begin to think in terms of the sequence of operations to be performed. For instance, at station 1, we will create two of the holes and use them as internal pilot holes. The holes being created at this station are shown shaded. Since these will be the only two features created at this station, all other features are removed (Figure 8.30).

Step 5: Although we could continue to create the remaining holes in subsequent workstations, let us first concentrate on notching the outer form shown shaded in Figure 8.31a. There are, obviously, several notching possibilities. We could consider the use of one complex notching punch as shown in Figure 8.31b, or three less complex punches as shown in Figure 8.31c. Since complex punches are costly to purchase, to regrind when worn, and to replace when broken, and punches with thin projections are susceptible to damage and would require replacement of the entire punch when damage occurs, the notching punch shown in Figure 8.31b is rejected.

To minimize the number of stations required, we settle on the compromise shown in Figure 8.32, namely, the use of two notching stations, one at station 2 and one at station 3.

Step 6: As seen in Figure 8.32, the notching punch at station 2 is close to the pilot punches at station 1, hence, station 2 is converted to an idle station and the two notching operations shown in Figure 8.32 are shifted to stations 3 and 4. We continue this procedure until we arrive at the process plan depicted in Figure 8.33.

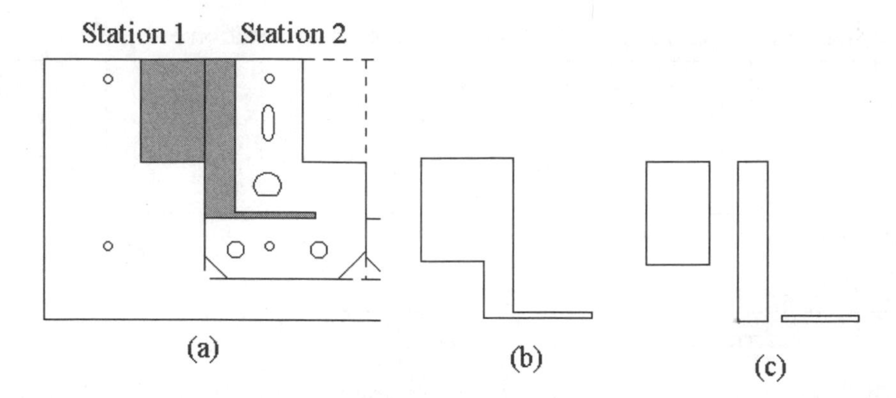

FIGURE 8.31 *Alternative notching punches for creation of the outer form shown in (a).*

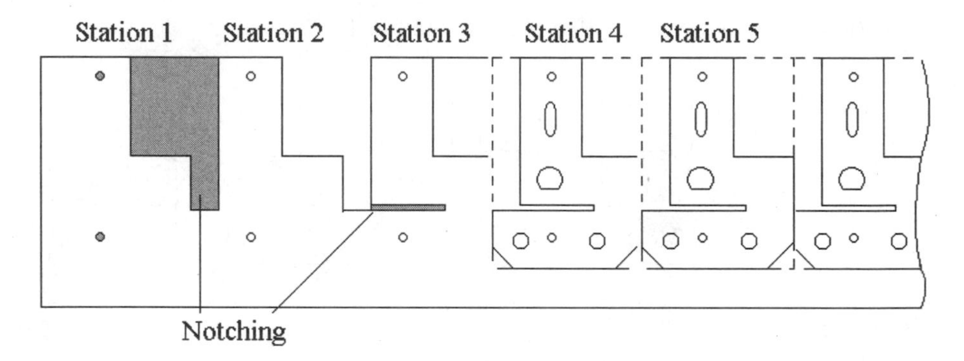

FIGURE 8.32 *Creation of workstations 2 and 3.*

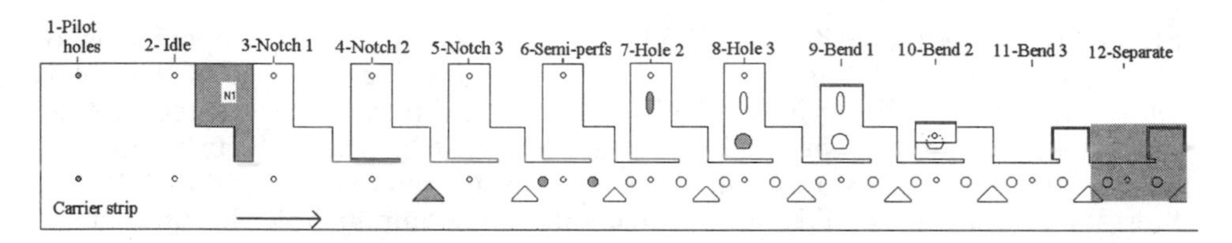

FIGURE 8.33 *One possible process plan for the part shown in Figure 8.26.*

There are obviously other plans that could be proposed. For example, why not punch all of the non-pilot holes at the same station and reduce the number of workstations from 12 to 10? (The reason for punching these holes at separate stations is explained in the next section.) Why not have an idle station between stations 4 and 5? The final process plan selected will depend on experience and the local practice of the stamping vendor used to produce the part.

8.8 DESIGN FOR MANUFACTURING GUIDELINES FOR STAMPED PARTS

As noted above, because of the very short cycle time per part, tooling and material costs are generally the dominating manufacturing cost factors when considering the design of stampings. For example, at production volumes less than 100,000, the proportion of part cost due to tooling is about 75%, but the proportion of part cost due to processing is less than 2%. Even at production volumes of about 2,000,000, the proportion of part cost due to processing still remains low (less than 5%). At such large production volumes, the greatest proportion of part cost (about 75%) is due to material cost.

At large production volumes, designing parts to reduce the amount of scrap can be important. However, in most cases it is the manufacturing process designer who designs the process plan to achieve minimum scrap. For these reasons, the discussion below is primarily about the relationship between part design and tooling cost.

As stated earlier, tooling cost is a function of tool construction cost and tool material costs. Both of these are functions of the size of the tool required. The size of the tool depends on the following issues.

Number of Features. In general, punches and dies wear at different rates. That is, a punch used to produce a hole wears at a rate different from a punch used to produce an extruded hole or a tab. Thus, these punches and dies must be removed and reground at different times. Hence, to facilitate their independent removal and maintenance from the die set, holes of different shapes are not produced at the same station; instead, one station is used to produce round holes and another station to produce extruded holes, for example. Obviously, as the number of distinct features increases, the number of die stations increases, and both die construction costs and die material costs increase. Therefore:

The number of distinct features in a part should be kept to a minimum.

This rule is violated if the tolerance requirement between features requires that the features be created at the same station or if the tool would become too large.

Spacing between Features. If features are spaced closely together (less than three sheet thicknesses), then there may be insufficient clearance for the punches. Even if space permits, however, the die sections become thin, making them susceptible to breakage. Punch breakage can occur due to metal deformation around any closely spaced features during piercing. As a result, two stations are required for each type of closely spaced feature; each station creates alternating features. Therefore:

Avoid closely spaced features when possible.

Narrow Cutouts and Narrow Projections. A link with a wide projection and a link with a wide cutout are shown in Figure 8.34a; a link with a narrow projection is shown in Figure 8.34b; and a link with a narrow cutout is shown in Figure 8.34c. As in the case of a plain link (Figure 8.18), a link containing either a wide cutout or a wide projection can be blanked out (separated from the strip) at a single station. If a narrow projection is present, however, then to separate the link from the strip at one station would require a blanking punch with a narrow cutout along with a die containing a narrow projection. In this situation the narrow section of the die would be easily susceptible to damage. In addition, damage to the narrow groove in the punch would require replacement of the entire blanking punch. For this reason the part is separated from the strip by first creating a notch to the left of the projection at one station (Figure 8.34b) and

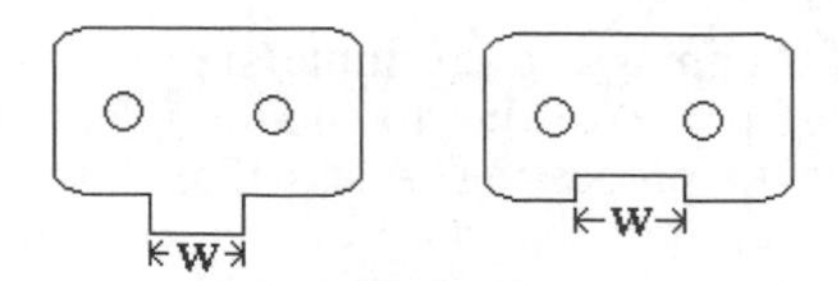

(a) Wide projection and/or wide cutout (w>3 sheet thicknesses) permit creation of the peripheral shape and separation from the strip at one station.

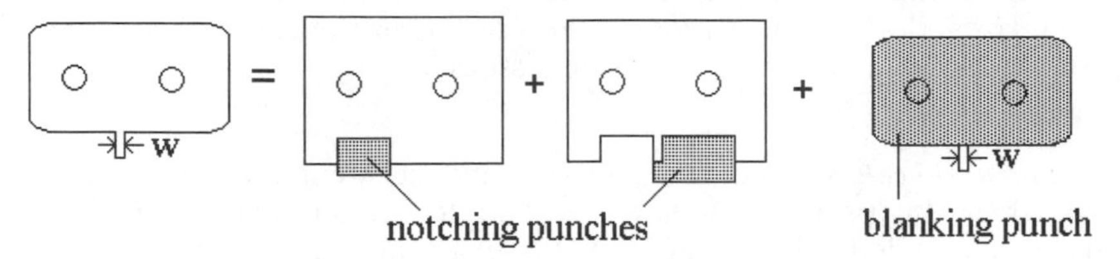

(b) Narrow projection requires the projection to be formed at two stations as shown and the part to be separated from the strip at a separate station.

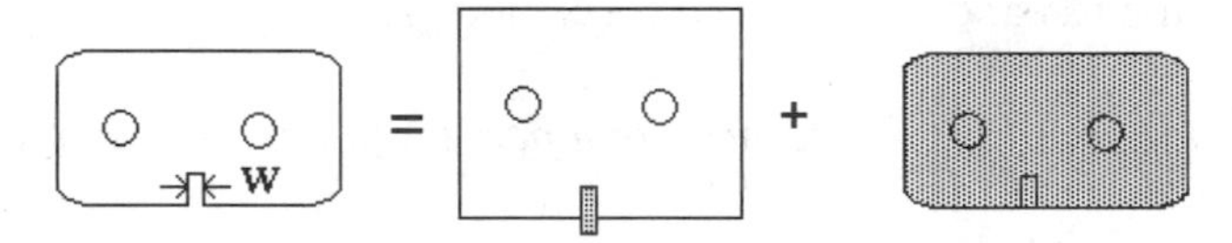

(c) Narrow cutout requires the cutout to be formed at one station and the part to be separated from the strip at another.

FIGURE 8.34 *Stamped part with a narrow projection or a narrow cutout.*

then creating the projection at the following station. Finally, the part is separated from the strip at a third station. Thus, in this case the addition of the narrow cutout to the link results in the addition of at least one and sometimes two additional stations.

In the case of the link with a narrow cutout, use of a single blanking tool would require the use of a narrow projection that is easily damaged. Damage to this tool requires replacement of the entire blanking tool. Thus, the narrow cutout is created at one station using a narrow punch. The link is then separated from the strip at the following station. Therefore:

The use of narrow cutouts and narrow projections should be avoided.

Bend Stages and Bend Directions. A U-shaped part, with both bends in the same direction, will have both bends created at the same time at one die station. An equivalent Z-shaped part, with bends in opposite directions, will require that each bend be separately created at two different die stations. Thus, bends in opposite directions create increased tool construction and die material costs. Therefore:

The number of bend stages in a part should be kept to a minimum.

In general, to keep the number of bend stages to a minimum, the number of bend axes should be kept to a minimum. A precise method for determining bend stages is discussed in Chapter 9, "Stamping: Relative Tooling Cost."

Overbends. To create a bend angle greater than 90° requires two die stations. At the first station the part is bent through 90°, and at the next station the part is bent past 90°. Clearly, then, parts with a bend angle greater than 90° require more costly tooling. Thus:

Bend angles greater than 90° should be avoided whenever possible.

Side-Action Features. Features, such as holes, whose shape or location from a bend line must be accurately located, must be created after bending. Once again, this requires that one or more additional die stations be added, thereby increasing tooling cost. Also, because the bends themselves cannot be closely controlled, they make poor reference points for important or close tolerances. Therefore:

Side-action features should be avoided or kept to a minimum.

In general, to keep side-action features to a minimum, the tolerances of feature dimensions that must be referred to bend lines should be "generous."

The effects of these rules on tooling costs are discussed in the next chapter.

8.9 SUMMARY

This chapter has described some of the most common methods used for the production of thin-walled sheet-metal parts. A discussion of the relationship between part attributes and the development of alternative process plans for the production of a stamped part were also included. The chapter concluded with a discussion of design for manufacturing issues as they apply to the production of stampings.

REFERENCES

American Society of Metals International. *American Society of Metals Handbook*, Vol. 14. "Forming and Forging." Metals Park, OH: ASM International, 1988.

Bralla, J. G., editor. *Handbook of Product Design for Manufacturing.* New York: McGraw-Hill, 1986.

Kalpakjian, S. Manu*facturing Engineering and Technology*. Reading, MA: Addison-Wesley Publishing, 1989.

Poli, C., Dastidar, P., and Graves, R. A. "Design Knowledge Acquisition for DFM Methodologies." *Research in Engineering Design*, Vol. 4, no. 3, 1992, pp. 131–145.

Wick, C., editor. *Tool and Manufacturing Engineers Handbook,* Vol. 2. "Forming, 9th edition." Dearborn, MI: Society of Manufacturing Engineers, 1984.

QUESTIONS AND PROBLEMS

8.1 In this chapter we have discussed the sheet-metal forming processes of stamping, stretch forming, and spinning. Name some products from the following product lines that would have parts made by these processes:

Consumer Products (Airplanes, cars, bikes, etc.)
Office equipment
Clothing
Hardware

8.2 In Chapter 2 the phenomenon of springback was introduced. What material properties affect springback? With regard to designing a process plan for a part that is to be wipe formed or stretch formed, discuss the implications of springback.

8.3 Imagine that you are faced with the problem of strengthening the link shown in Figure 8.18. One possibility of course is to change the material, but you are told that the material cannot be changed. What other possibilities exist?

8.4 Figure P8.4 shows the preliminary sketch of three proposed stamping designs. Based on the design for manufacturing guidelines discussed in Section 8.8, which of the designs requires the least costly tooling? Assume that the wall thickness is the same in all designs.

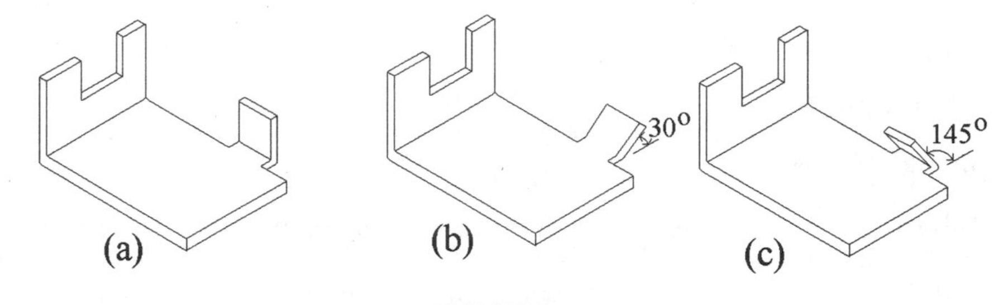

FIGURE P8.4

8.5 The proposed design of a stamping is shown in Figure P8.5. The sheet thickness of the part is constant and the location of hole A relative to the plate on which hole B is located is considered critical. Based on the design for manufacturing guidelines discussed in Section 8.8, what are the implications of this requirement with regard to the tooling required to produce the part?

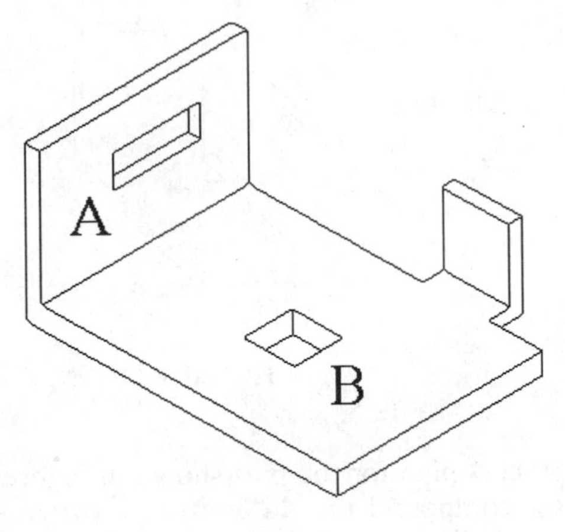

FIGURE P8.5

8.6 Repeat Exercise 5 for the stamping shown in Figure P8.6. Are there redesign suggestions that you can make to help reduce the cost to stamp the part?

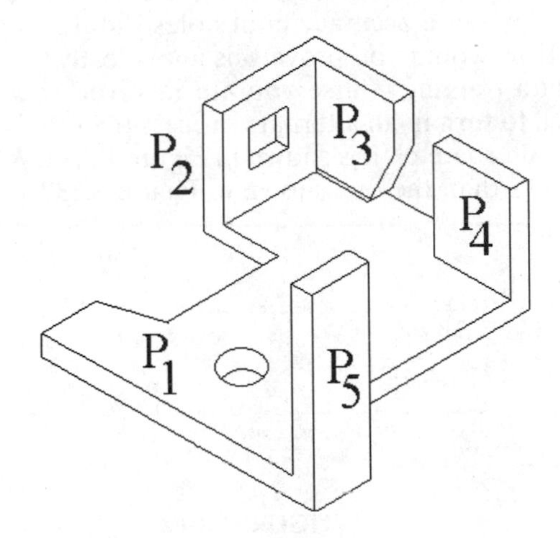

FIGURE P8.6

8.7 Three parts are shown in Figure P8.7. Assume that you are responsible for designing the process plan for these parts and have been given the task of proposing at least two alternative plans for each part. Which of the two plans proposed above would you recommend in each case? Explain your decision.

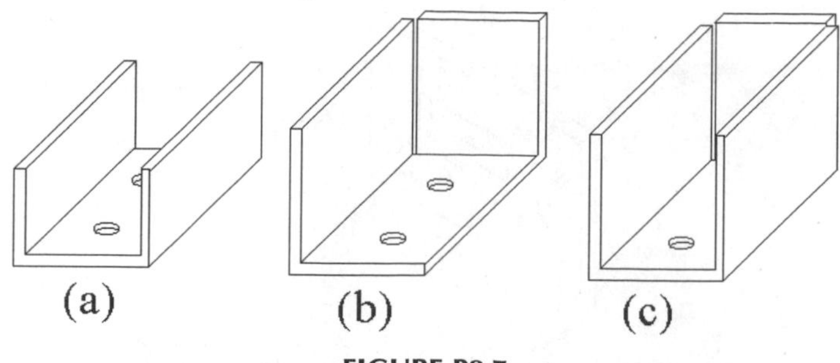

FIGURE P8.7

8.8 Design a process plan for the part shown in Figure P8.4c.

8.9 How does a compound die differ from a progressive die? Explain the advantages and disadvantages of each.

8.10 At low production volumes (less than 50,000 parts) which cost, tooling cost, processing cost, or material cost has the greatest overall affect on part cost? Explain. Does the same conclusion apply for very high production volumes.

8.11 As part of the same training program discussed in Exercise 4.1 of Chapter 4, you are in the process of explaining to some new hires at DFM.com that in order to increase accuracy pilot holes should be spaced as far apart as possible. How would you prove this analytically?

8.12 As part of a training course you are involved in at DFM.com you have been asked to turn in an alternate process plan to the one shown in Figure 8.33. Your alternate plan is shown in Figure P8.12. Why do you believe this plan is better than the one shown in Figure 8.33?

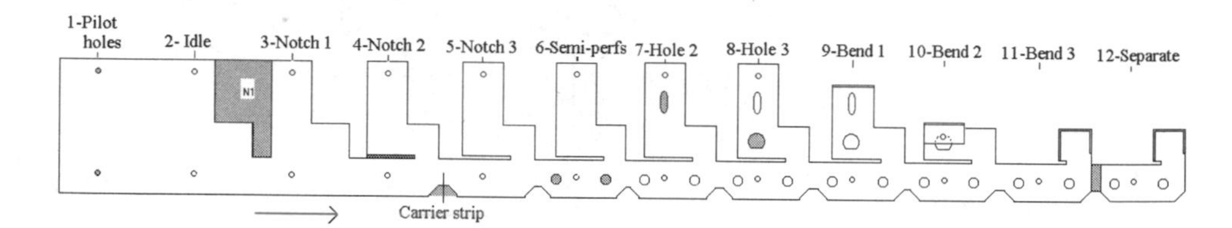

FIGURE P8.12

Chapter 9

Stamping: Relative Tooling Cost

9.1 INTRODUCTION

As indicated in the previous chapter, stamping is a process by which thin-walled metal parts are formed from sheets of metal. Parts are formed by a series of cutting (or shearing) and forming (e.g., bending) operations performed in combination punch and die sets mounted in mechanical or hydraulic presses (Figure 9.1). Although other sheet-metal forming processes exist, such as stretching, spinning, deep drawing, and others (see Wick, C., 1984) more component parts and products are made by stamping than by any other sheet-metal process.

As in the case of injection-molded and die-cast parts, stampings are generally designed by product designers working for large consumer products corporations, but the parts are produced by small (typically less than 100 employees) custom stampers. Unlike injection molding and die casting, however, most stampers also make the tooling necessary to produce the parts.

Short-run production of stamped parts is done using punch presses, steel rule dies (Figure 9.2), and press brakes (Figure 9.3). Punch presses are presses that are specifically designed to hold the tool and perform operations such as notching, piercing, perforating, slotting, and so on (Figure 9.4). A steel rule die (Figure 9.2) consists of a thin strip of steel, shaped like a ruler, formed to the outline of a part, set in plywood, and a thin steel punch set. (A steel rule die operates much as a cookie cutter.)

Long-run or conventional stamping is done using simple dies (one operation per stroke), compound dies (Figure 9.5), progressive dies, and transfer dies (similar to progressive dies but uses "fingers" to transfer to partially completed part from station to station).

One purpose of this chapter is to present a systematic approach for identifying, at the configuration design stage, those features of the part that significantly affect the cost of stampings. The goal is to learn how to design so as to minimize difficult-to-stamp features.

In this chapter we will also discuss a methodology for estimating the relative tooling cost of proposed stamped parts based on configuration information and on the assumption that progressive dies are used. Unlike injection molding and die casting, however, where the total relative tooling cost can be determined at the configuration stage of design, in stamping only the maximum and minimum relative tooling cost can be determined. The reason is that some of the details required to finalize the relative tooling cost can only be determined after parametric design. For example, in injection molding and die casting, any hole that is not parallel to the direction of mold closure is an undercut. In stamping, on the other hand, whether or not such a hole is an undercut (they're called side-action

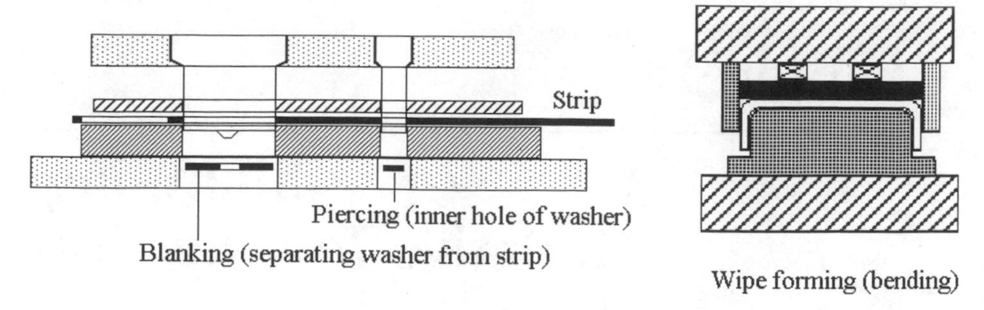

FIGURE 9.1 *Conventional stamping using a progressive die.*

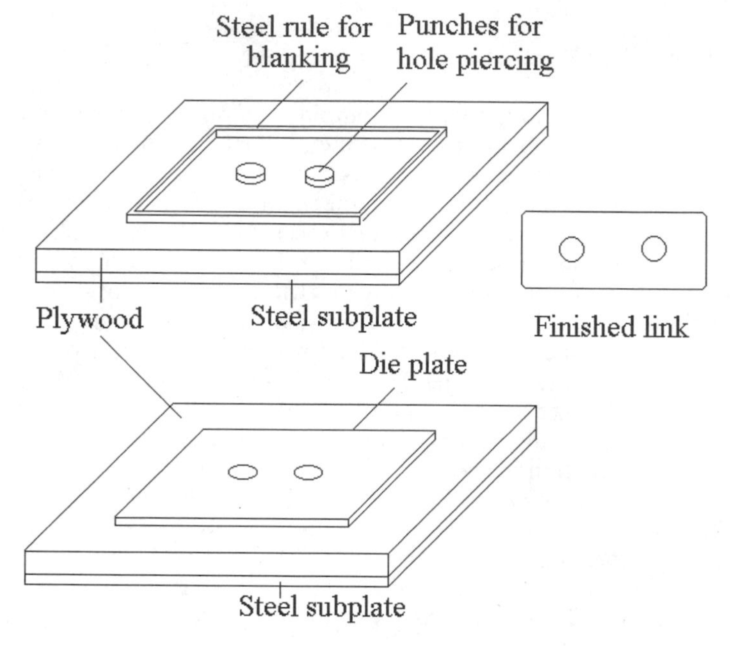

FIGURE 9.2 *Steel rule die. Used for short runs or to blank a part in order to begin production without waiting until the conventional die is delivered. Much less expensive than conventional dies.*

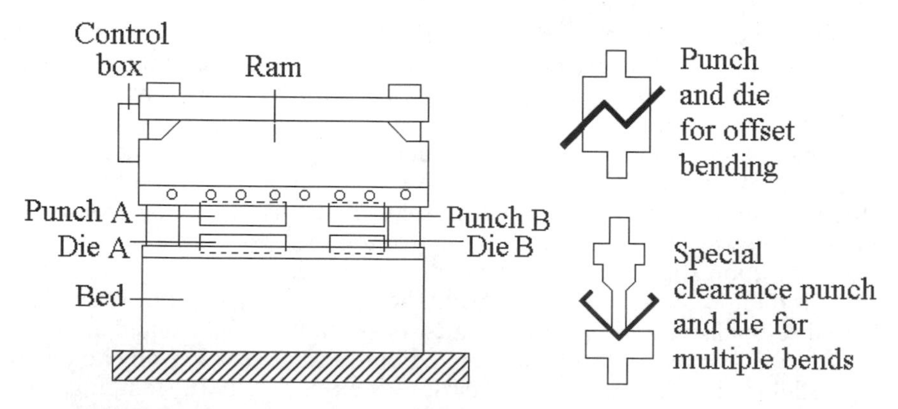

FIGURE 9.3 *A press brake and two of many possible dies used for bending.*

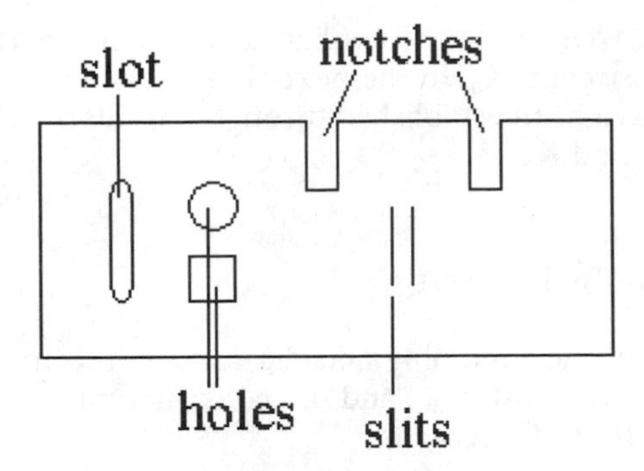

FIGURE 9.4 *A sample part illustrating piercing, notching, perforating, and slotting.*

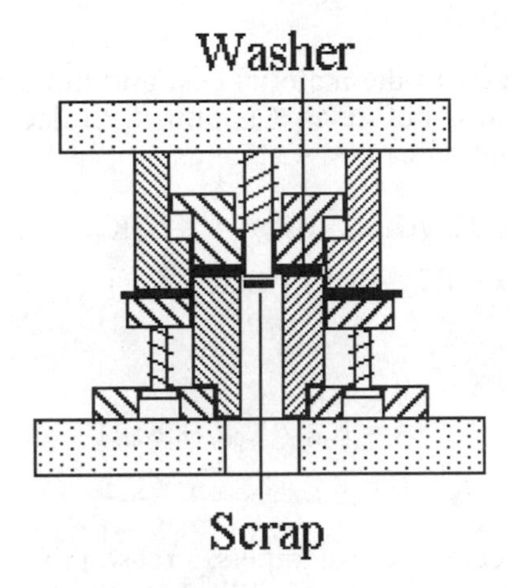

FIGURE 9.5 *A compound die used to produce the same washer shown in Figure 9.1.*

features in stamping) depends, not just on the presence of the feature, but also on its dimensions and tolerances. Nevertheless, a great deal of information concerning tooling cost can be determined at the configuration stage, and this information can and should be used to help guide the generation and selection of part configurations before parametric design is done.

9.2 ESTIMATING THE RELATIVE COST OF STAMPED PARTS

9.2.1 Total Cost

The total cost of a stamped part consists of: tooling cost, K_d/N; processing cost (or equipment operating cost), K_e; and material cost, K_m.

$$\text{Total Cost} = K_e + K_d/N + K_m \qquad \text{(Equation 9.1)}$$

Because of the high-speed presses used to produce stamped parts (24 strokes/minute to 700 strokes/minute) cycle times of less than 1 second are generally achievable. For this reason, even at high production volumes, the proportion of cost due to processing is low.

In this chapter, when only the configuration of the part is known, we can show only how to estimate K_d. In the next chapter, when we assume that more exact dimensions have been established through parametric design, we will show how to compute K_e and K_m.

9.2.2 Total Relative Tooling Cost

As in the case of injection molding and die casting, the total tooling costs are made up of die material cost, K_{dm}, and die construction cost, K_{dc}. Thus, relative to a reference part, total die costs are

$$C_d = \text{Cost of Tooling for Designed Part/Cost of Tooling for Reference Part}$$

$$C_d = (K_{dm} + K_{dc})/(K_{dmo} + K_{dco}) \qquad \text{(Equation 9.2)}$$

where K_{dmo} and K_{dco} refer to die material cost and die construction cost for the reference part. As shown before in Chapter 4, "Injection Molding: Relative Tooling Cost," this equation can be written as

$$\begin{aligned} C_d &= K_{dm}/(K_{dmo} + K_{dco}) + K_{dc}/(K_{dmo} + K_{dco}) \\ &= A(K_{dm}/K_{dmo}) + B(K_{dc}/K_{dco}) \qquad \text{(Equation 9.3)} \end{aligned}$$

where

$$A = K_{dmo}/(K_{dmo} + K_{dco})$$

$$B = K_{dco}/(K_{dmo} + K_{dco})$$

Based on data collected from stampers, a reasonable value for A is between 0.15 and 0.20, and a reasonable value for B is between 0.80 and 0.85. For our purposes we will take A and B to be 0.2 and 0.8, respectively. Hence, Equation 9.3 becomes

$$C_d = 0.8C_{dc} + 0.2C_{dm} \qquad \text{(Equation 9.4)}$$

where C_d is the total die cost of a part relative to the die cost of the reference part, C_{dc} is the die construction cost for the part relative to the die construction cost of the reference part, and C_{dm} is the die material cost for the part relative to the die material cost of the reference part.

9.3 DIE CONSTRUCTION COSTS

Die construction cost includes design cost, that is, the cost incurred in designing the die, and build cost, or the cost incurred in building and debugging the die. Both of these components are directly proportional to the number and types of features to be provided and to the hours spent in carrying out the respective activities.

To determine the die construction costs for a stamped part we could start by developing a detailed process plan as discussed in the previous chapter, Section 8.7, Processing Planning. Once the number of required stations is determined it is fairly straightforward to determine the die construction costs for a part. To create a strip layout, however, requires a decision concerning how the part will be oriented on the strip. This in turn depends upon such things as production volume, critical tolerances, the need to minimize scrap, and the presence of side-action features, for example (see Korneli, M., 1999). Often, a compromise must be made between orienting a part so as to minimize scrap and orienting to facilitate the creation of a side-action feature. Another consideration is the amount of lift that might be required to carry a strip through the die. One orientation may not require lifters, and another orientation may require either lifters imbedded in the die or external lifters. In addition, as pointed out in Chapter 8, "Sheet-Metal Forming," lifters can sometimes be eliminated by simply wiping up instead of down. Finally, decisions need to be made concerning the complexity of the notching punches to be used. For example, as we saw in Chapter 8, Section 8.7, Process Planning, we can use either one complex notching punch, or two to three less complex punches. Decisions such as these are often dependent on the size of equipment available to the process designer. Generally speaking, however, product designers are not sufficiently experienced to be able to produce such process plans, especially for complex parts.

A second approach we could use is a group technology approach similar to the ones used for injection molding and die casting. With this approach parts are grouped into various categories based on their manufacturing complexity. Coding of the part is equivalent to describing the presence or absence of features that significantly contribute to manufacturing costs. In the case of injection molding, for example, the tooling complexity, hence costs, required for parts that have the same six-digit part code, are about the same. This is the approach that was used in *Engineering Design and Design for Manufacturing* and works well in estimating the difference in relative costs between two or more alternative designs. When applied to stamping, however, extensive experience with this approach has shown that there are two principal drawbacks, namely,

1. The approach appears to inhibit the ability of users to visualize the relationship between part geometry and features, and tooling complexity. That is, it appears to hinder one's ability to develop a mental picture of the tooling required to produce a given stamped part. This does not appear to be the case when this approach is used for injection molding and dic casting.
2. At times the system grossly miscalculates the number of active stations required to produce the part, which in turn grossly under- or overestimates the relative tooling cost for the part.

A third approach, and the one we will use in this book, is to first calculate approximately the number of active stations required to produce a part by using an algorithm such as the one shown in Table 9.1. With the number of active stations estimated, the die design and build hours can be determined using an approach to be discussed later in this chapter. This approach is more precise than the group technology approach used in *Engineering Design and Design for Manufacturing* and avoids the need for development of a detailed process plan.

9.4 DETERMINATION OF THE NUMBER OF ACTIVE STATIONS FOR SHEARING AND LOCAL FEATURES

The algorithm presented in Table 9.1 is based on the following considerations.

9.4.1 Pilot Holes

Pilot holes are needed to accurately locate the partially completed part with respect to the punches in subsequent operations. Although many stampings have

Table 9.1 Algorithm for determination of the total number of active stations for shearing and local features (Chandrasekaran, S., 1993).

Features	*Number of Stations*
External pilot holes	1
Number of distinct feature types, n_f	0 1 2 3 4 5 6 7 8 9
Number of closely spaced feature types, n_{fc}	0 1 2 3 4 5 6 7 8 9
Number of feature types with features in opposite direction, n_{op}	0 1 2 3 4 5 6 7 8 9
Tabs? (y/n)	2/0
Embosses near part periphery? (y/n)	1/0
Curls or hems? (y/n)	2/0
Blanking out station	1
Total number of active stations, N_{a1}	

holes that are available for use as internal pilot holes, it is common practice to provide external pilots even when holes are available for use as internal pilots. Thus, Table 9.1 provides for 1 active station for pilot hole piercing. For especially simple parts such as washers and links (see Figures 8.15 and 8.18) where external pilots would not be necessary, this algorithm would obviously overestimate the number of active stations required by 1. For the most part, however, we are considering the use of this algorithm for more complex parts.

9.4.2 Number of Distinct Features and the Direction of Features

For ease of maintenance and construction, it is common practice among stampers to produce only one feature type at a given die station. Thus, features such as holes, extruded holes, tabs, and lance forms are each produced at separate die stations. Thus, Table 9.1 allocates 1 active station for each distinct feature type as explained below.

Figure 9.6 shows some of the features commonly found in stamped parts. Among those shown in Figure 9.6 that require only a single die station to create are extruded holes, lance forms, and semi-perfs.

An *extruded hole* is one that is generated at one station using a specially stepped punch that first shears a smaller hole and then follows through to deform the local area around the hole into a projection by limited forward extrusion. Alternately, the hole could be pre-pierced in a separate station, and then the edges extruded at a second station. This procedure creates, in effect, the equivalent of bosses that are often used for joining by tapping or thread-forming the projecting hole for use as a nut. Such bosses can also be used for alignment or reinforcement. Extruded holes are also referred to as flanged, collared, embossed, or drifted holes.

Lance forms are partial cuts in the stamping that are formed into projections. Lance forming involves a combination of shearing (or lance cutting) followed by local forming, to create the feature (which resembles a pocket-shaped opening). Lance forming is commonly used to create louvers for venting purposes, or to raise metal from the stamping surface so as to enclose a wire. Lance forming

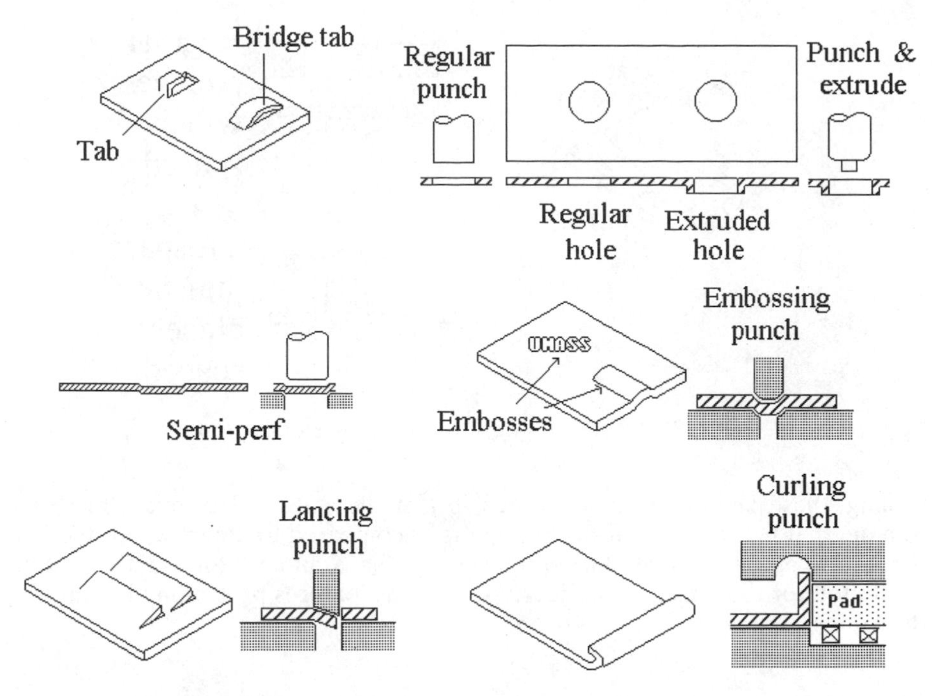

FIGURE 9.6 *Examples of some distinct feature types.*

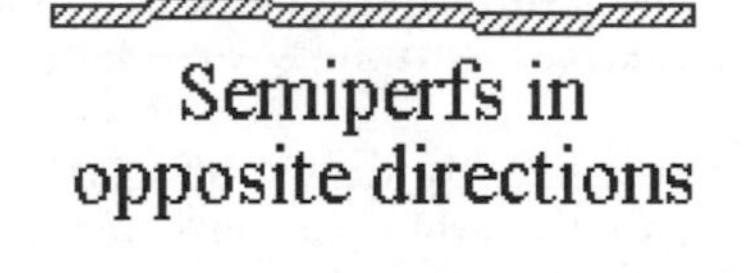

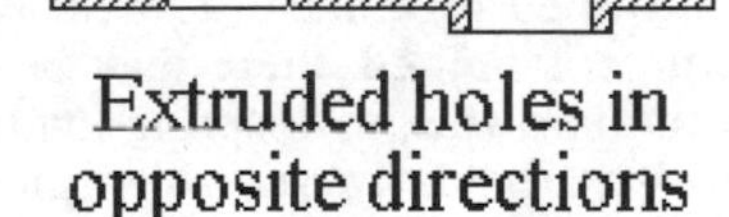

FIGURE 9.7 *Feature in opposite directions.*

punches have a cutting edge for shearing the lance cut, as well as a forming surface to form up the edge of the cut.

A *semi-perf* is a small—often about 3 mm diameter—button-like projection. It is frequently used as a locator during the assembly phase, or in subsequent fastening operations (such as when resistance spot-welding is used). Semi-perfs are also known as partial extrusions, rivet lug forms, dimples, bumps, or partial slugs.

In the case of protruding features, if two or more features are on the same surface of the stamping (and project out in opposite directions) then the features are said to be in opposite directions (Figure 9.7). Such combinations of features cause a torque on the die set at that station and may, therefore, need to be formed in separate stations or with separate secondary dies. Alternatively, by using pressure pads of varying stiffness together with specially sized

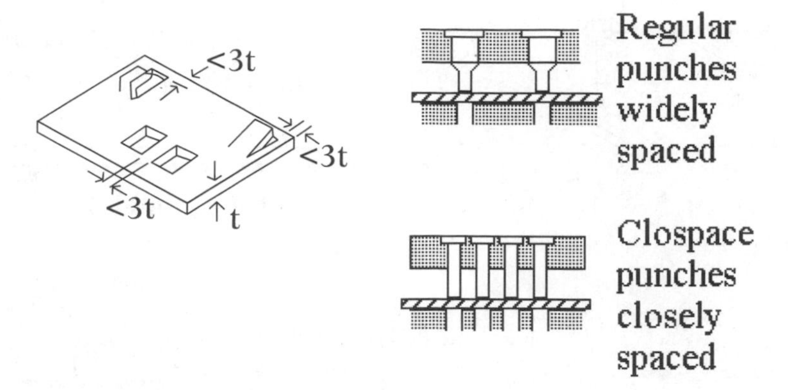

FIGURE 9.8 *Closely spaced nonperipheral features.*

tooling, these features can be formed in a single station. However, this leads to a die that needs more fabrication hours as compared to a part where the features were all on the same face of the surface. Thus, an additional active station is allotted for each distinct feature type that has features protruding in opposite directions.

9.4.3 Closely Spaced Features

Two nonperipheral features that are less than three times the stock thickness apart from each other, or any nonperipheral feature that is less than three times the stock thickness from the part periphery, are defined as closely spaced features (Figure 9.8).

In the case of compound and secondary tooling, if features are closely spaced with respect to each other or the edge, the part of the die block that supports the stamping between these features will have a thin section that will tend to crack or break in production. Moreover, the punches used to generate these features may have to be crowded together in too close a manner. In addition, if the space between features is limited, there may be distortion owing to insufficient space to clamp the part (i.e., with spring-loaded strippers or pads) in between the features (this is particularly true when pierced holes are near the edge that must be blanked). As a result of these constraints, closely spaced features necessitate the use of an additional die station. Thus, they add to tooling complexity. Hence, in Table 9.1, one additional active station is allocated for a closely spaced feature.

9.4.4 Narrow Cutouts and Narrow Projections

Any projections on the peripheral edge of the part that have a minimum width less than three times the stock thickness are considered narrow (Figure 9.9). Similarly, any cutout or notch in a cut edge whose width is less than three times the stock thickness is also considered narrow.

In general, a peripheral cutout (notch) or projection is imparted to the part when the part is blanked or separated from the strip at the final station. In the case of a narrow cutout or a narrow projection, these features would require a narrow die wall (to produce the narrow projection) or a narrow portion of the

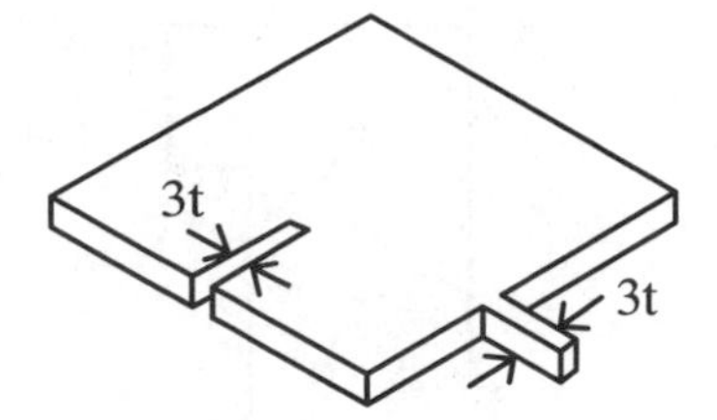

FIGURE 9.9 *Part with a narrow cutout and a narrow projection.*

blanking punch (to produce the narrow cutout). Both situations can result in damage to the blanking punch. Thus, the narrow cutout is imparted in a separate station using a narrow punch, and the narrow projection is imparted in two stages, first putting a notch on one side of the projection and then removing the remaining material at a separate station.

9.4.5 Tabs

The presence of tabs (Figure 9.6) necessitates a notching station, prior to tabbing, to relieve (i.e., remove) the portion of sheet metal around the section that will be bent in the next station. Thus, in addition to allocating 1 active station for producing the tab, Table 9.1 allocates another active station for notching.

9.4.6 Embosses

Embosses are small, shallow formed projections on the surface of stamped parts. Raised letters, beads, and ribs are examples of embossings. Embossings are produced by localized forming with the amount of deformation dependent on the design and depth of the projection. Sheet thickness remains unchanged, and embossings have smooth contours. Embossings call for complementary punch and die shapes, and these components may be incorporated into a die as inserts. The operation is not generally done simultaneously with any shearing operation, and it requires a separate station or secondary die. Embosses near the part periphery (Figure 9.6) can only be produced after the sheet metal on the periphery is notched. Thus, in addition to allocating one active embossing station, another active station is allocated for notching.

9.4.7 Curls and Hems

Curls (Figure 9.6) are features that result from rolling the edge of a stamped part. The curl diameter should be 10 to 20 times stock thickness. Curling is usually preceded by wiping (i.e., bending) the edge. A curling punch (with a groove cut into it corresponding to the curl) rolls the bent edge into the desired shape, while the part is gripped with a pressure pad. A curl requires at least two operations (stations) and, hence, is a complex feature from the tooling standpoint. Thus, the creation of curls or hems requires three stations. The first station is a notching station required to free metal from the strip to allow room to wipe form (bend) the sheet metal at the second station. The third station is used to curl or hem the part. Thus Table 9.1 allots two additional stations for the creation of curls and hems.

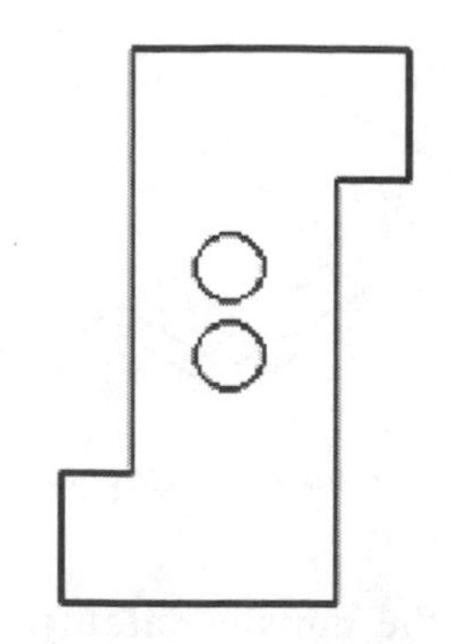

FIGURE 9.10 *Example 9.1.*

9.5 EXAMPLE 1—NUMBER OF ACTIVE STATIONS FOR A FLAT PART

9.5.1 The Part

As an example we consider the part shown in Figure 9.10. Although no dimensions are shown we will assume that the two circular holes are closely spaced and that the projections are wide projections.

9.5.2 Number of Active Stations

To determine the number of active stations required we complete Table 9.1. This completed table is shown below in Figure 9.11 and indicates that 4 active stations are required to stamp the part.

9.5.3 Strip Layout for Example 1

As a simple check against the results obtained by use of our algorithm, let us use the procedure outlined in Chapter 8, "Sheet-Metal Forming," to obtain one possible strip layout for the part shown in Figure 9.10.

We begin by enclosing the part in a rectangular flat envelope as shown in Figure 9.12. We then lay out two patterns, one with the largest dimension perpendicular to the direction of strip feed, and the other parallel to the direction of strip feed (Figure 9.12). At this point we need to add a carrier strip and begin to make decisions concerning which features should be stamped at each station. Figure 9.13 shows the result for the case where the largest dimension of the part is fed parallel to the direction of strip feed.

The layout presented in Figure 9.13 consists of a center carrier strip with two external pilot holes punched at the first station along with one of the circular holes. Since the holes are closely spaced, the second circular hole is punched at station 2. Station 3 is a notching station that provides the part with its peripheral contour. The part is separated from the strip at station 5. An idle station was added at station 4 in order to allow sufficient spacing between the blanking punch and the notching punches. Hence, this strip layout provides for 4 active stations, the number obtained by use of the algorithm.

If the part is a narrow one, then the pilot holes punched at station 1 would probably be too close to each other to be effective. In this case a process plan

Features	Number of Stations
External pilot holes	①
Number of distinct feature types, n_f	0 ① 2 3 4 5 6 7 8 9
Number of closely spaced feature types, n_{fc}	0 ① 2 3 4 5 6 7 8 9
Number of feature types with features in opposite direction, n_{op}	⓪ 1 2 3 4 5 6 7 8 9
Tabs? (y/n)	2 / ⓪
Embosses near part periphery? (y/n)	1 / ⓪
Curls or hems? (y/n)	2 / ⓪
Blanking out station	①
Total Number of active stations, N_{a1}	4

FIGURE 9.11 *Estimate of the number of active stations required to produce the part shown in Figure 9.10.*

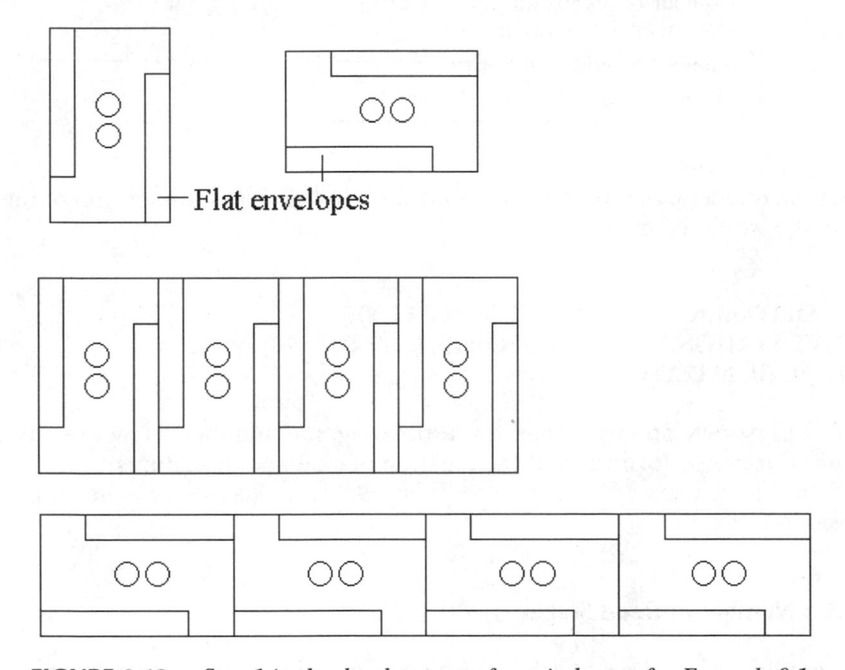

FIGURE 9.12 *Step 1 in the development of a strip layout for Example 9.1.*

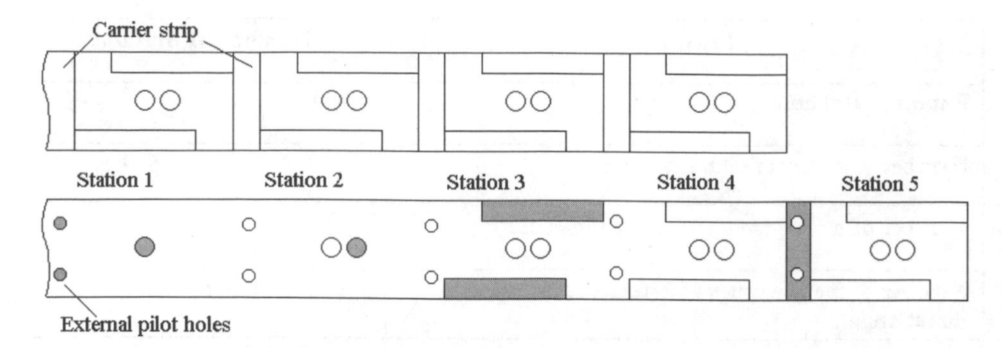

FIGURE 9.13 *One possible strip layout for Example 9.1.*

Table 9.2 Algorithm for determination of the total number of active stations for wipe forming and side-action features (Chandrasekaran, S., 1993).

Wipe Forming	*Number of Stations*
Number of bend stages 0 1 2 3 4 5	0 2 3 4 5 6
Bends in opposite directions? (applies only to parts with one bend stage) (y/n)	1/0
Number of overbends	0 1 2 3 4 5
Number of features in the primary plate near the bend line	0 1 2 3 4 5
Number of side-action features	0 1 2 3 4 5
Total Number of Active Stations, N_{a2}	

using the direction of strip feed perpendicular to the largest dimension of the flat envelope would be needed.

9.6 DETERMINATION OF THE NUMBER OF ACTIVE STATIONS FOR WIPE FORMING AND SIDE-ACTION FEATURES

Table 9.2 shows an algorithm for estimating the number of active stations required for wipe forming and the creation of side-action features.

The algorithms presented in Table 9.2 are based on the following considerations.

9.6.1 Number of Bend Stages

The number of stations at which bending operations are performed in a progressive die is defined as the number of bend stages. As explained in Section 8.7 in Chapter 8, "Sheet-Metal Forming," prior to wipe forming or bending, a notching operation is required to free metal from the strip and to generate the peripheral contour of the part. Thus, regardless of the number of bend stages, at least one notching operation is required. Hence, in Table 9.2 the number

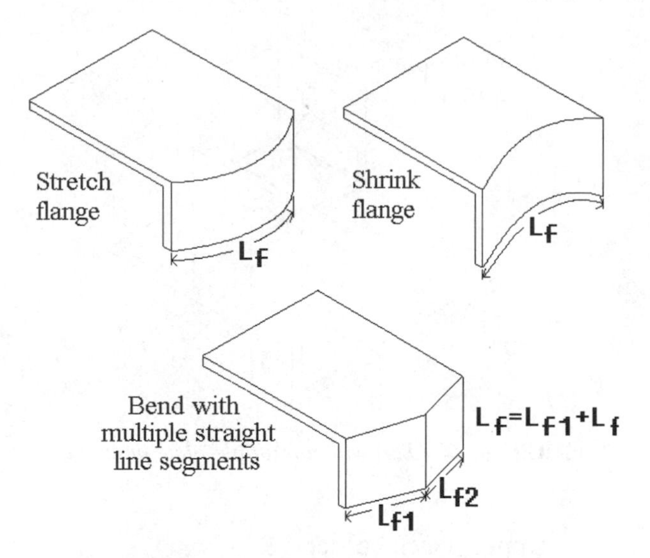

FIGURE 9.14 *Parts with nonstraight bends.*

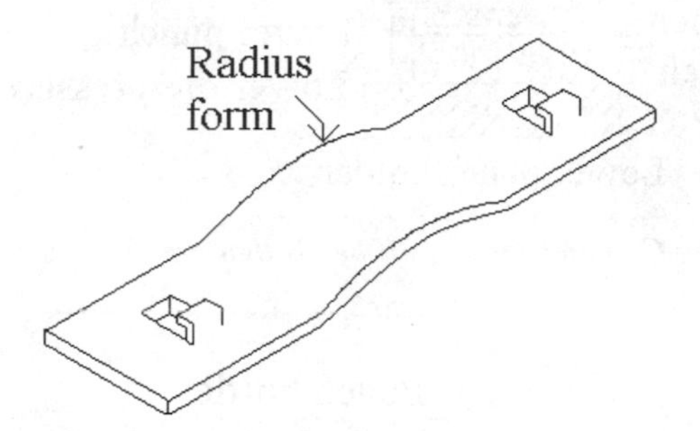

FIGURE 9.15 *Part with a radius form.*

of active stations provided for in bending is one greater than the number of bend stages.

A precise method for determining the number of required bend stages is presented in the next section. However, before determining the number of bend stages it is necessary to determine whether the bends can in fact be feasibly wipe formed. Parts without straight bends (Figure 9.14), parts containing radius forms (Figure 9.15), parts with a multiple-plate junction (Figure 9.16) are examples of bends that cannot be formed by wipe forming.

Assuming the part can be wipe formed, then the next step is determining the number of bend stages and, if there is only one bend stage, whether all the bends are or are not in the same direction.

9.6.2 Direction of Bends

For parts with two bends that can be formed at a single station (i.e., with only one bend stage), those parts with bends in the same direction require only one set of pressure pads. Parts with bends in opposite directions require two sets of pressure pads with different spring constants and specially sized dies—all of which increases die construction cost (Figure 9.17). Two stations are often used

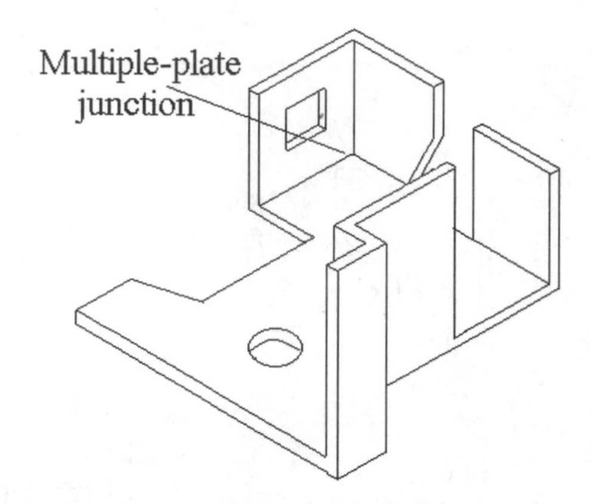

FIGURE 9.16 *Part with a multiple plate junction.*

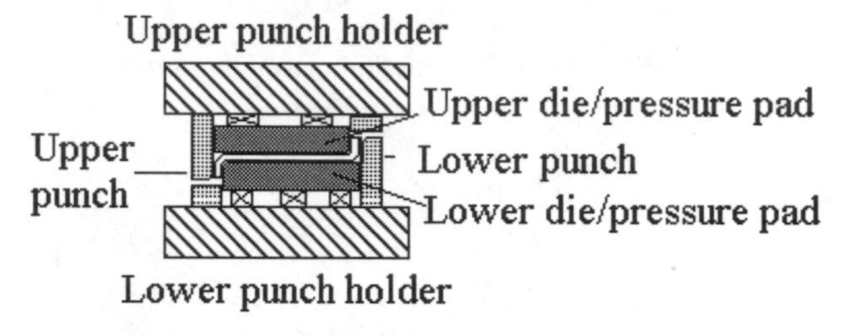

FIGURE 9.17 *Complex tooling for bends that are not in the same direction.*

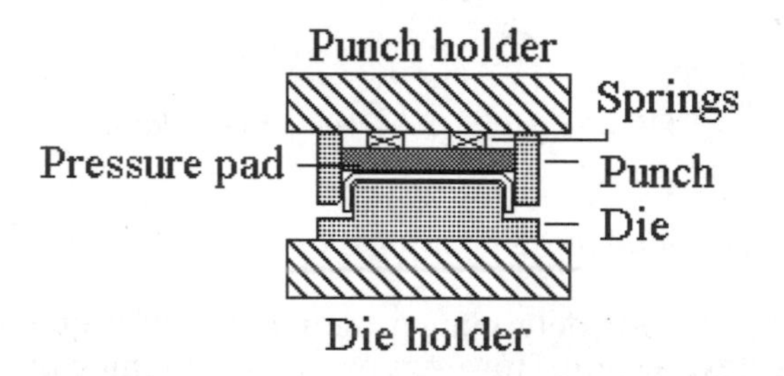

FIGURE 9.18 *A part wipe formed through 90°.*

to impart bends that occur in opposite directions. Table 9.2 assumes that two stations are used.

9.6.3 Bend Angle

For parts with bend angles between 90° and 105°, two bend stations are generally required. At the first station the part is bent through 90°, as shown in Figure 9.18, but at the following station it is overbent once more, as shown in Figure 9.19. Because of the limited overbend, relatively inexpensive tooling can be used to carry out the overbend.

If the bend angle is greater than 105°, once again two operations are required—though costs are higher. At the first station the part is bent through

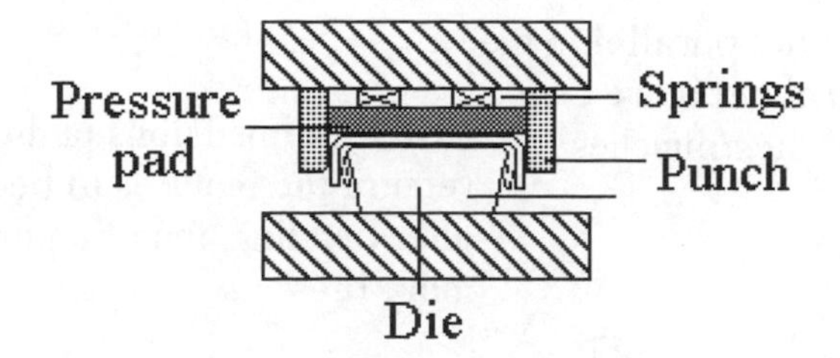

FIGURE 9.19 *Tooling for 90° < bend angles < 105°. The clearance between the punch and die must be less than the sheet thickness of the part.*

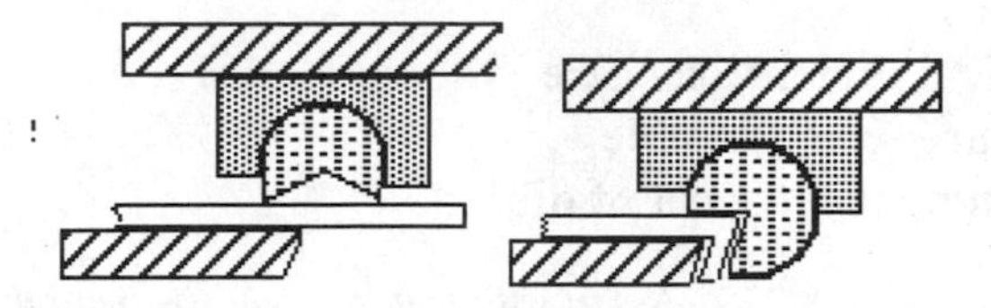

FIGURE 9.20 *A rotary bender for bending a part past 90° while using one workstation.*

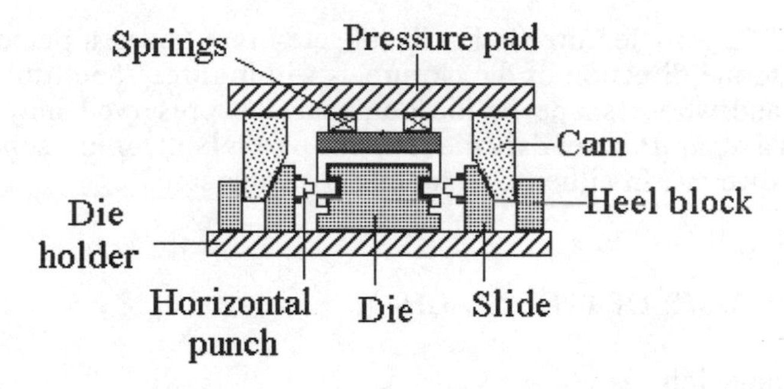

FIGURE 9.21 *A part with side-action features.*

90°, and at the second station more expensive cam actuated slides are required (Figure 9.21). Figure 9.20 shows a rotary bender being used at one station to achieve the bend in one operation. Most stampers, however, do not often use rotary benders. Hence, the algorithm presented in Table 9.2 provides for one additional workstation for each bend greater than 90°.

9.6.4 Side-Action Features

Features of a bent part, usually holes or slots, that do not lie in a plane perpendicular to the direction of die closure, and which must be accurately located or aligned (+/– 0.2 mm), must be created after bending by use of cam actuated tools (Figure 9.21). If such a feature is created before bending, then it may not be held to true position after bending. Such features are called side-action features and are equivalent to the undercuts that occur in injection molding and die casting. Hence, an additional die station is allocated for these types of features in Table 9.2.

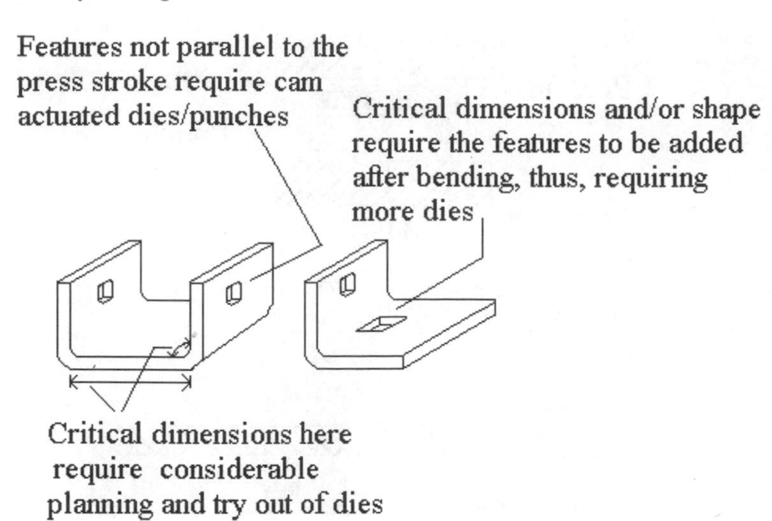

FIGURE 9.22 *A part with features near the bend line.*

9.6.5 Features Near the Bend Line

See Figure 9.22. Any feature (hole, rib, tab, etc.) that lies in a plane that is perpendicular to the direction of die closure, is within three sheet thicknesses of a bend line, and whose shape or location must be preserved may have to be imparted in a separate secondary die after the part is bent, or in a separate station in a progressive die. In either case, tooling costs increase.

9.7 THE NUMBER OF BEND STAGES

9.7.1 Introduction

One of the difficulties encountered in using the algorithm presented here is estimating the number of bend stations (or stages) required to produce the part. In this section, we present a method for estimating the number of bend stages that will be required to produce a given part. The method requires first partitioning the part into *elemental plates.*

9.7.2 Part Partitioning—Elemental Plates

A stamped part that is not flat can be decomposed into a set of elemental plates. Each plate is a part of the stamping that is bounded either by the part periphery, bend lines, radius forms, or a combination of these.

An elemental plate has a constant spatial orientation (all vectors normal to the surface point in the same direction), except in the case of a radius form. A radius form is an elemental plate that has a constant radius of curvature, and all surface normals point to the center of curvature of the form. Figure 9.23 shows two parts with elemental plates labeled.

Movement from one plate to another implies the crossing of a bend line or the boundary of a radius form. A flat part is a plate in itself.

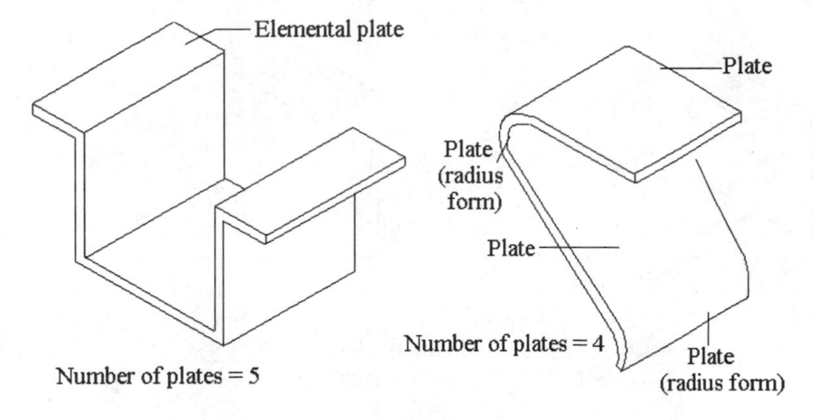

FIGURE 9.23 *Parts with elemental plates labeled.*

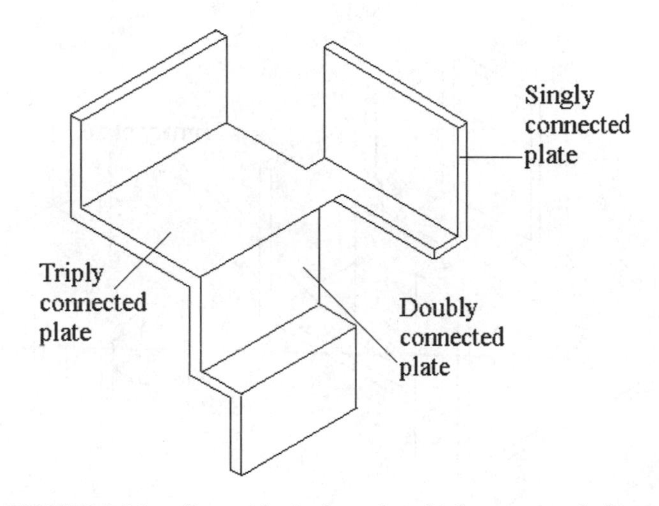

FIGURE 9.24 *Part with singly and multiply connected plates.*

An elemental plate that is connected (by a bend line or the boundary of a radius form) to only one other plate is called a *singly connected elemental plate*. Similarly, a plate that is connected to two other plates is called a *doubly connected plate*, and so on. Figure 9.24 shows a part with singly, doubly, and triply connected elemental plates.

The *primary plate* of a part is the plate that is most likely to be placed parallel to the plane of the die block (either in a progressive or secondary die). As noted earlier, the direction of die closure will always be normal to the plane of the die block.

Finding the Primary Plate

The primary plate of a part is determined by searching for an elemental plate with the following properties in this approximate order of significance:

1. The elemental plate with the largest area. Such an elemental plate is shown in Figure 9.25.

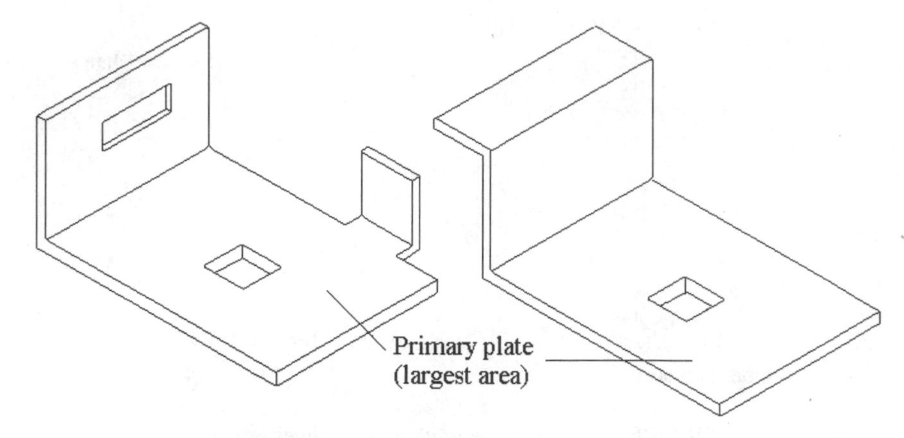

FIGURE 9.25 *Primary plate based on largest area.*

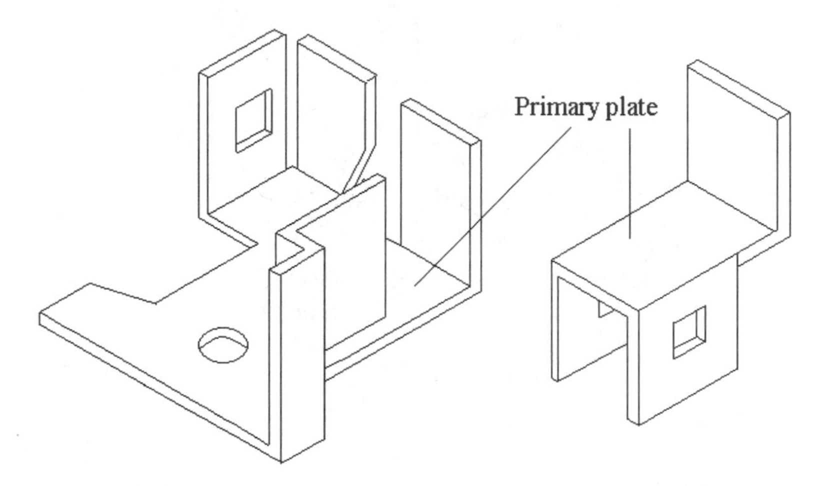

FIGURE 9.26 *Primary elemental plate based on maximum number of bend lines surrounding the plate.*

2. If the part has a complex formed profile with several bends, then the primary elemental plate is the elemental plate that is surrounded (on its boundary) by the largest number of bend lines. Figure 9.26 shows a part with such an elemental plate.
3. If no single elemental plate has a significantly larger area, and the part does not have a complex formed profile with several bends, then the primary elemental plate is that elemental plate with the maximum number of internal features (holes, slots, or protruding features). Figure 9.27 shows a part with such an elemental plate.

9.7.3 Approximate Number of Stages Required to Form the Bends

The part shown in Figure 9.28 will be used to illustrate a systematic procedure to determine the number of bend stages (Dastidar, P.G., 1991). The part in Figure 9.28 has all its elemental plates numbered as specified by the procedure.

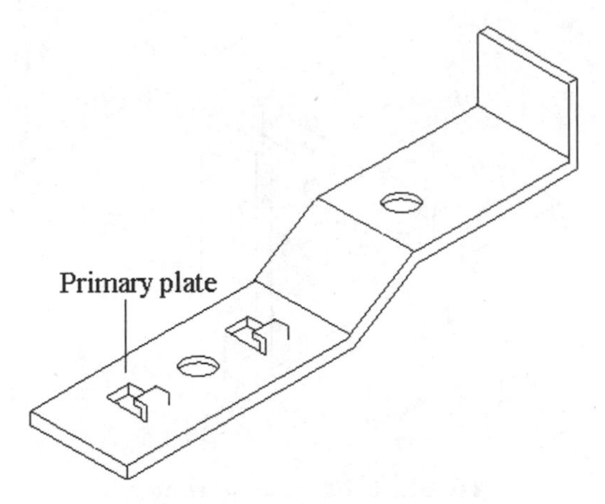

FIGURE 9.27 *Primary elemental plate based on maximum number of features.*

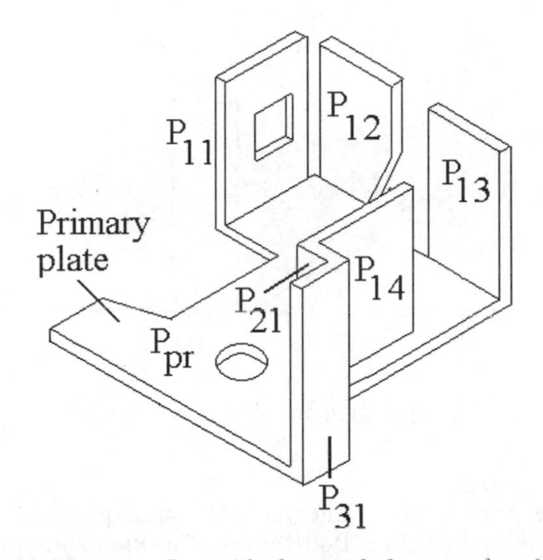

FIGURE 9.28 *Part with elemental plates numbered.*

Step 1: Remove All Singly Connected Plates.

Excluding the primary elemental plate, determine all the singly connected elemental plates of the stamping and number them as P_{ij}, where the subscript i is the current step number, and j is the number of the currently identified singly connected elemental plate.

Remove all the singly connected elemental plates identified above, namely, P_{11}, P_{12}, P_{13}, and P_{14}. The part after removing these elemental plates is shown in Figure 9.29, and has been modified by this step.

Step 2: Remove the Singly Connected Plates for the Modified Part.

Identify all the singly connected elemental plates in the modified part as shown in Figure 9.29. Once again, the primary elemental plate is not to be considered. The only elemental plate candidate is the elemental plate numbered P_{21}. Remove this elemental plate from the part, thus modifying it a second time. The truncated part is shown in Figure 9.30.

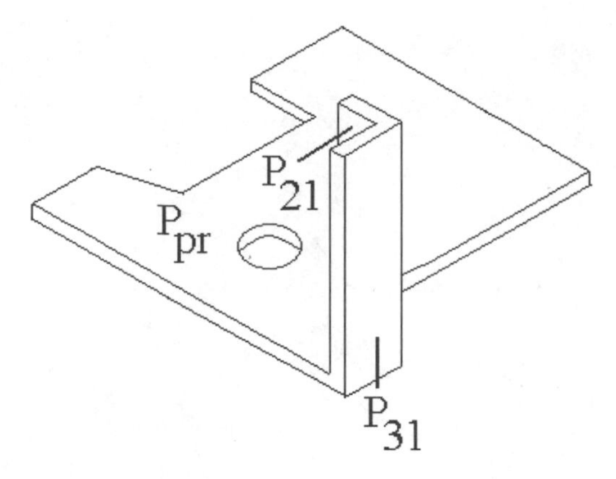

FIGURE 9.29 *Part after Step 1.*

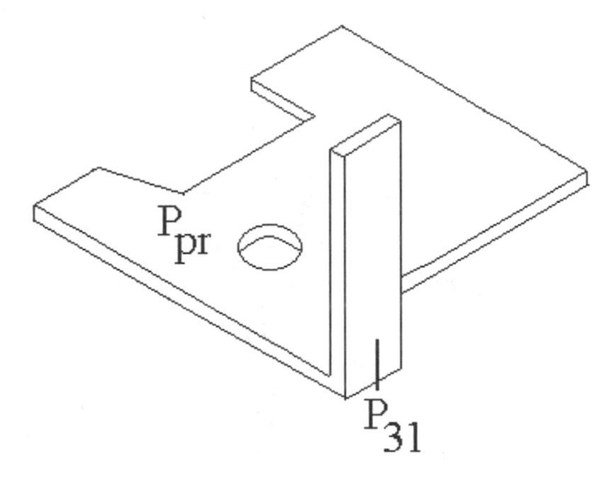

FIGURE 9.30 *Part after Step 2.*

Step 3: Repeat Step 2.

The only singly connected elemental plate in the modified part shown in Figure 9.30 is P_{31}. Remove this elemental plate. The truncated part is shown in Figure 9.31. After the third step, the truncated part comprises a single elemental plate that is the primary elemental plate. The procedure terminates when the primary elemental plate is the only remaining elemental plate. The number of steps required to reach this state is the number of bend stages. In this example, the number of bend stages is 3.

This process of identifying the singly connected elemental plates, removing them from the part, and repeating the process until the primary elemental plate is the only elemental plate left can be applied to any bent stamping to obtain a simple approximation to the number of required bend stages without considering tooling details. Figure 9.32 shows two parts with multiple bend stages.

9.7.4 Bend Directions

For a part with a single bend stage, if all the bend lines are in the same plane and all the singly connected elemental plates are on the same side of the primary plate, then the bends are in the same direction.

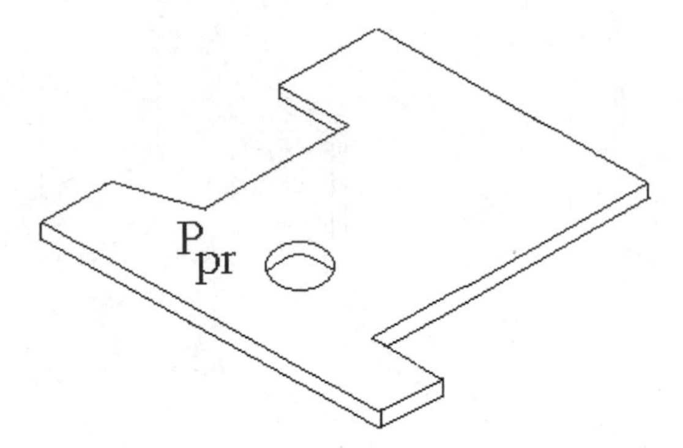

FIGURE 9.31 *Part after Step 3.*

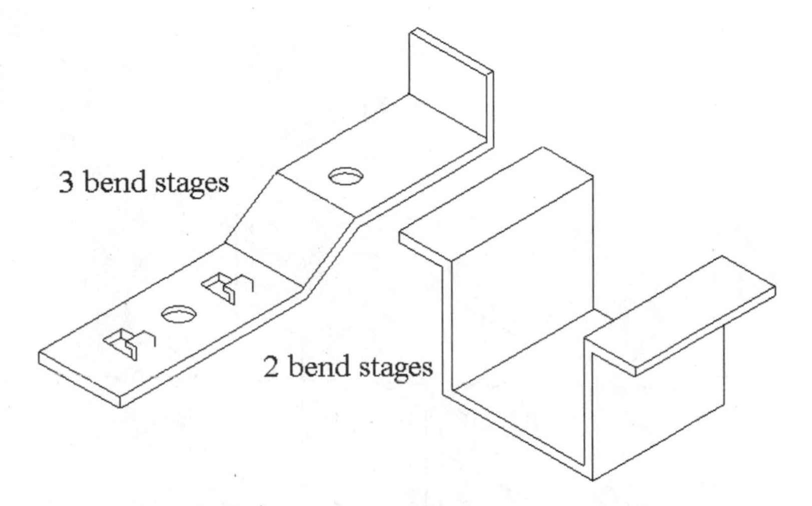

FIGURE 9.32 *Parts with multiple bend stages.*

Parts with all the bends in the same direction can be formed by wiping down all the singly connected elemental plates in a relatively easy to construct forming tool, with wiping punches, die block, and a pressure pad. The pressure pad is placed below the multiply connected elemental plate when wiping the singly connected elemental plates in a direction opposite to that of the press stroke (also known as "wiping up"), and placed above the multiply connected elemental plate when wiping the singly connected elemental plates in the same direction as the press stroke (also known as "wiping down"). Figure 9.33 shows parts that have all their bends in the same direction.

Parts with bends not in the same direction will have singly connected elemental plates on both sides of the primary plate. These parts require more complex form tooling in order to form them at a single progressive station or in a single secondary form die. Typically, such tooling would include specially sized punch and die steels and two pressure pads of differing spring constants.

As explained earlier in Section 9.6.2, Direction of Bends, another processing strategy is to form the part at more than one station (wiping down at one station, and wiping up at another) or using two secondary form dies (wiping down in both dies, but reorienting the part in between dies). Either way, such a bend configuration significantly raises fabrication hours for form tooling, and the part in Figure 9.34 is an example.

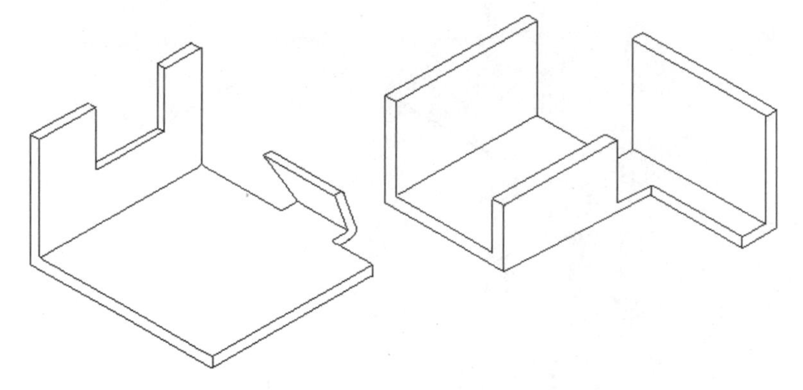

FIGURE 9.33 *Parts with all the bend lines in the same plane and all the bends in the same direction.*

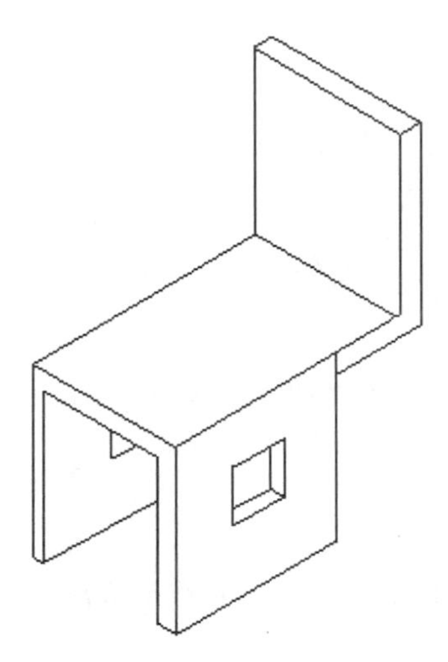

FIGURE 9.34 *Part with all bends not in the same direction.*

9.7.5 Determination of Bend Stages from Part Sketch

In general, during the initial stages of design we will probably not know the points of separation between the various elemental plates that form a bent part. For example, in place of the part sketch shown in Figure 9.28, one is more likely to encounter the part sketch shown in Figure 9.35. The part as shown in Figure 9.35 cannot be stamped. Instead, to produce the part with sheet metal as indicated in the drawing would require the use of deep drawing, or wipe forming followed perhaps by one or more welding operations. Deep drawing is a process that involves high tooling costs, due to extensive tryout, and high processing costs, due to the slow cycle time of hydraulic presses. To produce the part by stamping, then, some of the junctions of the plates that constitute the part must be separated from each other.

Figure 9.36 shows two possible alternatives. In the first alternative plates P_{11} and P_{12} are separated from each other but connected to the primary plate, Ppr.

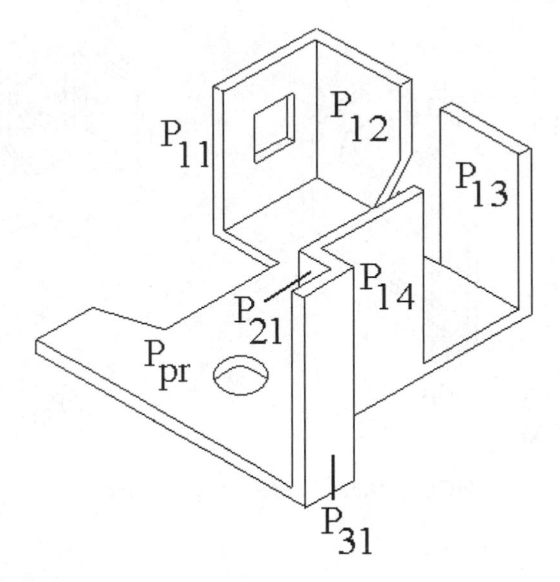

FIGURE 9.35 *Original sketch of part shown in Figure 9.28.*

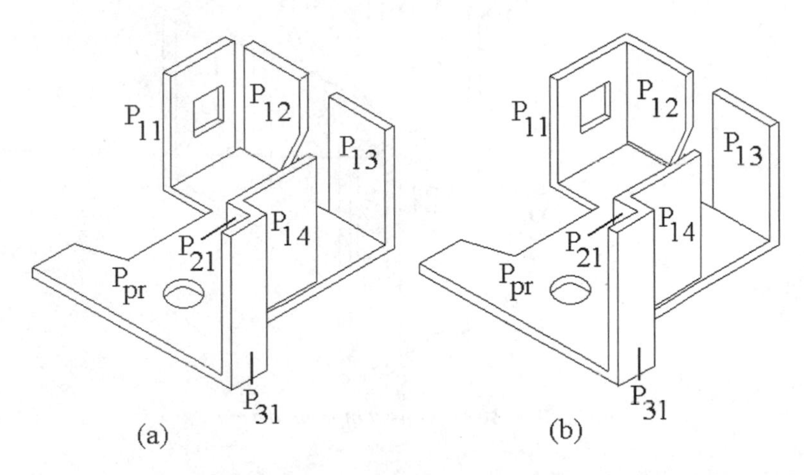

FIGURE 9.36 *Alternative designs to part shown in Figure 9.35.*

Plates P_{14}, P_{21}, and P_{31} are connected to each other. However, P_{14} and P_{21} are separated from the primary plate.

In the second alternative, P_{12} is connected to P_{11} but separated from plate Ppr. In both alternatives, three bend stages are required. Thus, from the standpoint of tooling, the two designs are approximately the same.

In some situations, the number of bend stages required for alternative designs will not be the same. For example, if the part did not contain plates P_{21} and P_{14}, then the design shown in Figure 9.37a would require one more bend stage than the one shown in Figure 9.37b. Except for situations where functionality does not permit, parts should be designed using the minimum number of bend stages. Thus, the design shown in Figure 9.37b should be selected.

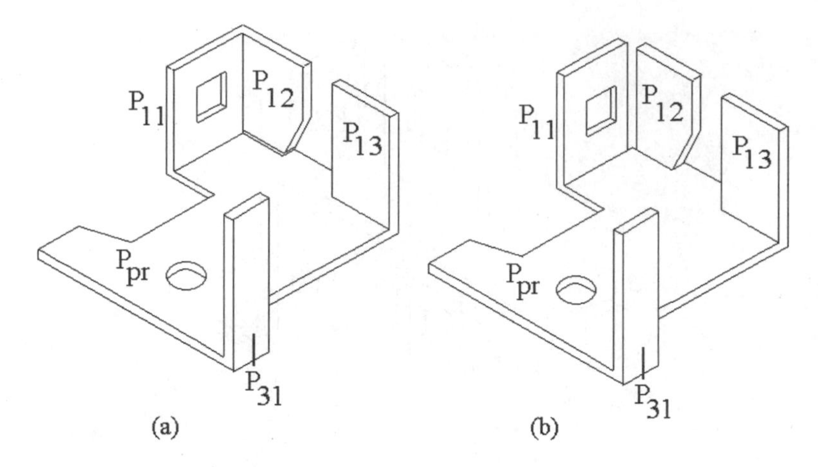

FIGURE 9.37 *Alternative designs.*

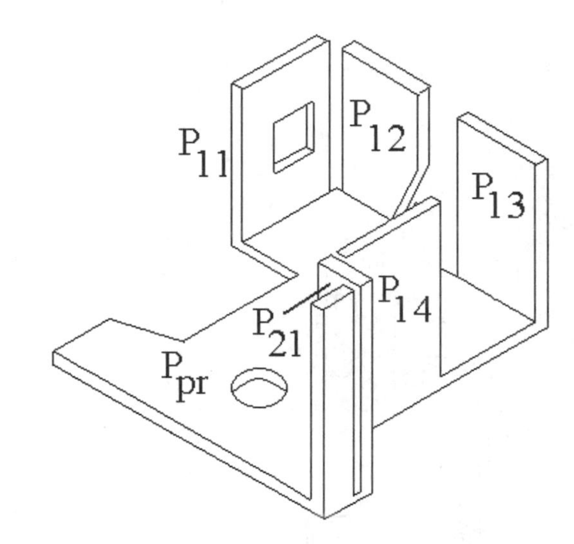

FIGURE 9.38 *Non-stampable design.*

Non-stampable Parts

In designing a part, one must take care not to design parts that cannot be stamped. Figure 9.38 shows plates P_{14}, P_{21}, and P_{31} separated from each other but connected to Ppr. This part cannot be produced as shown because of the interference between plates P_{21} and P_{14} when they are unfolded.

Other examples of non-stampable parts are

1. Parts that do not have a uniform sheet thickness,
2. Parts that have sheet thicknesses greater than 6.5 mm,
3. Parts whose nonperipheral features (holes, extruded holes, tabs, etc.) are not in a direction parallel to the sheet thickness,
4. Parts with hole diameters less than the sheet thickness,
5. Parts with projections, other than tabs or lance forms, that protrude a distance greater than four times the sheet thickness. Such features generally exceed the height that can be achieved by a local forming operation.

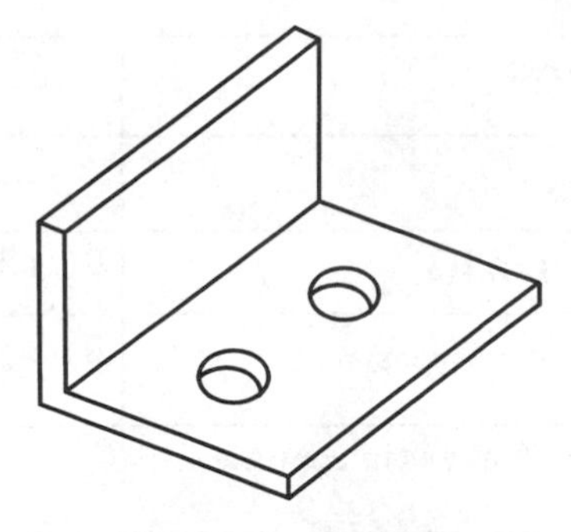

FIGURE 9.39 *Example 9.2.*

9.8 EXAMPLE 2—NUMBER OF ACTIVE STATIONS FOR A BENT PART

9.8.1 The Part

As our second example we consider the part shown in Figure 9.39. We'll assume that the two holes are widely spaced.

9.8.2 Number of Active Stations

To estimate the number of active stations required to produce this part we need to complete Tables 9.1 and 9.2. The completed tables are shown below in Figure 9.40. We see from Figure 9.40 that approximately 5 active stations will be required to produce this part.

As a check, let us compare our estimate with the number of active stations required to produce the part if we follow the process plan shown in Figure 8.21. For this example we see that if we do in fact use the layout shown in Figure 8.21 we overestimate the number of stations required by 1. The reason for this is that the process plan shown in Figure 8.21 uses the part holes as internal pilot holes.

Figures 8.24 and 8.25 show two alternative process plans for this same L-bracket. Once again the two part holes are used as internal pilot holes. However, in one case (Figure 8.24) 4 active stations are used to stamp the part, and in the other case (Figure 8.25) 5 stations are used. Thus, depending upon the actual process plan used our estimate may or may not be precise.

For the purpose of comparing two or more alternative designs and in order to avoid the need to produce a detailed process plan during the early stages of design, the algorithms proposed here have been found to be sufficiently accurate.

9.9 RELATIVE DIE CONSTRUCTION COSTS

9.9.1 Die Construction Hours

As indicated earlier in Section 9.3, die construction cost includes die design cost and die build cost. Design and build costs are related to the total die construction hours, t_{dc}, which is equal to the sum of the build hours, t_b, and design hours, t_d. That is:

Features	Number of Stations
External pilot holes	①
Number of distinct feature types, n_f	0 ① 2 3 4 5 6 7 8 9
Number of closely spaced feature types, n_{fc}	⓪ 1 2 3 4 5 6 7 8 9
Number of feature types with features in opposite direction, n_{op}	⓪ 1 2 3 4 5 6 7 8 9
Tabs? (y/n)	2/⓪
Embosses near part periphery? (y/n)	1 /⓪
Curls or hems? (y/n)	2/⓪
Blanking out station	①
Total Number of active stations, N_{a1}	**3**

Wipe Forming	Number of Stations
Number of bend stages 0 ① 2 3 4 5	0 ② 3 4 5 6
Bends in opposite directions? (applies only to parts with one bend stage) (y/n)	1 /⓪
Number of overbends	⓪ 1 2 3 4 5
Number of features in the primary plate near the bend line	⓪ 1 2 3 4 5
Number of side action features	⓪ 1 2 3 4 5
Total Number of Active Stations, N_{a2}	**2**

FIGURE 9.40 *Estimate of the number of active stations required to produce the part shown in Figure 9.39.*

$$t_{dc} = t_b + t_d \qquad \text{(Equation 9.5)}$$

The build hours for each station are determined using the following equation (9.6), namely,

$$t_i = \frac{(n+1)}{2} t_{basic} \qquad \text{(Equation 9.6)}$$

where t_i is the hours spent in building the tool for the ith station, n is the number of identical features to be imparted in the ith station and t_{basic} is the number of

Table 9.3 Basic hours required to produce various features using a medium-grade tool. For a high-grade tool add 10 hours. For a low-grade tool, subtract 10 hours. (Note: Data based on information provided by collaborating stampers.)

Operation	*Medium Grade*
Bending	40
Blanking	40
Piercing	30
Standard Hole	30
Nonstandard Hole	45
Extruded Hole	50
Lancing	40
Notching	40
Embossing	40
Lettering	40
Semi-perf	25
Tab	65
(notch)	(40)
(form)	(25)
Drawing	55
Forming	40
Coining	40
Curl	120
(notch)	(40)
(bend)	(40)
(form)	(40)
Hem	120
(notch)	(40)
(bend)	(40)
(form)	(40)
Side-Action Feature	80

hours required for building a medium-grade tool to produce a single feature. Build hours include the time spent in creating the die block and stripper plate cavities, machining the punches, punch plate, die holder and punch holder, and assembling all the components (see Figure 8.18).

The total build hours for the die is the sum of the build hours required for each station, namely,

$$t_b = \sum t_i \quad \text{(Equation 9.7)}$$

Table 9.3 gives typical values of $t_{basic.}$

Design hours are a function of the part complexity. The more complex the part and the greater the number of features in it, the greater the number of workstations, hence, the greater the number of hours spent in designing the tool. It has been found that design hours are directly proportional to the number of active stations (Mahajan, P.V., 1991) and can be estimated from the following equation:

$$t_d = 18.33N_a - 3.33 \quad \text{(Equation 9.8)}$$

The total die construction time, t_{dc}, is obtained by summing together the times to both design and build the tool, that is,

$$t_{dc} = t_b + t_d \qquad \text{(Equation 9.9)}$$

9.9.2 Die Construction Costs Relative to a Reference Part

The relative die construction cost for a given part was defined earlier (see Section 9.2.2) as the die construction cost for a designed part divided by the die construction cost for a reference part. That is:

$$C_{dc} = \frac{K_{dc}}{K_{dco}} \qquad \text{(Equation 9.10)}$$

where K_{dc} and K_{dco} refer to the die construction costs for the designed part and reference part, respectively. The die construction costs are obtained by multiplying the die construction time by the hourly rate, C_h. Thus,

$$K_{dc} = t_{dc}C_h \qquad \text{(Equation 9.11)}$$

and

$$K_{dco} = t_{dco}C_{ho} \qquad \text{(Equation 9.12)}$$

Thus, if $C_h = C_{ho}$ then

$$C_{dc} = \frac{t_{dc}}{t_{dco}} \qquad \text{(Equation 9.13)}$$

If the reference part is taken as a stamped washer (OD = 50 mm; ID = 10 mm; t = 1.5 mm) made of low-carbon cold-rolled steel (CRS), then the tool construction time for the washer is about 138 hours.

9.10 EXAMPLE 3—RELATIVE DIE CONSTRUCTION COST FOR A FLAT PART

As an example we will consider the part shown in Figure 9.10 and first introduced in Example 1 in Section 9.5. It was shown in Example 1 that 4 active stations are required to produce this part. Table 9.4 lists the particular operation to be carried out at each workstation along with a calculation of the build time required for the die. In this case 145 hours are required to build the die.

The number of hours required for designing the die for the part is given by substituting into Equation 9.8. Thus,

Table 9.4 Build hours required for Example 3.

Station	*Operation*	t_{basic}	t_i
1	Pierce 2 pilot holes	30	45
2	Pierce 1 standard hole	30	30
3	Pierce 1 standard hole	30	30
4	Blank out part	40	40
	t_b		145

Table 9.5 Build hours required for Example 4.

Station	*Operation*	t_{basic}	t_i
1	Pierce 2 pilot holes	30	45
2	Pierce 2 standard holes	30	45
3	Notch	40	40
4	Wipe form	40	40
5	Blank out part	40	40
	t_b		210

$$t_d = 18.33N_a - 3.33 = 18.33(4) - 3.33 = 70 \text{ hours}$$

The total die construction time in this case is, therefore,

$$t_{dc} = 145 + 70 = 215 \text{ hours}$$

and the relative die construction cost is

$$C_{dc} = (215)/138 = 1.56$$

9.11 EXAMPLE 4—RELATIVE DIE CONSTRUCTION COST FOR A BENT PART

As our next example we will consider the part shown in Figure 9.39 that was first considered in Example 2 (in Section 9.8). It was shown in Example 2 that 5 active stations are required to produce this part. Three of these stations are required for shearing and local forming, and the other 2 stations are required to carry out the wipe forming operation. Table 9.5 lists the particular operation to be carried out at each workstation along with a calculation of the build time required for the die. In this case it is seen that 210 hours are required to build the die.

The design hours for this die are

$$t_d = 18.33N_a - 3.33 = 18.33(5) - 3.33 = 88 \text{ hours}$$

Thus, the total die construction time is,

$$t_{dc} = 210 + 88 = 298 \text{ hours}$$

The relative die construction cost is

$$C_{dc} = (298)/138 = 2.16$$

9.12 EFFECTS OF PART MATERIAL AND SHEET THICKNESS ON DIE CONSTRUCTION COSTS

Unlike the processes of injection molding and die casting where maintenance costs are charged as needed, in stamping maintenance charges, learning and nervous factors are charged up front as part of tool construction cost.

Harder workpiece materials require more tool maintenance (especially tool regrind). In addition, some materials exhibit more springback and as a result

Table 9.6 Factors to account for the effects of sheet thickness on die construction and die material costs.

Sheet thickness (mm)	F_t	F_{dm}
0.125–3.00	1.0	1.0
<0.125	1.7	1.0
>3	1.0	1.4

more experimentation (called tryout) and learning is required to obtain satisfactory results. Thus, in those cases where "harder" workpiece materials such as

- Cold rolled red brass (C23000)
- Cold rolled cartridge brass (C26000)
- Hard aluminum alloys (1100-H18; 5052-H18)
- Stainless steel (302 annealed, 304 cold rolled)

the design and build hours for bend stations should be increased by 10%.

Die construction costs are also affected by the sheet thickness of the part. Sheet thickness is the thickness of the stock metal material used in the stamping. For shearing operations, the clearance between the punch and die must be less than the sheet thickness. For sheet thickness less than 0.005 inches (0.125 mm), adequate tool clearances between the punch and die are difficult to achieve, causing tool fabrication costs to increase. In addition, thin material can buckle when being fed in a progressive die. Hence, such a stock thickness is considered critically thin.

Although a stock thickness greater than 0.125 inches (3.00 mm) has only a negligible effect on tool fabrication costs, it does significantly increase tool material costs. In addition, thick stock, which is coiled for use on progressive dies, has undesirable wrinkles. Hence, such a stock thickness is often considered critically thick.

The factors F_t and F_{dm} account for the effects of sheet thickness on the die construction costs and the die material cost, respectively. Thus, the relative die construction cost, C_{dc}, is given by:

$$C_{dc} = C'_{dc}F_t \qquad \text{(Equation 9.14)}$$

where C'_{dc} is defined as the value of C_{dc} as obtained from Equation 9.13. Values for F_t and F_{dm} are given in Table 9.6.

The method for estimating the die material cost, C_{dm}, and the role that F_{dm} plays in the determination of C_{dm} is discussed in the next section.

9.13 RELATIVE DIE MATERIAL COST FOR PROGRESSIVE DIES

Die material costs consist of the cost of the die set (punch holder, die holder, guide posts, bushings—Figure 9.41), tool steel (die block, punches, die inserts, cams, back gage, finger stop, etc.), and soft or machine steel (punch plate for holding individual punches, stripper plates, and noncritical parts). As mentioned earlier, some items are standard and can be catalog purchased. Such items include the die set itself, punches, die buttons, pilots, springs, and other products. In general, 25% of the total cost of the die material is due to the die set itself. The remaining 75% is due to tool steel and soft material. Hence, a good estimate of

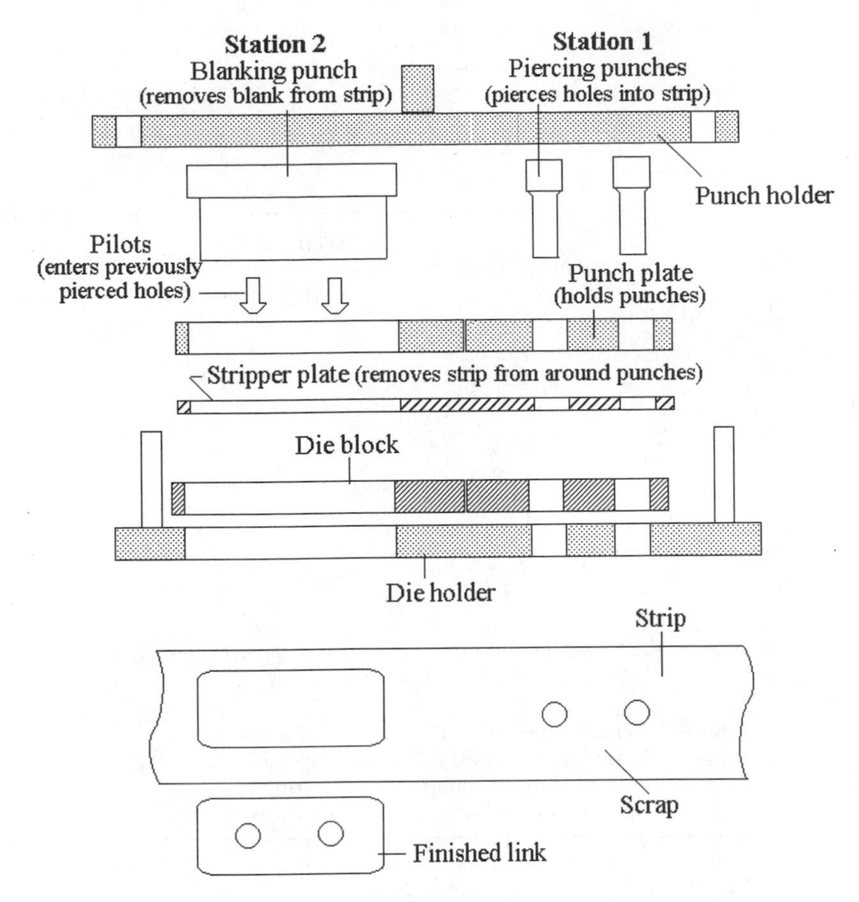

FIGURE 9.41 *Some components included in die material cost.*

the total die material cost can be taken as four times the cost of the die set required.

Assuming that a "one-up" tool is being used (i.e., that only one part at a stroke is being made), then the die material cost is a function of the size of the die set. The size of the die set is, in turn, a function of the number of stations required to produce the part. The number of active stations required is obtained from Tables 9.1 and 9.2.

As discussed before in Chapter 8, Section 8.7.2, Strip Development, in addition to active stations, most progressive dies also make use of idle stations. This is done in order to strengthen the die, to avoid closely spaced punches, and to distribute the cutting load over a larger area. The number of idle stations required depends on the unbent length of the part, L_{ub}, in the direction of strip feed (see Figure 9.42). When L_{ub} is small, more idle stations are added in order to avoid closely spaced punches. Table 9.7 can be used to determine the total number of stations required.

Table 9.8 can be used to determine the size of the die set, S_{ds}. Substituting the value of S_{ds} into the following equation (Mahajan, P.V., 1991) gives the die material cost for the part relative to the die material cost of the reference washer:

$$C_{dm} = F_{dm}[2.7S_{ds}/(25.4)^2 + 136]/260.2 \qquad \text{(Equation 9.15)}$$

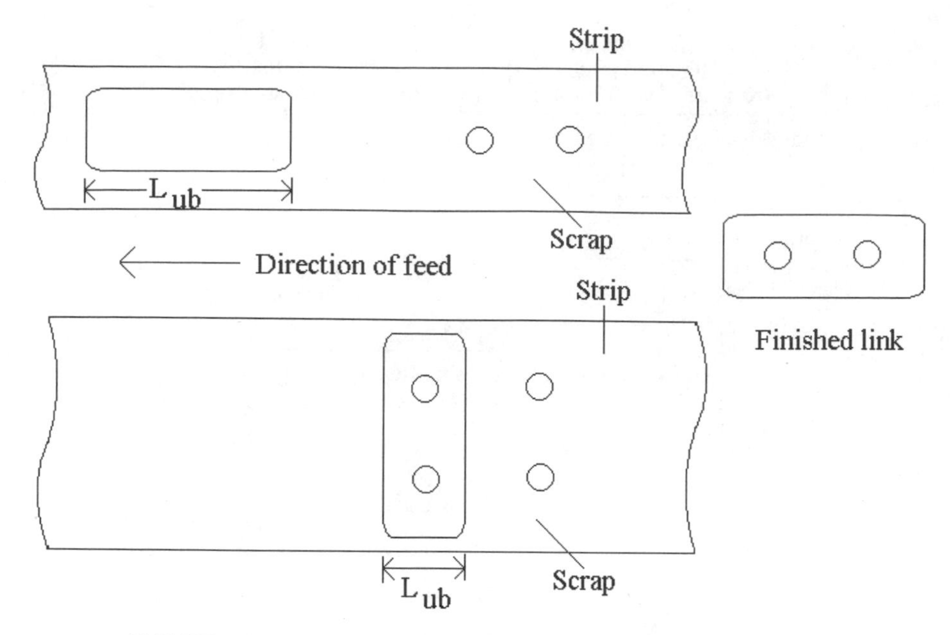

FIGURE 9.42 *Link laid out on strip in two different orientations.*

Table 9.7 Determination of the number of stations required (Mahajan, P.V., 1991). The subscripts 1 and 2 refer to the number of stations required for shearing and local forming, and bending and side-action features, respectively.

Number of Active Stations: $N_a = N_{a1} + N_{a2} =$
Number of Idle Stations: $N_i = N_{i1} + N_{i2} =$

(a) For $L_{ub} \leq 25\,mm$
$N_{ij} = 1.5(N_{aj} - 1), j = 1, 2$

(b) For $25\,mm < L_{ub} < 125\,mm$
$N_{i1} = 0$ (parts without curls)
$N_{i1} = 2$ (parts with curls)
$N_{i2} = (N_{a2} - 1)$

(c) For $L_{ub} \geq 125\,mm$, $N_{i1} = N_{i2} = 0$

Total Number of Stations: $N_S = N_a + N_i =$

The value of F_{dm} accounts for the effect of sheet thickness on die material cost and is obtained from Table 9.6.

9.14 EXAMPLE 5—RELATIVE TOOLING COST

9.14.1 The Part

As our first example for calculating the total relative tooling cost for a part let's consider the part, made of cold rolled steel, shown in Figure 9.43. Assume that we've sketched the part as shown so that the separation between the various elemental plates reflects what is indicated in Figure 9.43. Let's assume that we are

Table 9.8 Die block and die set size (Mahajan, P.V., 1991) with bottom plate of die set shown.

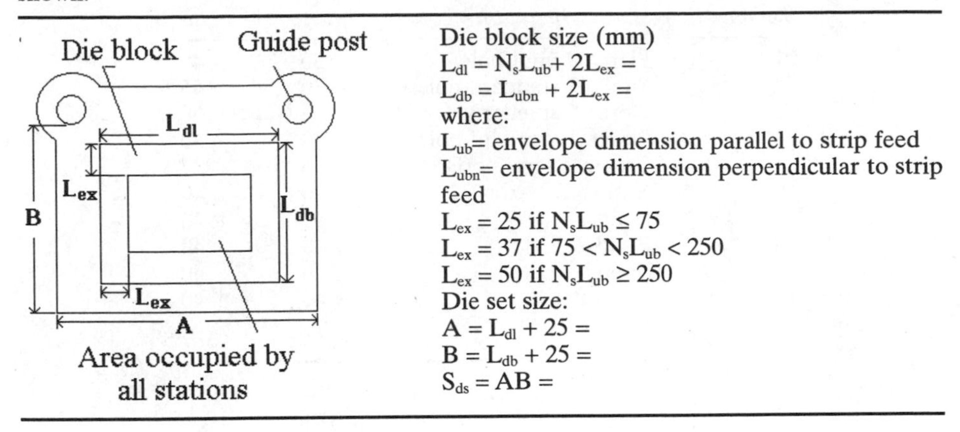

Die block size (mm)

$L_{dl} = N_s L_{ub} + 2L_{ex} =$

$L_{db} = L_{ubn} + 2L_{ex} =$

where:

L_{ub}= envelope dimension parallel to strip feed

L_{ubn}= envelope dimension perpendicular to strip feed

$L_{ex} = 25$ if $N_s L_{ub} \leq 75$

$L_{ex} = 37$ if $75 < N_s L_{ub} < 250$

$L_{ex} = 50$ if $N_s L_{ub} \geq 250$

Die set size:

$A = L_{dl} + 25 =$

$B = L_{db} + 25 =$

$S_{ds} = AB =$

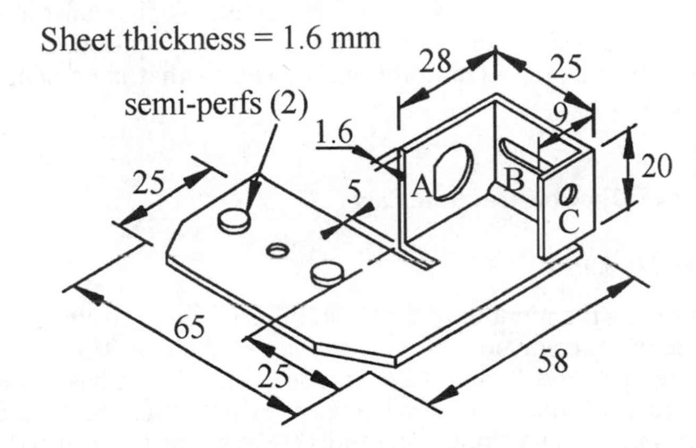

FIGURE 9.43 *Example 5—Original design. All dimensions in mm.*

at the configuration design stage and have several alternative approaches we could use to analyze the part.

In one approach, we could assume the worst possible conditions. In this case, we would treat holes A, B, and C as side-action features and dimension the cutout so that it is narrow. The advantage to this approach is that we could obtain a comparison between the most costly design and one that is less costly.

Alternatively, we could assume the best possible conditions. In this case, we could assume that we do not know as yet whether holes A, B, and C are to be side-action features. We would also assume that the dimensions of the cutout are not yet known, and consequently that it is not to be treated as narrow. With these assumptions, we obtain a lower bound with regard to tooling cost, and investigate the possibility of redesign suggestions that will lower the cost to produce the part still further.

Finally, we could take a middle approach. In this case, we might assume that we know that the location of holes A and C are not critical, but that we're uncertain about hole B. We also assume that, since we are not certain about the precise dimensions of the cutout, we will treat it as narrow. Thus, we will obtain a comparison between a design that treats the cutout as narrow and hole B as critical, with one that does not treat them as narrow and critical, respectively.

Table 9.9 Number of build hours required for Example 5.

Station	*Operation*	t_{basic}	t_i
1	Pierce 2 pilot holes	30	45
2	Pierce 2 standard holes	30	45
3	Pierce 1 standard slot	30	30
4	Pierce 1 standard hole	30	30
5	Create 2 semi-perfs	25	37.5
6	Notch	40	40
7	Bend 1	40	40
8	Bend 2	40	40
9	Bend 3	40	40
10	Separate part	40	40
	t_b		387.5

In this example we will assume the best possible conditions and obtain the minimum relative tooling cost. Thus we are assuming that all dimensions are approximate and that none of the features are closely spaced, near a bend, or result in side-action features. In addition, the cutout that is present will not be assumed to be narrow.

9.14.2 Relative Die Construction Cost

Number of Active Stations

Before determining the number of active stations required to stamp the part let's first determine whether or not the part is even stampable. The part is certainly unfoldable (see Figures 8.26 and 8.27). In addition, the part has a uniform sheet thickness less than 6.5 mm, and hole diameters greater than the 1.6 mm thickness of the sheet. Also, the nonperipheral features are in a direction normal to the sheet thickness, and the part contains only straight bends. Hence, the part is stampable.

There are 4 distinct feature types, 3 distinct types of holes, and semi-perfs. There are also 3 bend stages. Thus, from Figure 9.44 we see that 10 active stations are needed.

Relative Die Construction Cost

The number of build hours required is 387.5 (see Table 9.9). The die design hours are

$$t_d = 18.33N_a - 3.33 = 18.33(10) - 3.33 = 180 \text{ hours}$$

hence, the total die construction time is

$$t_{dc} = 387.5 + 180 = 567.5 \text{ hours}$$

Thus, the relative die construction cost is

$$C_{dc} = (567)/138 = 4.11$$

Since the part has a sheet thickness between 0.125 and 3.00 mm and is made of cold rolled steel, F_t is 1.0 and no correction needs to be made to the calculation of C_{dc}.

Features	Number of Stations
External pilot holes	(1)
Number of distinct feature types, n_f	0 1 2 3 (4) 5 6 7 8 9
Number of closely spaced feature types, n_{fc}	(0) 1 2 3 4 5 6 7 8 9
Number of feature types with features in opposite direction, n_{op}	(0) 1 2 3 4 5 6 7 8 9
Tabs? (y/n)	2 /(0)
Embosses near part periphery? (y/n)	1 /(0)
Curls or hems? (y/n)	2 /(0)
Blanking out station	(1)
Total Number of active stations, N_{a1}	6

Wipe Forming	Number of Stations
Number of bend stages 0 1 2 (3) 4 5	0 2 3 (4) 5 6
Bends in opposite directions? (applies only to parts with one bend stage) (y/n)	1 /(0)
Number of overbends	(0) 1 2 3 4 5
Number of features in the primary plate near the bend line	(0) 1 2 3 4 5
Number of side action features	(0) 1 2 3 4 5
Total Number of Active Stations, N_{a2}	4

FIGURE 9.44 *Number of active stations required for Example 5.*

9.14.3 Relative Die Material Cost

The length of the flat envelope of the part before bending is approximately 92 mm, and the width of the flat envelope is about 65 mm. We assume that the part is fed with the width of the envelope parallel to the direction of strip feed. Thus, $L_{ub} = 65$ mm and $L_{ubn} = 92$ mm.

Using Table 9.7, since $L_{ub} = 65$ mm, and since no curls are present, then no idle stations, N_{i1}, are required for shearing and forming.

Since $N_{a2} = 4$, then the number of idle stations required during the wipe forming stages of the part is

$$N_{i2} = (N_{a2} - 1) = 3$$

Thus, the total number of stations required, N_s, is 13.

From Table 9.8, the die block size required is

$$L_{dl} = N_s L_{ub} + 2L_{ex} = 13(65) + 2(50) = 945\,\text{mm}$$
$$L_{db} = L_{ubn} + 2Lex = 92 + 2(50) = 192\,\text{mm}$$

Hence, the die set size, S_{ds} is

$$S_{ds} = (L_{dl} + 25)(L_{db} + 25) = 970(217) = 210{,}490\,\text{mm}^2$$

Since $F_{dm} = 1$, the relative die material cost is, from Equation 9.15,

$$C_{dm} = [2.7S_{ds}/(25.4)^2 + 136]/260.2 = 3.91$$

9.14.4 Total Relative Die Cost

The total relative die cost is obtained from Equation 9.4, thus,

$$C_d = 0.8C_{dc} + 0.2C_{dm} = 0.8(4.11) + 0.2(3.91) = 4.07$$

9.14.5 Redesign Suggestions

The major cost drivers in this case are the number of stations required to impart the various shearing and local forming operations and the number of bend stages required. The number of shearing stations could be reduced if all the holes were identical. Thus, instead of requiring 3 stations to pierce the holes only 1 would be required. Reducing the number of multiply connected plates can reduce the number of bend stages. Redesigning the part as shown in Figure 9.45 can reduce the number of multiply connected plates.

Redesign Alternative 2 has no multiply connected plates, and hence requires only one bend stage. In addition, all of the bends are in the same direction, hence, the total number of active stations is 8: 6 for shearing and local forming, and 2 for bending. The build hours are reduced to 307.5 hours and the design hours are reduced to 143 hours. Hence, the total die construction time for alternative redesign 2 is 450.5 hours, and C_{dc} for this design becomes 3.26. This results in a reduction of die construction cost of approximately 21%.

The reduction in the number of active stations from 10 to 8 will also result in a reduction in the die material cost. In this case, N_{i1} is still 0 and N_{i2} is reduced to 1. Hence the total number of stations, N_S, required is reduced from 13 to 9. The new die block set is then approximately 865 mm (L_{dl}) by 185 (L_{db}) and the die set size, S_{ds}, becomes 186,900 mm^2. The die material cost, C_{dm}, for Alternative 2 is, therefore, 3.52. This is a 10% savings in die material cost.

The overall die cost for the redesigned part is

$$C_d = 0.8(3.26) + 0.2(3.52) = 3.31$$

This is an overall savings in tooling cost of 19%.

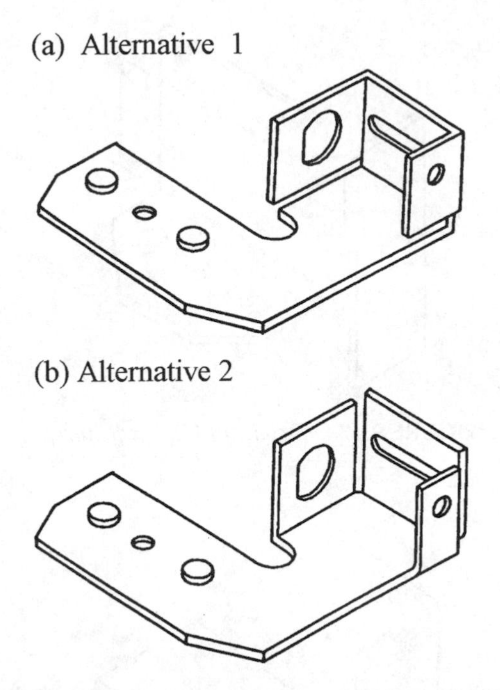

FIGURE 9.45 *Alternative redesigns for Example 5.*

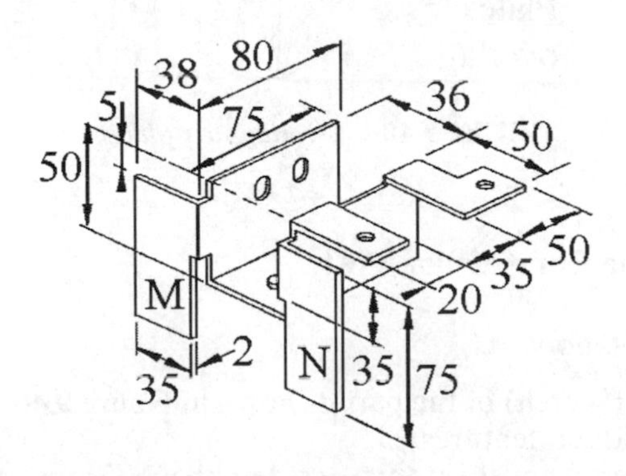

FIGURE 9.46 *Original design.*

9.15 EXAMPLE 6—RELATIVE TOOLING COST

9.15.1 The Part

As our final example we consider the part shown in Figure 9.46. To analyze this part, we will take a middle-of-the-road approach. That is, except for holes A (see Figure 9.47), we will assume that the other holes are not critical and they will, therefore, not be treated as side-action features. Although the details of holes A are not as yet known, we will assume that we know enough about this design that even at this configuration stage we know that their location is critical and, thus, they will be treated initially as side-action features.

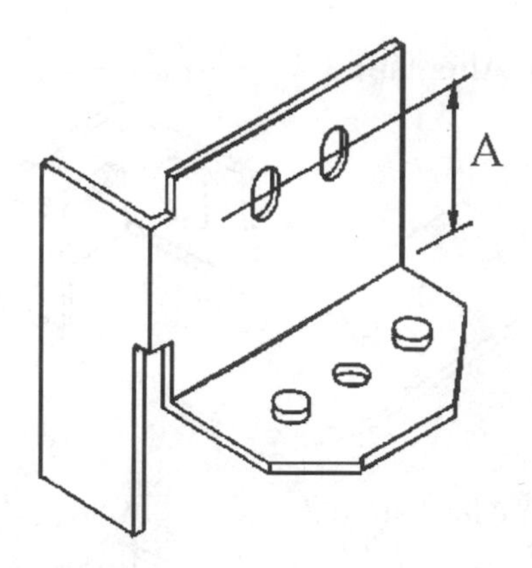

FIGURE 9.47 *Details of hidden features.*

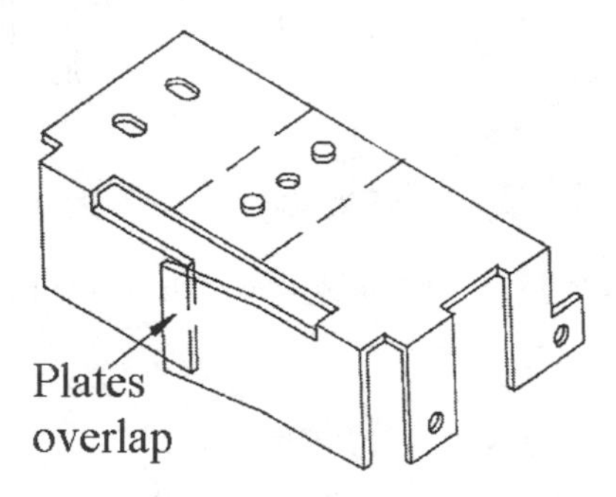

FIGURE 9.48 *Overlapping plates.*

9.15.2 Relative Die Construction Cost

Number of Active Stations

The original design (sketch) of the part is shown in Figure 9.46. Figure 9.47 shows the details of the hidden features.

The part has a uniform sheet thickness less than 6.5 mm, and hole diameters greater than the 2.0 mm thickness of the sheet. In addition, the nonperipheral features are in a direction normal to the sheet thickness. However, upon unfolding the part, in order to determine the size of the flat envelope, it is seen that elemental plates M and N overlap (Figure 9.48), hence, the part is not stampable!

Redesign Suggestions

If the length of the plates M and N are reduced as shown in Figure 9.49, the plate is now stampable. This redesigned part will now be analyzed.

Number of Active Stations for Redesigned Part

In order to determine the amount of die detail present, it is necessary to check for the presence of side-action features. In this case the location of holes A was specified to have a tight location tolerance from the bend line. Hence, holes A

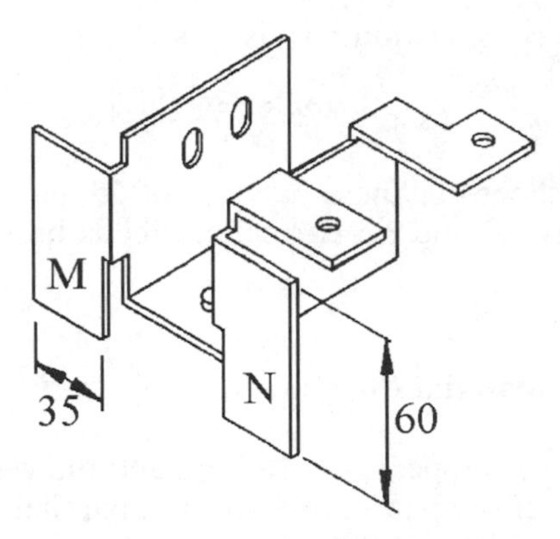

FIGURE 9.49 *Redesigned part.*

Table 9.10 Number of build hours required for Example 6.

Station	*Operation*	t_{basic}	t_i
1	Pierce 2 pilot holes	30	45
2	Pierce 3 standard holes	30	60
3	Create 2 semi-perfs	25	37.5
4	Notch	40	40
5	Bend 1	40	40
6	Bend 2	40	40
7	Side-action feature	80	80
8	Separate part	40	40
	t_b		382.5

must be formed after bending and are, therefore, a side-action feature. Thus, holes A are not included in the determination of the number of active stations required for shearing and local forming.

Using Table 9.1, it is seen that since the part contains 3 standard holes and 2 semi-perfs in the same direction, the total number of active stations required for shearing and local forming, N_{a1} is 4. Using Table 9.2, it is seen that since the part contains 2 bend stages and has 1 side-action feature, the number of active stations required for bending and side-side action features, N_{a2}, is also 4. Thus, a total of 8 active stations is required to produce this part.

Relative Die Construction Cost

The calculations carried out in Table 9.10 show that the number of build hours required is 382.5.

The die design hours are

$$t_d = 18.33N_a - 3.33 = 18.33(8) - 3.33 = 143.3 \text{ hours}$$

hence, the total die construction time is

$$t_{dc} = 382.5 + 143.3 = 525.8 \text{ hours}$$

Thus, the relative die construction cost is

$$C_{dc} = 525.5/138 = 3.81$$

Since the part has a sheet thickness between 0.125 and 3.00 mm and is made of cold rolled steel, F_t is 1.0 and no correction needs be made to the calculation of C_{dc}.

9.15.3 Relative Die Material Cost

The length of the flat envelope, L_{ul}, is 186 mm and the width of the flat envelope of the part is 148 mm. If the part is fed with L_{uw} parallel to the direction of strip feed, then $L_{ub} = 143$ and $L_{ubn} = 186$ mm.

Using Table 9.7, since $L_{ub} = 148$ mm, no idle stations, N_{il}, are required for shearing and forming.

Since $N_{a2} = 4$, then the number of idle stations required during the wipe forming stages of the part are

$$N_{i2} = (N_{a2} - 1) = 3$$

Thus, the total number of stations required, N_s, is 11.

From Table 9.8, the die block size required is

$$L_{dl} = N_s L_{ub} + 2L_{ex} = 11(148) + 2(50) = 1728\,\text{mm}$$

$$L_{db} = L_{ubn} + 2L_{ex} = 186 + 2(50) = 286\,\text{mm}$$

Hence, the die set size, S_{ds}, is

$$S_{ds} = (L_{dl} + 25)(L_{db} + 25) = 1753(311) = 545{,}183\,\text{mm}^2$$

Since $F_{dm} = 1$, the relative die material cost is, from Equation 9.15,

$$C_{dm} = [2.7S_{ds}/(25.4)^2 + 136]/260.2 = 9.29$$

9.15.4 Total Relative Die Cost

The total relative die cost is obtained from Equation 9.4, thus,

$$C_d = 0.8C_{dc} + 0.2C_{dm} = 0.8(3.81) + 0.2(9.29) = 4.91$$

9.16 WORKSHEET FOR RELATIVE TOOLING COST—STAMPING

To facilitate the calculation of the relative tool costs for stamping, the worksheet shown below can be used. The worksheet may be reproduced for use with this book.

Worksheet for Relative Tooling Costs—Stamping

Original Design

	Yes	No
Unfoldable?		
Uniform sheet thickness?		
Sheet thickness < 6.5 mm?		
Hole diameter > sheet thickness?		
Features normal to sheet thickness?		
Are bends straight bends?		
Without multiple plate junctions?		
Primary plate without overbends on all sides?		
IF ALL RESPONSES ARE YES—PART IS STAMPABLE!		

Number of Active Stations

N_{a1} (see Table 9.1) =	N_{a2} (see Table 9.2) =
$N_a = N_{a1} + N_{a2}$ =	

Relative Die Construction Cost

t_b (see Table 9.3) =	$t_d = 18.33N_a - 3.33$ =
$t_{dc} = t_b + t_d$ =	$C_{dc} = t_{dc}/138$ =
F_t (see Table 9.6) =	$C_{dc} = C'_{dc}F_t$ =

Relative Die Material Cost

Flat Envelope:	L_{ul} =	L_{uw} =
Direction of strip feed:	L_{ub} =	L_{ubn} =
Idle Stations	N_{i1} (see Table 9.7) =	N_{i2} (see Table 9.7) =
Total Stations	$N_s = N_a + N_{i1} + N_{i2}$ =	F_{dm} (see Table 9.6) =

$L_{dl} = N_sL_{ub} + 2L_{ex}$ =	$L_{db} = L_{ubn} + 2L_{ex}$ =
$A = L_{dl} + 25$ =	$B = L_{db} + 25$ =
$S_{ds} = AB$ =	

$C_{dm} = F_{dm}[2.7S_{ds}/(25.4)^2 + 136]/260.2$ =
$C_d = 0.2C_{dm} + 0.8C_{dc}$ =

Redesign Suggestions (or % savings if a redesign):

9.17 SUMMARY

The purpose of this chapter was to present a systematic approach for identifying, at the configuration design stage, those features of the part that significantly affect the cost of stampings. In addition, the goal was to learn how to design so as to minimize difficult-to-stamp features. Therefore, a methodology for estimating the relative tooling cost of proposed stamped parts based on configuration information was presented. Unlike injection molding and die casting, however, where the total relative tooling cost can be determined at the configuration stage of design, we found that in stamping only a range of relative tooling costs can be determined. Nevertheless, a great deal of information concerning tooling cost can be determined at the configuration stage, and this information can be used to help guide the generation and selection of part configurations before parametric design is done.

REFERENCES

Chandrasekaran, Srinivasan. "A PC-Based Design for Manufacturability Model for Economical Stamping Design." M.S. Thesis, Mechanical Engineering Department, University of Massachusetts Amherst, Amherst, MA, 1993.

Dastidar, P. G. "A Knowledge-Based Manufacturing Advisory System for the Economical Design of Metal Stampings." Ph.D. Dissertation, Mechanical Engineering Department, University of Massachusetts at Amherst, Amherst, MA, 1991.

Dixon, John R., and Poli, Corrado. *Engineering Design and Design for Manufacturing: A Structured Approach.* Conway, MA: Field Stone Publishers, 1995.

Korneli, M. "Designing Progressive Dies." *Stamping Journal*, Vol. 11, No. 1, January/February 1999, pp. 28–31.

Mahajan, P. V. "Design for Stamping: Estimation of Relative Die Cost for Stamped Parts." M.S. Final Project Report, Mechanical Engineering Department, University of Massachusetts at Amherst, Amherst, MA, 1991.

Poli, C., Dastidar, P. G., and Mahajan, P. V. "Design for Stamping—Analysis of Part Attributes that Impact Die Construction Costs for Metal Stampings." Transactions of the ASME, Journal of Mechanical Design, Vol. 115, 1993, pp. 735–743.

Wick, C., editor. *Tool and Manufacturing Engineers Handbook,* Vol. 2. "Forming, 9th edition." Dearborn, MI: Society of Manufacturing Engineering, 1984.

QUESTIONS AND PROBLEMS

9.1 As part of Example 1 (see Figure 9.10), a process plan was obtained for the case with the largest dimension of the flat envelope of the part fed parallel to the direction of strip feed. Develop a process plan for this same part with the largest dimension of the rectangular flat envelope of the part fed perpendicular to the direction of strip feed. How do these results compare with the results obtained in Example 1, Section 9.5?

9.2 In determining the relative die material cost for the part shown in Figure 9.43 (Example 5, Original Design), it was assumed that the strip layout was such that the longest flat dimension of the part, L_{ul}, was perpendicular to the direction of strip feed. Determine the relative die material cost for this same part if L_{ul} is laid out parallel to the direction of strip feed. Do you anticipate any difficulties in implementing such a process plan?

9.3 In determining the relative die material cost for the Alternative 2 design (Figure 9.45b) of the part considered in Problem 9.2 above, it was assumed that the strip layout was such that the longest flat dimension of the part, L_{ul}, was perpendicular to the direction of strip feed. Determine the relative die material cost for this same part if L_{ul} is laid out parallel to the direction of strip feed. What is the total relative die cost in this case? Which of the two layouts (i.e., L_{ul} parallel to the direction of strip feed, or L_{ul} perpendicular to the direction of strip feed) seems more feasible?

9.4 As part of the same training course discussed earlier in Problem 4.1 in Chapter 4, for example, assume that you have been teaching some new hires the approach discussed in this chapter for comparing two alternative designs. One of the trainees seems skeptical that this approach works. In order to convince the trainee of the viability of this method:
(a) Develop a detailed process plan for the Alternative 2 design of Example 5.
(b) Estimate the relative die construction costs for Alternative 2 design using the process plan just developed.
(c) Using the process plan developed in Chapter 8, "Sheet-Metal Forming" (see Section 8.7.3), estimate the relative die construction costs for the original design.
(d) Using the results obtained in parts b and c of this problem, estimate the percentage savings in relative die construction costs if Alternative 2 design is used.
(e) Compare the results obtained in part d above with the results obtained in Section 9.14.

9.5 The part shown in Figure P9.5 is made of a soft cold rolled steel. The location of holes A and C are not critical. Distortion of hole B is not permitted. Determine:
(a) The relative die construction cost for the part.
(b) The relative die material cost for the part.
(c) The overall die cost for the part.

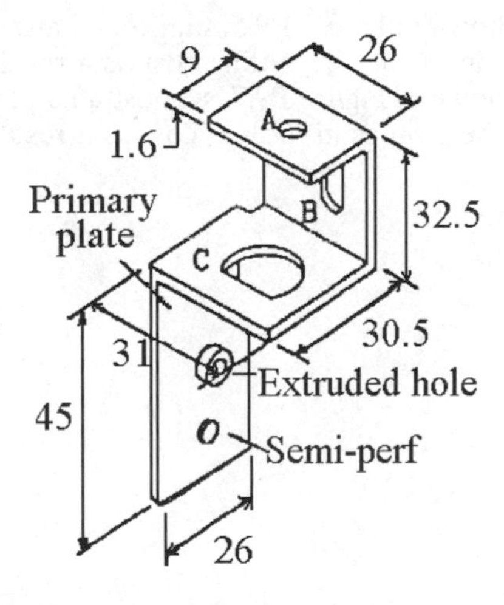

FIGURE P9.5 *(All dimensions are in mm.)*

9.6 Develop a detailed process plan for the part shown in Figure P9.5. Compare the number of active and idle stations, N_a and N_I, respectively, that you used in your plan with the number of active and idle stations used in your estimate of the relative die constructions costs in Problem 9.5 above.

9.7 The part shown in Figure P9.7 is made of a soft cold rolled steel. The dimension A is 9 mm. It is an important dimension that must be held to within +/– 0.02 mm. Determine:

(a) The relative die construction cost for the part.
(b) The relative die material cost for the part.
(c) The overall die cost for the part.

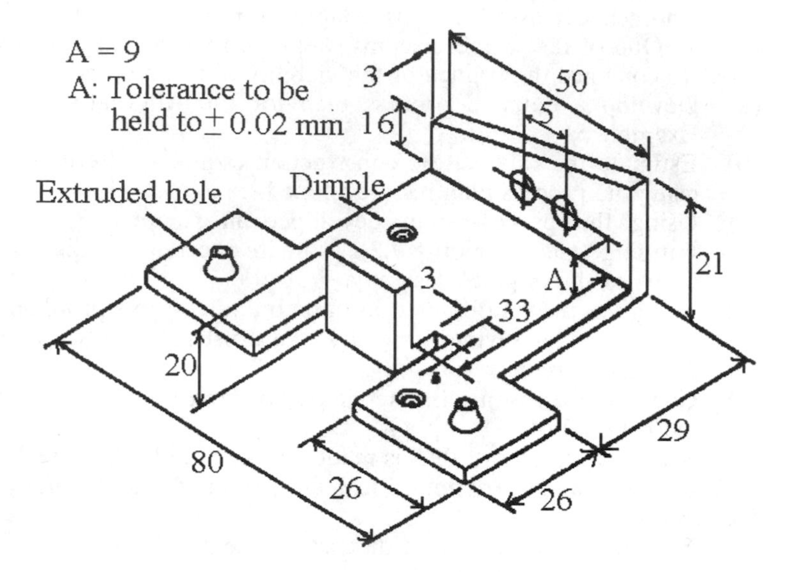

FIGURE P9.7 *(All dimensions are in mm.)*

9.8 For the part shown in Figure P9.5, suggest an alternative design for the part and estimate the savings in tooling cost as a result of the redesign.

9.9 For the part shown in Figure P9.7, suggest an alternative design for the part and estimate the savings in tooling cost as a result of the redesign.

Chapter 10

Stamping: Total Relative Part Cost

10.1 INTRODUCTION

In Chapter 9, "Stamping: Relative Tooling Cost," we expressed the total cost of a stamped part in terms of: tooling cost per part, K_d/N; plus processing cost, K_e; plus material cost, K_m. At the configuration stage, since the final dimensions and tolerances of the part are not yet established, only tooling cost can be estimated. In the case of stamping, it may be possible to obtain only a lower boundary on the tooling cost at the configuration stage.

Once we arrive at the parametric design stage, however, the final dimensions and tolerances are known. Thus the total relative tooling cost can now be more accurately obtained by the methods presented in Chapter 9. In addition, at the parametric stage the relative processing cost, the relative material cost, and thus the overall relative part cost can be obtained.

In this chapter we concentrate on methods for evaluating the relative processing cost and the overall relative part cost for stamped parts made on progressive dies. In general, progressive dies are used for parts whose unfolded length, L_{ul}, is less than 100 mm. For values of L_{ul} between 100 mm and 200 mm, both progressive dies and die lines of compound and single dies are commonly used. For parts larger than 200 mm, die lines are more common than progressive dies. This is due primarily to press capacity constraints that arise as the size of progressive dies increases.

The analysis that follows is restricted to the use of progressive dies. It has been shown that for production volumes in excess of about 100,000, parts are most economically produced using progressive dies (Dastidar, P.G., 1991). In fact, in many cases even for production volumes down to about 10,000, the use of progressive dies can result in a lower production cost.

10.2 RELATIVE PROCESSING COST, C_E

If t_{cy} represents the effective cycle time of the press, and C_h represents the machine hourly rate, then the relative processing cost, C_e, is:

$$C_e = \frac{t_{cy} C_h}{t_o C_{ho}} = t_r C_{hr} \qquad \text{(Equation 10.1)}$$

where t_o and C_{ho} represent the cycle time and the machine hourly rate for the press used to produce the reference washer introduced in Chapter 9 (Section 9.9.2).

As Equation 10.1 indicates, in order to estimate the processing cost, we must first estimate the relative machine hourly rate, C_{hr}, and the relative cycle time, t_r. Since both these parameters require an estimate of the total force required of the press to perform the stamping operations, we present next a procedure for estimating the required press tonnage.

10.3 DETERMINING PRESS TONNAGE, F_P

The press size required depends on the length of the press bed, A (which must be large enough to accommodate the die set), and the force required to stamp the part. The force required is a function of the number and type of features being formed, the size of the part, and the workpiece material.

The required (or rated) press tonnage, F_p, is generally taken as approximately 50% larger than the computed force, F, in tons, required to form the part. The added 50% is an approximate factor to account for uncertainties in the material properties and other estimates. That is:

$$F_p = 1.5F \qquad \text{(Equation 10.2)}$$

where F is the estimated total force in tons required to form the part.

The required force, F, is the sum of several factors:

$$F = F_s + F_b \qquad \text{(Equation 10.3)}$$

where

F_s = Forces for Shearing
F_b = Forces for Bending

The shearing force required, F_s, is the sum of stripper forces, F_{st}, required to remove the strip from around the shearing and blanking punches, and the separation force, F_{out}. The separation force F_{out} is the force required to separate the part from the strip at the last station of the progressive die, and the force required to shear and form the individual nonperipheral features. The force required to create the individual nonperipheral features can be approximated by multiplying the sum of F_{st} and F_{out} by a factor, X_{dd}, which depends on die detail. Values for X_{dd} can be obtained from Table 10.1. Thus, the required shearing force is given by

$$F_s = (F_{out} + F_{st})(1 + X_{dd}) \qquad \text{(Equation 10.4)}$$

where

$$F_{out} = 1.5 L_{out} \sigma_s t \qquad \text{(Equation 10.5)}$$

$$F_{st} = 21 L_{out} t \qquad \text{(Equation 10.6)}$$

F_{out} and F_{st} = the force in Newtons

σ_s = the shear strength of the material in MPa

L_{out} = the length of the periphery of the part in mm

t = sheet thickness in mm

The 1.5 factor in Equation 10.5 is used to account for any peripheral features (cutouts, notches, etc.) that occur on the periphery of the part and are not accounted for by L_{out}. If no notches are present, then Equation

Table 10.1 Values of X_{dd} for use in Equation 10.4.

Feature	Number of Features (n)	Opposite Directions? (Y/N)	Penalty
Standard Hole		n / n	
Non-standard Hole		2n / 2n	
Coin		3n / 3n	
Standard Emboss		(n+1) / n	
Non-standard Emboss		2(n+1) / 2n	
Extruded Hole		2(n+1) / 2n	
Lance Form		3(n+1) / 3n	
Curl		3(n+1) / 3n	
Curl		3(n+1) / 3n	
Hem		4(n+1) / 4n	
Semi-Perf		(n+1) / n	
Tab		3(n+1) / 3n	
Lettering		2(n+1) / 2n	
		Total Penalty	

Small Parts (L ≤ 100 mm)

Total Penalty ≤ 4 => Low Die Detail
4 < Total Penalty ≤ 8 => Medium Die Detail
Total Penalty > 8 => High Die Detail

Medium Parts (100 mm < L ≤ 200 mm)

Total Penalty ≤ 6 => Low Die Detail
6 < Total Penalty ≤12=> Medium Die Detail
Total Penalty >12=> High Die Detail

Large Parts (L > 200 mm)

Total Penalty ≤10=> Low Die Detail
10 < Total Penalty ≤17=> Medium Die Detail
Total Penalty >17=> High Die Detail

Die Detail	X_{dd}
Low	0
Medium	0.22
High	0.28

10.5 will overestimate the force required. However, given the approximate nature of the calculations performed here, this is not considered a serious limitation.

The bending force in Equation 10.3 above is computed as follows:

$$F_b = \sigma_t L_{bt} t/18 \qquad \text{(Equation 10.7)}$$

where

F_b is the bending force in Newtons

σ_t is the tensile strength of the material in MPa,

L_{bt} is the sum of all the straight bend lengths.

Values for the shear strength and tensile strength of some common materials can be determined with the help of Table 13A.2 of Chapter 13, "Selecting Materials and Processes for Special Purpose Parts." When shear strength is unknown but tensile strength is known, an approximate value for the shear strength is obtained by multiplying the tensile strength by a value between 0.65 and 0.85. For high values of tensile strength one should use 0.85 as the multiplying factor. For low values of tensile strength one should use 0.65. When in doubt, use 0.70.

10.4 PRESS SELECTION

Given the length of a die set, A, and the total force, F, required to stamp a part, a press can be selected from Figure 10.1. Then Figure 10.2 can be used to determine the relative hourly rate, C_{hr}, for the selected press.

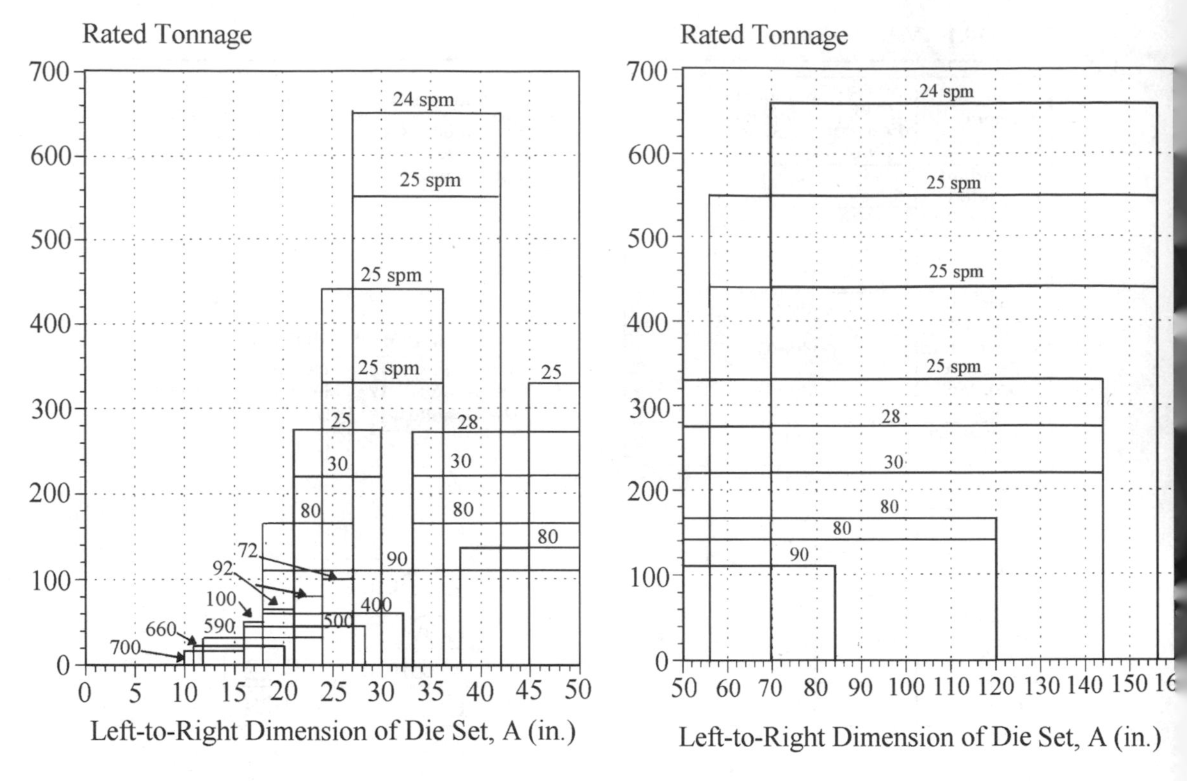

FIGURE 10.1 *Rated tonnage of presses.*

Figures 10.1 and 10.2 are used as follows. Enter the Figure at the value of A along the abscissa. Then move vertically upward at this value until the first horizontal line is reached that has a rated tonnage equal to or greater than the required force F_p. This horizontal line will correspond to a press with a speed expressed in strokes per minute (SPM) shown along the horizontal line. The rated tonnage of the press can be read along the ordinate at the level of the horizontal line. The rated tonnage, SPM value, and length A determine a press whose hourly rate is given in Figure 10.2. The vast majority of stampers using progressive dies uses coiled stock that is automatically fed to the die. Occasionally, the stock is manually fed.

By continuing to move upward in the Figure 10.1 at the value of A, other presses—larger, slower (usually), and more expensive—can also be selected. Generally speaking, however, the smallest press capable of producing the forces required will result in the lowest overall part cost.

10.5 DETERMINING THE RELATIVE CYCLE TIME

The previous section describes how a press can be selected and its hourly rate determined. To establish part cost, however, we must also know the rate at which useful parts can be produced. The press has a speed (SPM), but this is not the same as the rate at which useful parts are produced, even though it is assumed in what follows that one part is completed with each stroke of the press. (This is done with what are called "one-up" coil fed progressive dies.) The reasons are (1) not all parts produced are in fact useful; and (2) it takes time to set up the

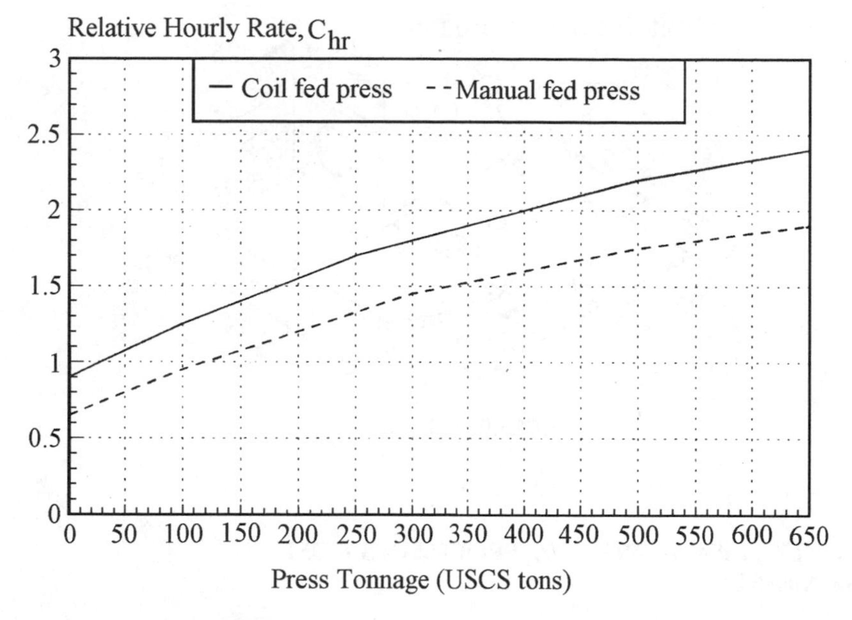

FIGURE 10.2 *Relative machine hourly rate, C_{hr}.*

progressive dies for actual production, during which time no parts are produced at all.

We define the cycle time, t_{cy}, as the average time required to produce a useful part (or the effective cycle time). The relative cycle time, t_r, is the average cycle time of a part compared with the average time to produce the standard part. Thus, relative cycle time is given by:

$$t_r = \frac{t_{cy}}{t_o}$$

where t_o is the cycle time for the reference part and t_{cy} is the effective cycle time for the part to be produced. In the case of stamping, the cycle time required to produce the reference washer is 0.234 s. (See Table 10.3.)

The effective cycle time, t_{cy}, in seconds, for a given press is given by (Dastidar, P.G., 1991)

$$t_{cy} = 3600\left[\frac{1}{(F_{eff}\,60SPM)} + \frac{t_{setup}}{N}\right] \qquad \text{(Equation 10.8)}$$

where F_{eff} is the press efficiency (i.e., the proportion of time the press is up) and accounts for the downtime of the press due both to routine maintenance of the press and dies. SPM is the number of strokes per minute achievable by the press; t_{setup} is the approximate setup time in hours (approximately 1 hour in many cases); and N is the production volume.

As noted in Section 10.2 above, the relative part cost is given by

$$C_e = t_r C_{hr}$$

With the methods described above, both t_r and C_{hr} can be computed.

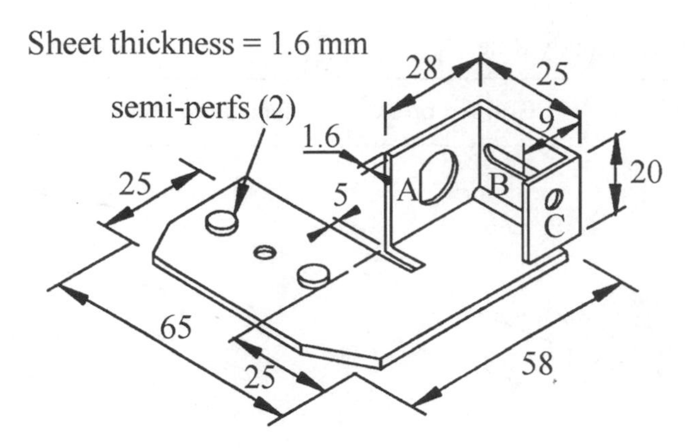

FIGURE 10.3 *Example 1.*

10.6 EXAMPLE 1—RELATIVE PROCESSING COST FOR A PART

The part shown in Figure 10.3 is made of a commercial quality, cold rolled steel. The relative tool construction cost for this part was determined earlier in Chapter 9, Section 9.14. It was assumed there that holes A, B, and C were not side-action features and that the cutout shown was not narrow.

Let us assume that the dimensions and tolerances of the part have been finalized such that holes A, B, and C remain as non-side-action features and that the cutout remains wide. Let us now determine the relative processing cost for the part.

The length of the perimeter of the unfolded part, L_{out}, is about 350 mm. From Table 13A.2 of Chapter 13, the approximate shear strength of a commercial quality carbon steel sheet is about 306 MPa (i.e., 0.85 times tensile strength). Since the sheet thickness is 1.6 mm, then from Equation 10.5, F_{out} is

$$F_{out} = 1.5L_{out}\sigma_s t = 1.5(350)(306)(1.6) = 257\,\text{kN}$$

From Equation 10.6, the stripping force, F_{st}, is

$$F_{st} = 21L_{out}t = 21(350)(1.6) = 11.8\,\text{kN}$$

Previously, in Chapter 9, we found that the largest dimension of the unfolded length is 92 mm. Thus, From Table 10.1, it is seen that since the part has 4 standard holes (4 penalty points), and 2 semi-perfs in the same direction (2 penalty points), the part is considered to have medium die detail, and $X_{dd} = 0.22$. Thus, the total force required for shearing, F_s, is

$$F_s = (F_{out} + F_{st})(1 + X_{dd}) = (257 + 11.8)(1.22) = 328\text{ kN}$$

The total length of all the bends, L_{bt}, is 65 mm. Since the tensile strength of a commercial quality carbon steel sheet (Table 13A.2 of Chapter 13) is about 360 MPa, then, from Equation 10.7,

$$F_b = \sigma_t L_{bt} t/18 = (360)(65)(1.6)/18 = 2.1\,\text{kN}$$

Thus, the total force requirement is

$$F = F_s + F_b = 328 + 2.1 = 330.1\,\text{kN}$$

Thus, the minimum press force required, F_p, is obtained from Equation 10.2, that is,

$$F_p = 1.5F = 495\,\text{kN} = 56 \text{ tons}$$

The die set length A was determined earlier (Chapter 9) to be given by

$$A = L_{dl} + 25\,\text{mm} = 970\,\text{mm} = 48.2\,\text{in}$$

Hence, from Figure 10.1, the smallest press that can satisfy both the required tonnage and the required die set length, is a 110-ton press with a speed of 90 strokes per minute.

From Figure 10.2, the relative hourly rate for this press, C_{hr}, is about 1.3.

Production Volume = 10,000

A reasonable value for the press efficiency, F_{eff}, is 0.75. Thus, from Equation 10.8, the effective cycle time for the press, at a production volume of 10,000, is

$$t_{cy} = 3600\left[\frac{1}{0.75(60)(90)} + \frac{1}{10,000}\right] = 1.25\,\text{s}$$

Therefore, since the cycle time required to produce the reference washer is 0.234 s (see Table 10.3),

$$t_r = \frac{t_{cy}}{t_o} = \frac{1.25}{0.234} = 5.34$$

The relative processing cost for the part is, thus,

$$C_e = t_r C_{hr} = 5.34(1.3) = 6.94$$

Production Volume = 100,000

For a production volume of 100,000,

$$t_{cy} = 0.924\,\text{s}$$
$$t_r = 3.95$$
$$C_e = 3.95(1.3) = 5.13$$

(As an exercise you should confirm the values shown for a production volume of 100,000.)

10.7 EXAMPLE 2—RELATIVE PROCESSING COST FOR A PART

Figure 10.4 shows the redesigned version of the part shown in Figure 10.3. Because L_{out} and L_{bt} do not significantly change, and since the amount of die detail remains at medium, the minimum required press force is still about 56 tons.

Because the number of stations required to produce this redesigned part has been reduced (see Chapter 9, Section 9.14), the value of A is about (865 + 25)mm

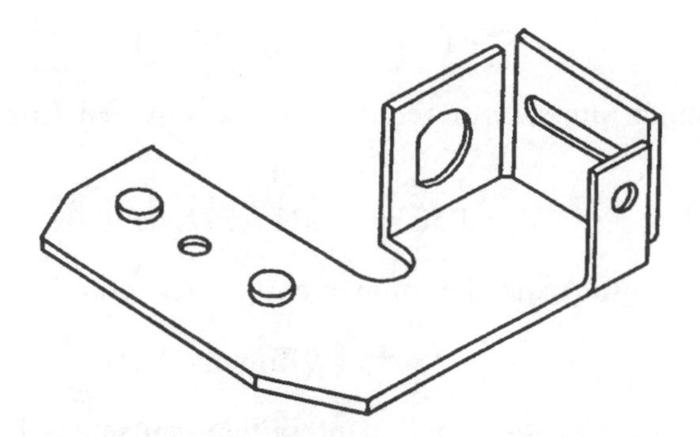

FIGURE 10.4 *Redesigned part.*

or 35 inches. In this case it is seen, from Figure 10.1, that we still require a 110-ton press with a speed of 90 SPM. Thus, in this case there would be no savings in processing cost. On the other hand, if the number of stations required to produce the part can be reduced still further such that A becomes about 32 inches, then from Figure 10.1 we see that a 60-ton press with a speed of 400 SPM can be used. The relative hourly rate for this machine is about 1.2.

Punching all of the holes at 1 station instead of 3 stations could reduce the number of active stations. This can be done by either making all of the holes identical or by ignoring the guideline that distinct features should be stamped at separate stations. For purposes of illustration we will assume that a 60-ton press can be used for this part.

Production Volume = 10,000

From Equation 10.8 the effective cycle time for this press, at a production volume of 10,000, is

$$t_{cy} = 3600\left[\frac{1}{0.75(60)(400)} + \frac{1}{10,000}\right] = 0.560\,s$$

Therefore,

$$t_r = \frac{t_{cy}}{t_o} = \frac{0.560}{0.234} = 2.39$$

The relative processing cost for the part is, thus,

$$C_e = t_r C_{hr} = 2.39(1.2) = 2.87$$

This is a 57% reduction in processing cost.

Production Volume = 100,000

For a production volume of 100,000, show that

$$t_{cy} = 0.236\,s$$

$$t_r = 1.01$$

$$C_e = 1.01(1.2) = 1.21$$

This is about an 83% reduction in processing cost.

Table 10.2 Some relative material prices, C_{mr}, for sheet metal.

Material	C_{mr}
Hot rolled, oiled steel	0.84
Low carbon CRS, killed	0.97
High carbon CRS, spring	0.98
Low carbon CRS, half hard	0.98
Low carbon CRS	1.00
Low carbon CRS, galvanized	1.39
Low carbon CRS, zinc coated	1.41
Aluminum alloy	1.79
Annealed CRS, spring	1.86
Stainless steel	5.09
Cu sheet, electrolytic tough	6.04
Phosphor bronze, spring	8.41

10.8 RELATIVE MATERIAL COST

The expression for the relative material cost for a stamped part, C_m, is identical to Equation 5.5, namely,

$$C_m = \frac{K_m}{K_{mo}} = \frac{V}{V_o} C_{mr} \qquad \text{(Equation 10.9)}$$

where V and V_o are the volume of the part and the reference part, respectively, and C_{mr} is the price of the material relative to low-carbon cold rolled steel (CRS). The value of V_o, from Table 10.3, is $3750\,mm^3$. Table 10.2 contains the relative material prices for some of the most commonly stamped materials.

10.9 TOTAL RELATIVE PART COST

As in the case of injection molding, the total production cost of a stamped part, K_t, can be expressed as the sum of the material cost of the part, K_m, the tooling cost, K_d/N, and the equipment operating cost (processing cost), K_e. Thus,

$$K_t = K_m + \frac{K_d}{N} + K_e \qquad \text{(Equation 10.10)}$$

where K_d represents the total cost of the tool and N represents the number of parts, or production volume.

If K_{mo}, K_{do}, and K_{eo} represent the material cost, tooling cost, and equipment operating cost for the reference part, then as shown in Section 5.17 of Chapter 5, the relative part cost, C_r, can be written as

$$C_r = C_m f_m + (C_d/N) f_d + C_e f_e \qquad \text{(Equation 10.11)}$$

where f_m, f_d, and f_e represent the ratio of the material cost, tooling cost, and processing cost of the reference part to the total manufacturing cost of the reference part, that is

$$f_m = K_{mo}/K_o,$$

Table 10.3 Relevant data for reference part.

Part Material	*CRS*
Material Cost (K_{po})	$5.72 \times 10\text{–}4$ cents/mm$^{3(1)}$
Part Vol (V_o)$^{(2)}$	3750 mm^3
Die Material Cost (K_{dmo})	\$1,041$^{(3)}$
Die Construction Time (Includes design and build hours)	138 hours$^{(3)}$
Labor Rate (Die Construction)	\$40/hr$^{(3)}$
Cycle Time (t_o)	0.234 s
Press Hourly Rate (C_{ho})	\$27.20$^{(3)}$

(1) Material Price Update, Xerox Corp., Jan. 1991; (2) Based on the total material required to make the part, i.e., L_{ul} L_{uw} t; (3) From collaborating companies.

$$f_d = K_{do}/K_o,$$

$$f_e = K_{eo}/K_o$$

C_d, C_m, and C_e are, respectively, the tooling cost, material cost, and equipment operating cost of a part relative to the material cost, tooling cost, and equipment operating cost of the reference part. Values for C_d can be obtained from Equation 9.4, which is repeated below for convenience, and values for C_m and C_e can be obtained from Equations 10.9 and 10.1, respectively:

$$C_d = 0.8C_{dc} + 0.2C_{dm} \qquad \text{(Equation 9.4)}$$

If we assume that a "one-up tool" is being used, that is, only one part per stroke is produced, then from the data given in Table 10.3 for the reference part, the cost of the reference part in dollars is given by

$$K_o = 0.0233 + 6561/N_o \text{ (dollars)}$$

where the first term is the sum of the material and processing costs and the second term is the tooling cost for the reference part. As in the case of injection molding and die casting, it is seen that at low production volumes most of the cost of the part is due to tooling cost.

At a production volume $N_o = 6717$ the cost of the reference part is \$1.00. Since once again the main concern is a comparison of alternative designs for a given part, the actual cost of the reference part is not vital. All that is really of interest is a comparison in relative costs between two competing designs. For this reason it becomes once again convenient to obtain the relative cost of a part with respect to the standard part when K_o is \$1.00. In this case the values of f_m, f_e, and f_d become

$$f_m = K_{mo}/K_o = 0.0215$$

$$f_e = K_{eo}/K_o = 0.00177$$

$$f_d = K_{do}/K_o = 6561$$

and

$$C_r = 0.0215C_m + (6561/N)C_d + 0.00177C_e \quad \text{(Equation 10.12)}$$

A comparison of Equations 10.12 and 5.12 show that, unlike injection molding and die casting, processing costs in stamping do not make up a significant proportion of part costs. This is primarily due to the high speed (short cycle time) of the presses used.

10.10 EXAMPLE 3—TOTAL RELATIVE PART COST FOR A PART

As originally designed, the total relative die construction cost, C_d, for the part shown in Figure 10.3 (Example 5 in Chapter 9, Section 9.14) was found to be 3.49.

For a production volume of 10,000, the relative processing cost, C_e, for the part was found (Example 1, Section 10.6) to be 6.94, while for a production volume of 100,000, C_e was found to be 5.13.

The volume of the part material needed to make this part is

$$V = L_{ul}L_{uw}t = 92(65)(1.6) = 9568\,\text{mm}^3$$

Thus, using the values of V_o and C_{mr} found in Table 10.3,

$$C_m = (V/V_o)C_{mr} = (9568)/(3750)(1.0) = 2.55$$

Production Volume = 10,000

At a production volume of 10,000, the total relative cost of the part is, from Equation 10.13,

$$C_r = 0.0215(2.55) + (6561/10{,}000)(3.49) + 0.00177(5.13) = 2.35$$

For the redesigned version of this part (Figure 10.4), $C_d = 2.84$, and, from Example 2, $C_e = 2.87$. Thus,

$$C_r = 0.0215(2.55) + (6561/10{,}000)(2.84) + 0.00177(2.87) = 1.92$$

This is an 18% savings in overall part cost. This is essentially equal to the savings in tooling cost found in Example 5 of Chapter 9 and is due to the fact that at this production volume tooling cost is the dominant factor.

Production Volume = 100,000

As an exercise shows that even at a production volume of 100,000 the savings in part cost achieved by redesigning the part are still almost entirely due to the savings in tooling cost.

10.11 WORKSHEET FOR RELATIVE PROCESSING COST—STAMPING

To facilitate the calculation of the relative processing cost and the overall relative part cost, the worksheet can be used. This copy may be reproduced for use with this book.

Worksheet for Relative Processing Costs and Total Relative Cost—Stamping

Original Design/Redesign

Tonnage required (linear dimensions in mm)

Material =	σ_s = MPa	σ_t = MPa	L_{out} =
t =	Detail = low/med/high	X_{dd} =	L_{bt} =

$F_{out} = 1.5L_{out}\sigma_s t =$ kN	$F_{st} = 21L_{out}t =$ kN
$F_s = (F_{out} + F_{st})(1 + X_{dd}) =$	$F_b = \sigma_t L_{bt} t/18 =$ kN
A = inches (from die material calculation)	$F_p = 1.5F = 1.5(F_s+F_b) =$ (Note: 1 ton = 8.89 kN)

Press Selection (See Figure 10.1)

Tons =	SPM =	C_{hr} = (See Figure 10.2)

Relative Processing Cost

N	$t_{cy} = 3600[1/F_{eff}(60SPM) + 1/N]$	$t_r = t_{cy}/0.234$	$C_e = t_r C_{hr}$

Relative Material Cost

C_{mr} = (See Table 10.2)	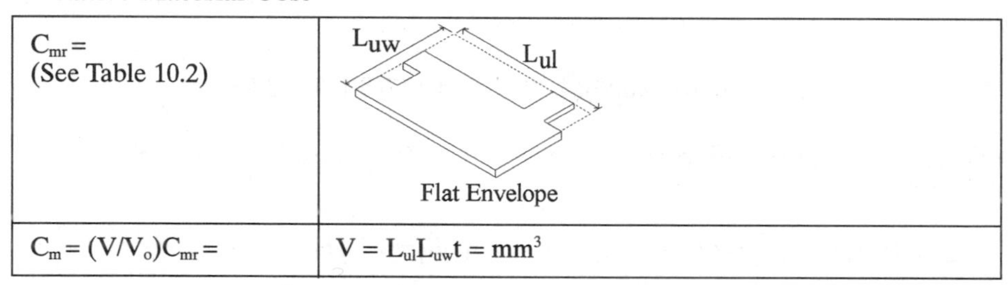
$C_m = (V/V_o)C_{mr} =$	$V = L_{ul}L_{uw}t =$ mm^3

Total Relative Part Cost ($C_r = 0.0215C_m + (6561/N)C_d + 0.00177C_e$)

N	C_r

Redesign Suggestions (or % savings if a redesign):

10.12 SUMMARY

In this chapter we concentrated on methods for evaluating the relative processing cost and the overall relative part cost for stamped parts made on progressive dies. It was shown that the processing cost of a stamped part represents a very small percentage of the overall part cost. This is due to the high-speed presses that are used and that result in very short cycle times. For this reason for production volumes less than 100,000 most of the savings achievable in the design of stampings occurs due to the reduction in tooling costs. At very high production volumes (2 million, say), material costs dominate and part design and strip layout design become vital so as to reduce material consumption.

REFERENCES

Dallas, Daniel B. "Metricating the Pressworking Equations, Part 1." *Pressworking: Stampings and Dies.* Dearborn, MI: Society of Manufacturing Engineers, 1980.

Dastidar, P. G. "A Knowledge-Based Manufacturing Advisory System for the Economical Design of Metal Stampings." Ph.D. Dissertation, Mechanical Engineering Department, University of Massachusetts at Amherst, Amherst, MA, 1991.

Poli, C., Dastidar, P. G., and Mahajan, P. V. "Design for Stamping—Analysis of Part Attributes that Impact Die Construction Costs for Metal Stampings." Transactions of the ASME, *Journal of Mechanical Design*, Vol. 115, 1993, pp. 735–743.

Poli, C., Mahajan, P. V., and Dastidar, P. G. "Design for Stamping, Part II: Quantification of the Tooling Cost for Small Parts." Proceedings of the ASME Design Automation Conference, Vol. 42, ASME, 1992.

QUESTIONS AND PROBLEMS

10.1 Imagine that you work for a local stamping company, ABC Tool and Die, situated near a large university. As part of the company's outreach program you've been invited to give a talk to a mechanical engineering design class that is involved in the design of a product that contains a substantial number of stamped parts. As part of your talk you decide to focus on the design of the parts as it affects tooling cost.

(a) Explain why you decided to focus on tooling cost as opposed to material or processing cost.

(b) Explain in detail exactly which features of a part affect tooling cost.

10.2 As part of the same talk given in Exercise 10.1 above, show a sketch of two parts that can easily be injection molded or die cast but would be difficult if not impossible to stamp. Be sure to explain why.

10.3 For a production volume of 50,000 parts, determine the relative processing costs for the part shown in Figure P10.3. Assume that the part is made of soft cold rolled steel. (Note: The relative tooling costs for this part was determined in Problem 9.5 of Chapter 9).

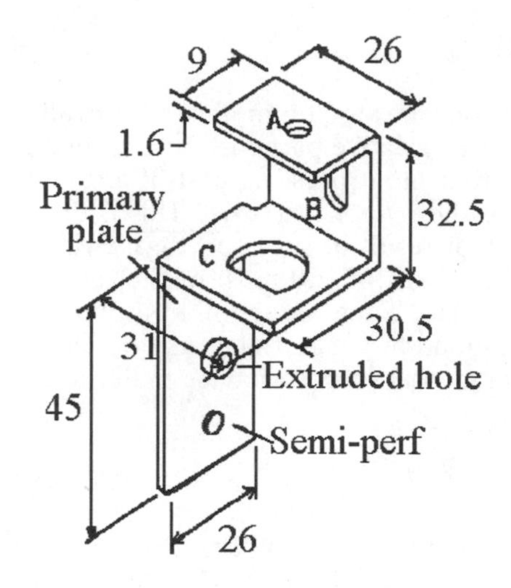

FIGURE P10.3 *(All dimensions are in mm.)*

10.4 Repeat the calculations of Problem 10.3 for a production volume of 1,000,000.

10.5 For a production volume of 50,000 parts, determine the relative processing costs for the part shown in Figure P10.5. Assume that the part is made of soft cold rolled steel. Note: The relative tooling cost for this part was determined in Problem 9.7 of Chapter 9.

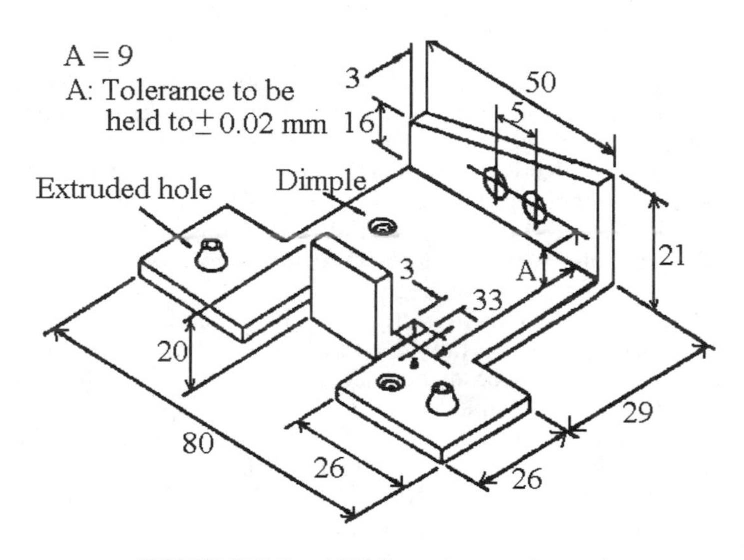

FIGURE P10.5 *(All dimensions are in mm.)*

10.6 Repeat the calculations of Problem 10.5 for a production volume of 1,000,000.

10.7 For the redesigned version of the part shown in Figure P10.3 (Problem 9.8 of Chapter 9), what are the relative processing costs for the part at production volumes of 50,000 and 1,000,000?

10.8 Determine the overall relative part cost for the parts considered in Problems 10.3 and 10.5.

10.9 Determine the savings in overall part costs for the part considered in Problem 10.3 if it is redesigned as discussed in Problem 9.8 of Chapter 9.

10.10 Determine the savings in overall part costs for the part considered in Problem 10.5 if it is redesigned as discussed in Problem 9.9 of Chapter 9.

Chapter 11

Other Metal Shaping Processes

11.1 INTRODUCTION

The vast majority of piece-parts found in consumer products is formed by those processes emphasized in the previous chapters, in particular, injection molding, die casting, and stamping. This is the reason so much time was spent discussing these particular processes. These are the processes that you, as a designer or manufacturing engineer, will most likely use or encounter while engaged in the practice of mechanical or industrial engineering. There are, however, other processes that are important, and these processes will be briefly discussed in this chapter. More detailed information concerning the processes can be found in the list of references for this chapter.

11.2 METALS AND PROCESSES

There are three broad categories of metal shaping processes, namely, *casting, forming,* and *machining*. Casting is used to produce thinned-walled parts with intricate shapes or hollowed areas. Casting is usually selected as the manufacturing process of choice based on a part's complex geometry and lower final part cost. A decision to use casting is not typically based on the necessity of providing a part with particular mechanical properties.

With forming, a part shape is obtained by deforming the workpiece while the volume remains constant. There are two broad categories of forming processes, namely, bulk deformation processes and localized deformation processes. In bulk deformation we change the thickness of the cross-section in order to obtain the part shape. Rolling (Figure 11.1), drawing (Figure 11.2), extruding (Figures 11.3 and 11.4), and forging (Figure 11.5) are examples of bulk deformation processes. Stamping, which was discussed in Chapter 9, "Stamping: Relative Tooling Cost," and Chapter 10, "Stamping: Total Relative Part Cost," is an example of a localized deformation process. With localized deformation the thickness remains essentially constant.

Machining, which is used primarily for low-production volume parts, for the production of prototypes, and for the production of the tooling used in processes such as stamping, injection molding, and other processes, generates a part shape by changing the volume of the workpiece through the removal of metal.

From your previous study of materials you are probably aware that the properties of metal alloys are accompanied by such terms as cast and wrought. Cast metals are those metals that have been formed by casting. Wrought metals are those which have been formed by rolling, drawing, extruding, or forging.

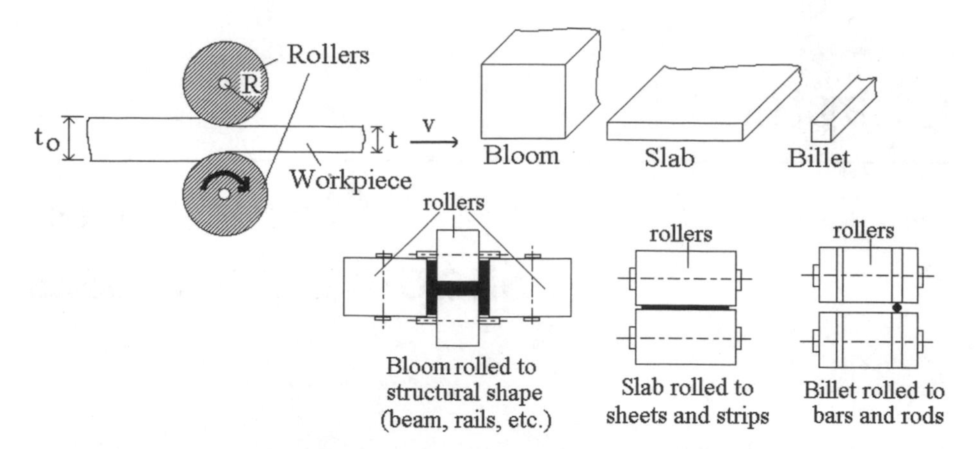

FIGURE 11.1 *Reducing the thickness of a cast ingot via rolling to produce blooms, slabs, and billets. Also shown are blooms, slabs, and billets converted to structural shapes, sheet metal, and bars.*

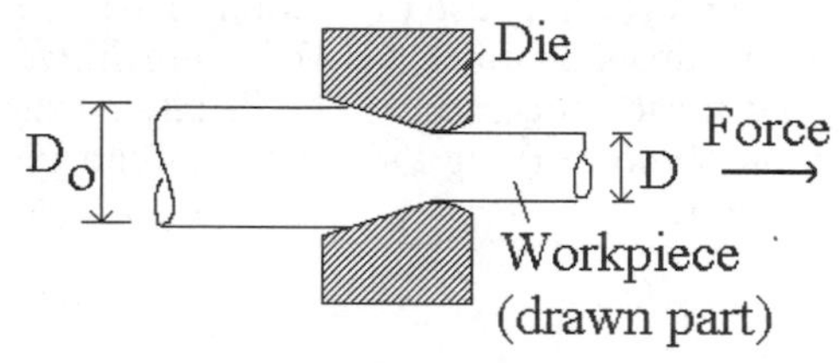

FIGURE 11.2 *Reducing the diameter of a rod using drawing.*

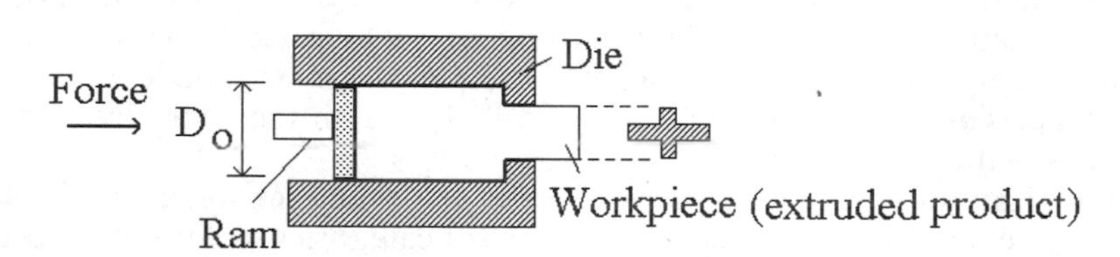

FIGURE 11.3 *An example of extrusion using flat dies.*

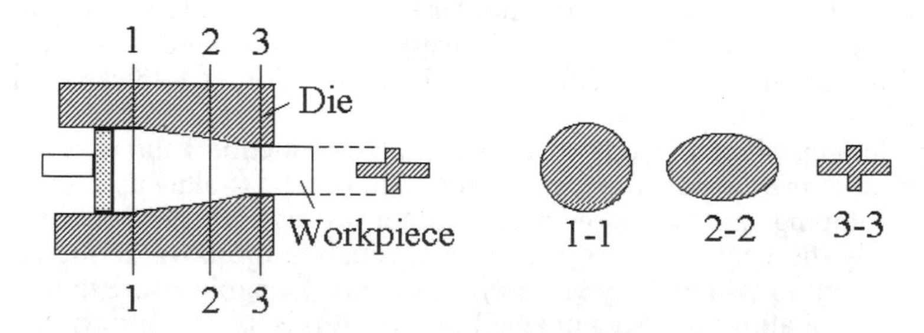

FIGURE 11.4 *An example of extrusion using gradually shaped dies.*

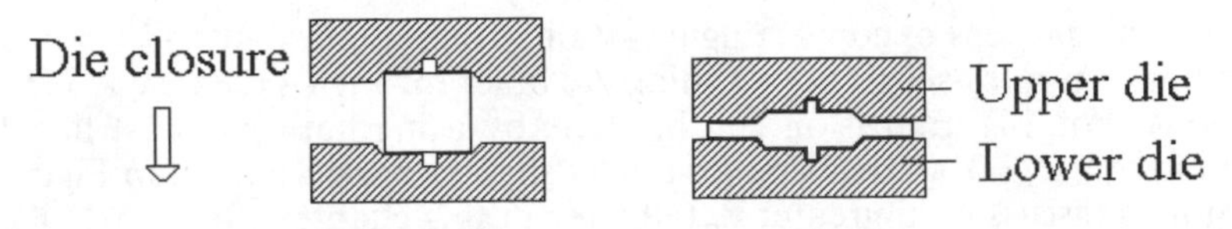

FIGURE 11.5 *An example of forging.*

In Chapter 6, "Metal Casting Processes," we discussed various casting processes. In Chapter 8, "Sheet-Metal Forming," we discussed several sheet-metal forming processes in which wrought metals, formed by rolling, are used to produce thin-walled parts. In Sections 11.3 and 11.4 we will discuss forging and extruding. In Section 11.5 we will discuss machining.

11.2.1 Rolling

Rolling (Figure 11.1) is a process of producing large reductions of thickness and of forming metal strips into profiles by passing the metal workpiece through rollers. The process can be done either hot or cold. The large reductions are usually done hot, and the final reductions are generally done cold in order to produce a better surface finish and closer tolerances (see *ASM Metals Handbook*, Vol. 14).

Rolling is usually the first process used after casting an ingot. Cast ingots are rolled to form slabs (thick flat plates, say 40 mm thick), billets (long thick rods with square, rectangular, or circular cross-sections), and blooms. Slabs are then rolled into sheets, plates, and welded pipes, and billets are rolled and drawn into bars, rods, pipes, and wires. Blooms are roll formed into structural shapes such as I-beams and rails.

Cold roll forming is currently used to produce a variety of shapes and products and is presently employed to produce parts that were previously formed by extrusion (see Section 11.2.3 and Problem 11.2). Unlike extrusion, however, roll-formed parts must be designed to have a constant wall thickness. Metal gutters, tubes, and moldings are just three examples of parts that are currently roll-formed.

11.2.2 Drawing

Drawing (Figure 11.2) is a process of reducing the diameter of a wire, bar, or tube by pulling it through a die of similar cross-section (see *ASM Metals Handbook*, Vol. 14). Drawing is usually done at room temperature; as a result strain hardening occurs. Since several passes are usually needed to reduce the thickness to the desired diameter, annealing is often required after two or three passes. Rods up to 6 inches (150 mm) and wires down to 0.001 inch (0.025 mm) are drawn. Rods with diameters greater than 5 mm are done on a draw bench. Rods and wires with diameters less than 5 mm are drawn in coiled form.

In order to initiate the process, the workpiece must be pointed so that it can be slipped into the die. Pointing can be done by machining, dipping in acid, or swaging. Occasionally the workpiece is simply heated then pushed through the die.

11.2.3 Extrusion and Forging

Extrusion is a process of converting a cast or wrought billet into a long prismatic part of uniform cross-section by pushing the billet through a die (see *ASM Metals Handbook*, Vol. 14). Extrusion can be done by using flat or almost flat dies as shown in Figure 11.3, or by using gradually shaped dies as shown in Figure 11.4. Extrusion is discussed in greater detail later in this chapter (Section 11.4).

Forging is a process of plastically deforming metal, usually hot, into desired shapes by compressing the workpiece between two dies (Figure 11.5). Like injection molding, die casting, and stamping, forging presents the designer with the possibility of designing a multiplicity of semicomplex (geometrically) special purpose parts. The next section is devoted to a more detailed discussion of forging.

11.3 FORGING

Forging is a bulk deformation process in which the geometry of a part is shaped by squeezing (with a mechanical or hydraulic press) or hammering (with a gravity or power-assisted hammer) a hot workpiece between two die halves attached to a press or hammer. Because the workpiece is plastically deformed, its cast structure is refined and the grains or fibers align in the direction of flow. The resulting directional alignment of the fibers makes a forging stronger and more ductile than castings, and enables a forging to exhibit greater resistance to shock and fatigue. Thus forging is used to produce some of the most critically stressed parts found in aircraft, automobiles, and tools.

11.3.1 Types of Forging

There are two broad categories of forging: *open die* and *closed die*. Closed-die forgings are also referred to as *impression forgings* since the forging dies partially enclose the workpiece material and restrict the flow of metal.

Open-die forging is a process where a hot metal workpiece is squeezed or hammered between flat, circular, or v-shaped dies. The workpiece is not enclosed so the metal flow is not completely restricted. Figure 11.6 shows an example of dies used for open-die forging.

Open-die forgings are used when the part is too large to be produced in closed dies, or when the quantity involved is too small to justify the use of closed dies, or when a short lead-time exists.

Although complex shapes can be produced by the use of open dies, their production requires skilled operators and is time-consuming. For these reasons, most open-die forgings are restricted to the production of bars and shafts (round, square, rectangular, or hexagonal cross-sections) and flat pancake-type parts.

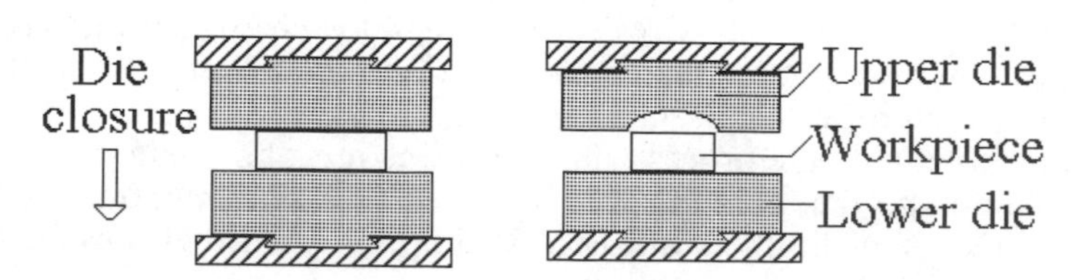

FIGURE 11.6 *Some open die sets for forging.*

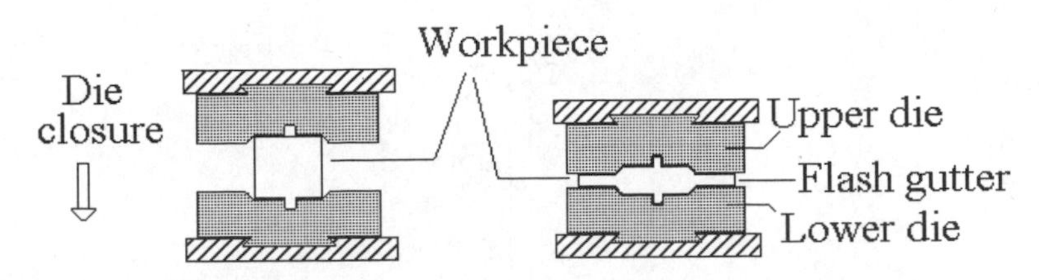

FIGURE 11.7 *An example of closed-die forging using a parallel flash gutter. Other types of gutter designs exist that permit the dies to touch upon closure. The type of gutter used is a function of the forging equipment used and the workpiece material (see* ASM Metals Handbook, Vol. 14*).*

Although similarly shaped shafts and bars may be available from off-the-shelf rolled stock, forged shafts are used when superior mechanical properties are required.

In closed-die forging the workpiece is squeezed or hammered between one or a series of dies that enclose the workpiece on all sides. Often a series of dies is used because solid metal is difficult to form in a single stage.

Closed-die forgings are subdivided into blocker-type forgings, conventional forgings, and precision (low-draft) forgings.

Blocker-type forgings only roughly approximate the general shape of the final part, with relatively generous contours, large radii, large draft angles, and liberal finish allowances. A blocker-type forging requires considerable machining to satisfy the final dimensions, tolerances, and surface finish of the part.

Conventional forgings, which represent the vast majority of forgings produced, more accurately achieve the final dimensions with closer tolerances and smaller radii. Because dies must produce a more accurate part, more dies are required. In fact, most conventional forgings are preformed using one or two blocker dies prior to the use of the final or conventional die.

Conventional forgings generally have portions of the part machined and other portions that remain "as forged." Both blocker-type forgings and conventional forgings are produced in dies that allow for excess material (flash) to escape from the die cavity (Figure 11.7).

Precision forgings have little if any draft and generally require little if any machining. Precision dies are more costly than conventional dies, and the forgings produced with them require higher forging pressures than conventional forgings.

Because the vast majority of special-purpose forgings are produced as conventional closed die forgings, the discussion of forging machines that follows is restricted to the discussion of conventional forgings.

11.3.2 Forging Machines

The two main types of forging "machines" used in commercial forging plants in the United States are *hammers* and *presses*.

Hammers

Hammers are either gravity type or power assisted. With a gravity drop hammer, the upper die is attached to a ram and is raised by either a board, belt, or air (Figure 11.8). It is then allowed to fall freely to strike the workpiece. In power-assisted drop hammers, air or steam is used against a piston to supplement the

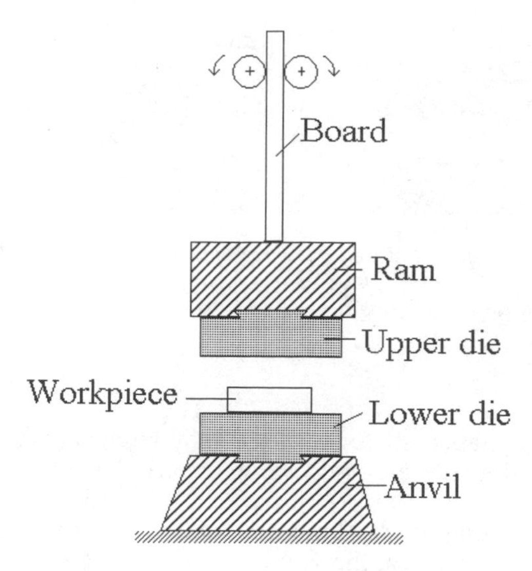

FIGURE 11.8 *A gravity drop hammer.*

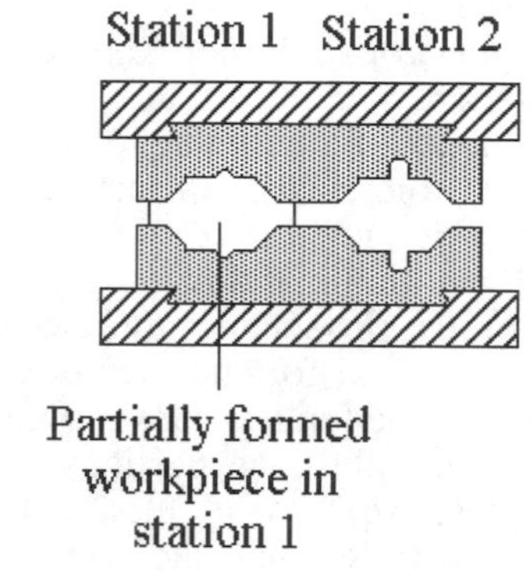

FIGURE 11.9 *A closed die with two die stations. Although forging dies are often depicted as separate blocker and conventional dies, this is not usually the case. Generally, a forging die is divided into two or more sections or stations, and the workpiece is moved from one station to the other in order to forge the part.*

force of gravity during the downward stroke. The energy used to deform the workpiece is obtained from the kinetic energy of the moving ram and die. Because hammers are energy-restricted machines, multiple blows (usually three) are required at each stage (i.e., for each die) during the forging process.

In addition to regular drop hammers, there are also counterblow hammers. In a counterblow hammer, two rams are activated simultaneously and driven toward each other. The two die halves strike the workpiece at some midway point between the two rams. Vertical counterblow hammers are used in order to avoid the need for the heavy anvil and foundation weights required with gravity drop hammers.

Horizontal counterblow hammers (also called impacters) also exist. In this case, two rams of equal weight are driven toward each other by compressed air in a horizontal plane. Horizontal counterblow hammers are used for the automatic forging of workpieces by utilizing a transfer unit to move the partially completed forging from die station to die station (Figure 11.9).

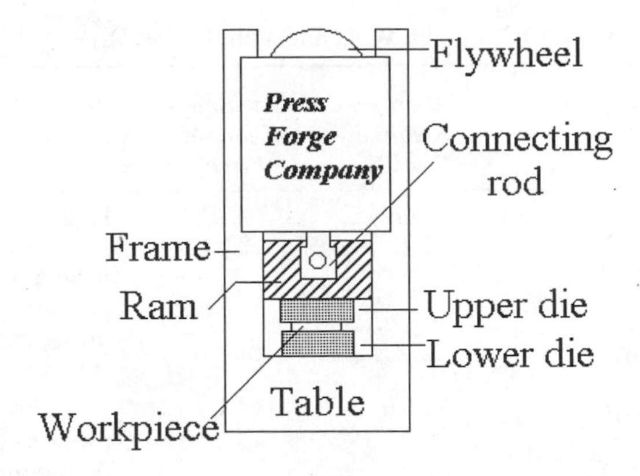

FIGURE 11.10 *A mechanically driven forging press.*

Presses

Forging presses (Figure 11.10) can be either mechanically driven or hydraulically driven. Although mechanical presses contain the same general components as stamping presses, forging presses tend to be stiffer and more robust. Unlike the blow delivered by a hammer, a mechanical press squeezes the metal between the two die halves; one squeeze is used for each stage (die) of the process. While hammers are energy restricted, mechanical presses are stroke limited. The largest force is delivered at bottom dead center.

The general components of a hydraulic forging press are similar to those of hydraulic presses used for stamping. Hydraulic presses are load-limited since the maximum load can be delivered at any position within the press stroke range. Multiple ram hydraulic presses are available, and can be used to forge cavities that are equivalent to external undercuts found in die castings.

Presses tend to be more expensive than hammers. For this reason, hammers are used whenever possible. However, presses can produce all of the types of forgings produced by hammers and, in addition, can forge some low ductile alloys that might fragment under hammer blows. In addition, for rate-sensitive materials, hydraulic presses are preferred since the load can be slowly applied.

11.3.3 Factors Influencing Design for Forging

Materials

Forging difficulty (and hence cost) is a function of both part material and part shape. These two factors are strongly interrelated; thus, a shape that is relatively easy to forge in one material (aluminum, for example) may be difficult or impossible to forge in another (say a nickel based superalloy).

Table 11.1, taken from Knight, W.A. and Poli, C., 1982, contains a list of materials and alloys that can be forged. The materials have been divided into six groups and have been ranked in general order of forging difficulty. Details concerning the actual material alloys allocated to each group, together with recommendations for forging conditions and other issues, can be found in Knight, W.A. and Poli, C., 1982. Materials that are not generally forged (e.g., cast iron) have not been included.

Table 11.1 Materials for forged parts (Knight, W.A. and Poli, C., 1982).

Group	Material	*Average Relative Density*[1]	*Average Relative Price*[1]	*Comments*
0	Aluminum Magnesium	0.346 0.228	4.70 22.84	High ductility; readily forged into precise, intricate shapes; low forging pressures; can be forged on both presses and hammers, but presses preferred when deformation is severe.
1	Copper and Copper Alloys (Brass, bronze, etc.)	1.076	13.10	Readily forged into intricate shapes; requires generally less pressure than equivalent shapes in low-carbon steels; presses preferred. With good process design can be forged with small draft angles down to zero degrees.
2	Carbon and Alloy Steels	1.0	1.04	Most widely forged materials; readily forged into wide variety of shapes using conventional methods and standard equipment; hammers and presses both used. Oxidizes considerably during heating and scale may present a production problem, as does decarburization unless steps are taken to control it. Forged with a variety of lubricants, to reduce die wear and oxidation, including glass coatings, and colloidal graphite.
3	Martensitic and Ferritic Stainless Steel Maraging Steels Tool Steels	0.992 1.018 1.014	5.38 34.24 12.96	Forged by conventional methods, but require higher pressures than carbon or low alloy steels; hammers and presses both used.
4	Austenitic Stainless Austenitic Nickel Alloys of Iron	1.012 1.019	7.37 16.16	More difficult to forge than carbon and alloy steels; requires greater pressure; hammers preferred to presses. Two to three times as much energy is required for forging as for carbon and alloy steels for otherwise similar parts.
5	Titanium and Titanium Alloys	0.577	89.53	Forging pressure increases rapidly with decreasing temperature; close tolerance forgings can be produced but only at great expense.

[1]Relative to low-carbon steel

Shapes

There are basically three types of forged parts produced: compact (or chunky) parts in which the length, width, and height of the part are approximately equal; flat (disk-like) parts; and long parts where the part length is significantly greater than the part width.

Compact shapes can generally be made from billets or blanks, and usually have a simple forging sequence. In general, they can be formed using one blocker die and one conventional die. Thus, if formed on a press, two operations are required. If formed using a hammer, generally the workpiece will be struck three times with each die station. A significant proportion of parts in this category contain external undercuts that require multiple-action forging machines.

Flat parts are also generally produced from billets or blanks, and upsetting (counterflow) type material flow predominates. Flat parts are usually produced using two or three die stations. For simple flat parts with uniformly thick walls, a blocker die is used to distribute the material properly, and then an impression die is used to finish the forging. For flat parts with thin sections, two separate blocker die stations are often required to properly distribute the workpiece material.

Long parts are produced directly from bar stock and in general require elongation and drawing stages prior to the impression forging sequence. For example, in the case of long slender parts with two or more heavy sections separated by light sections, some preliminary preforming operations are used to thin down the metal in the center section and provide more mass at the ends for later operations. In general, three or four die stations are required to produce the final forged part.

From this discussion it should be apparent that the basic shape of a forged part affects the number of dies required, and hence the tooling costs.

As in die casting and injection molding, die costs increase if the part requires a non-planar parting surface, and if multiple action dies are required to produce external undercuts. Unlike injection molding and die casting, however, die costs are also a function of the part material as noted above. Material that is difficult to move requires more preforming stages and decreases die life.

Although part shape, part complexity, and part material affect the number of dies or die stages required (hence die cost), as the number of dies or die stages increase the sequence of operations necessary for processing the part also increases; hence, processing costs increase too. In addition, multiple-action machines are more costly, so parts with external undercuts are also more costly to process.

11.3.4 Design for Manufacturing Guidelines for Forged Parts

The nature of the forging process in which solid metal is squeezed and moved within a die set to form a part leads to the following broad DFM guidelines:

1. Because all the pre-forming operations required to forge a part result in long cycle times, and because the robustness required of the dies, hammers, and presses result in high die and equipment cost, when compared to stamping and die casting, forging is an expensive operation. Thus, if possible, forging should be avoided.

   Of course, there are times when functionality dictates a forged part, or when other processes are even more costly. In these cases:
2. Select materials that are relatively easy to deform. These materials will require fewer dies, shorten the processing cycle, and require a smaller hammer or press.
3. Because of the need for the metal to deform, part shapes that provide relatively smooth and easy external flow paths are desirable. Thus, corners with generous radii are desirable. In addition, tall thin projections should be avoided since such projections require large forces (hence large presses and/or hammers), more pre-forming stages (hence more dies), cause rapid die wear, and result in increased processing cycle time.

4. For ease of producibility, ribs should be widely spaced (spacing between longitudinal ribs should be greater than the rib height; spacing between radial ribs should be greater than 30 degrees). Closely spaced ribs can result in greater die wear and an increase in the number of dies required to produce the part.
5. Internal undercuts, and external undercuts caused by projections, must be avoided since they are impossible to form by the movement of solid metal. External undercuts that are the result of holes should be avoided since they increase both die costs and processing costs.

A more quantitative guide to design for forging, complete with classification systems and relative cost models, can be found in Knight, W.A. and Poli, C., 1982; Gokler, M.I., et al., 1981.

11.4 ALUMINUM EXTRUSION

11.4.1 The Process

Like forging, aluminum extrusion is a solid metal flow process, but the process is very different. In aluminum extrusion, a round aluminum billet is heated to 700–900°F; the billet remains solid. It is then inserted into the cylindrical cavity of a large "container" and, by means of a hydraulically powered ram, the metal is forced out through a die hole or holes of the desired shape (Figures 11.3 and 11.4).

After extrusion, the metal cools slowly, is given a slight mechanical stretching, and then cut to desired lengths. Almost always, extrusions are given additional heat treating to add strength and hardness, and sometimes a surface treatment called anodizing is done to improve appearance and provide better weatherability.

Aluminum extrusions can be formed into an amazingly wide variety of shapes, and anodized into many attractive colors. Examples of extrusions are found in residential window casements, commercial storefront structures, picture frames, chalk trays, and many standard structural shapes.

There are a number of aluminum alloys (most containing small amounts of magnesium, silicon, and other elements) that can be readily extruded and then heat treated (see *ASM Metals Handbook*, Vol. 14).

Extrusion dies for aluminum are relatively inexpensive. If there is to be a reasonable quantity of production, die costs are usually essentially negligible on a per pound basis.

11.4.2 Qualitative Reasoning on Design for Manufacturing for Aluminum Extrusion

Most of the DFM considerations for aluminum extrusions evolve from the difficulty of forcing the metal to flow uniformly from a large round billet out through small complex die openings. This issue leads to the following "rules" that can help guide the generation of more easily extrudable aluminum parts:

1. Sections with both thick and thin sections are to be avoided. Metal tends to flow faster where thicker sections occur, thereby giving rise to distortions in the extruded shape. Use designs that will function with all the walls as

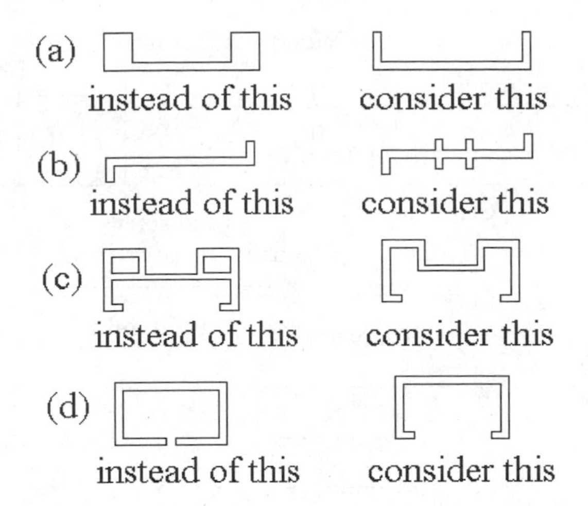

FIGURE 11.11 *Some examples of design for extrusion rules.*

uniform in thickness as possible. Avoid designs that require "slugs" of material (Figure 11.11a).
2. Long, thin wall sections should be avoided, since such shapes are difficult to keep straight and flat. If such sections are absolutely necessary then the addition of ribs to the walls will help distribute the flow evenly (Figure 11.11b).
3. Hollow sections are quite feasible, though they cost about 10% more per pound produced. The added cost is often compensated for in the added torsional stiffness that the hollow shape provides. It is best if hollow sections can have a longitudinal plane of symmetry (Figure 11.11c).
4. "Semi-hollow" features should be avoided. A semi-hollow feature is one that requires the die to contain a very thin—and hence relatively weak—neck. Figure 11.11d shows the applications of this rule.

For additional information on design guidelines for aluminum extrusion, see *The Aluminum Extrusion Manual.*

11.5 MACHINING

11.5.1 The Process and the Tools

Machining is a process that produces parts of desired size and shape by removing material in the form of small chips from a solid workpiece using a single or multiple-edged cutting tool. Since the process removes material already paid for, machining is not an economical process and is not generally used to produce special purpose parts for consumer products. However, it is often used to improve the tolerances or surface finish of parts made via other processes (e.g., sand casting, forging, etc.) by accurately removing small amounts of material from selected portions of the surface. In addition, machining is used to produce the tools, punches, and dies used in most processes such as injection molding, die casting, stamping, and other processes.

There are several kinds of machine tools, though not all of them will be discussed here. Among the most common are the following.

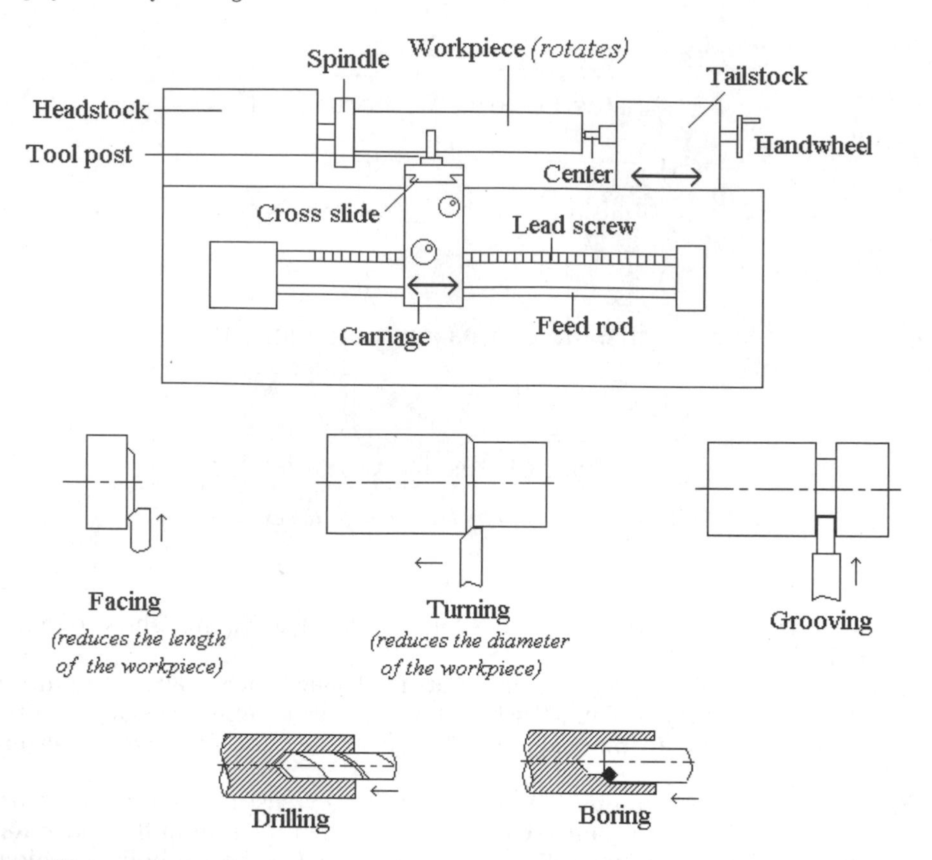

FIGURE 11.12 *Principal components and movements of a lathe and the basic operations that can be performed on it.*

Lathes

Lathes are used primarily for the production of cylindrical or conical exterior and interior surfaces, via turning, facing, boring, and drilling (see Figure 11.12). Lathes are also used for the production of screw threads. In a lathe, the workpiece is rotated while the cutting tool is moved ("fed") into the workpiece in a direction parallel and/or perpendicular to the axis of rotation of the workpiece.

Vertical and Horizontal Boring Machines

These machines are used in place of lathes for the machining of large workpieces. They can be used to perform turning, facing, and boring. Boring machines are also used to form grooves and for increasing the diameters of existing holes.

Vertical and Horizontal Milling Machines

These machines (Figure 11.13) are used to form slots, pockets, recesses, holes, and other features. In this case the cutting tool is rotated and the workpiece is fed.

Planing and Shaping Machines

These machines are used primarily for reducing the thickness of blocks and plates and for "squaring up" blocks and plates. Shapers are also used to machine notches and keyways and to form flat surfaced on parts formed by other processes such as casting and forging. In the shaping machine shown in Figure 11.14 the

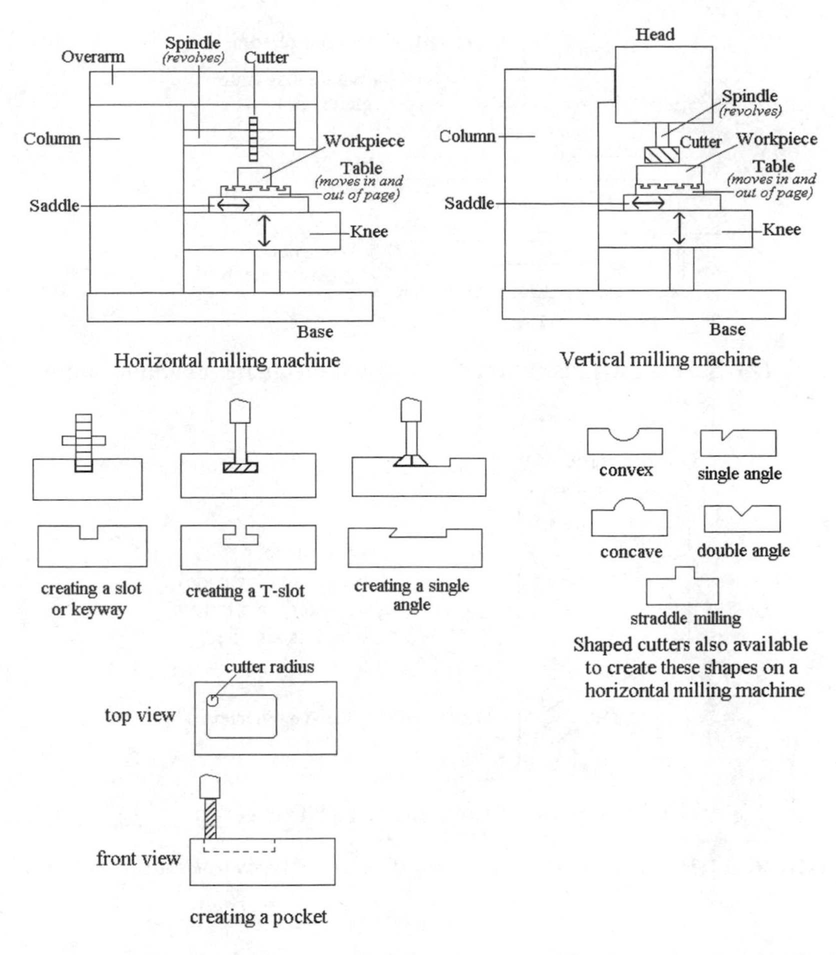

FIGURE 11.13 *Principal components and movements of horizontal and vertical milling machines and the basic operations that can be performed on them.*

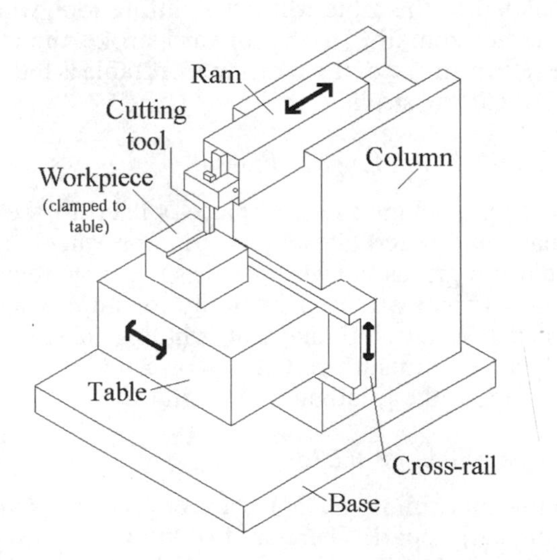

FIGURE 11.14 *Horizontal shaper, used primarily to reduce the thickness of blocks and plates.*

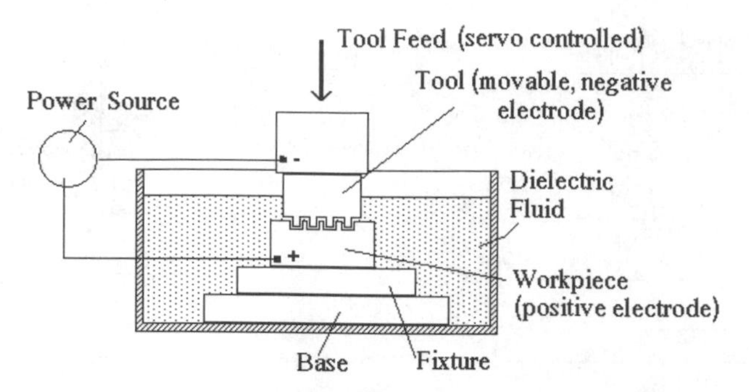

(a) Schematic illustration of electrical discharge machine setup

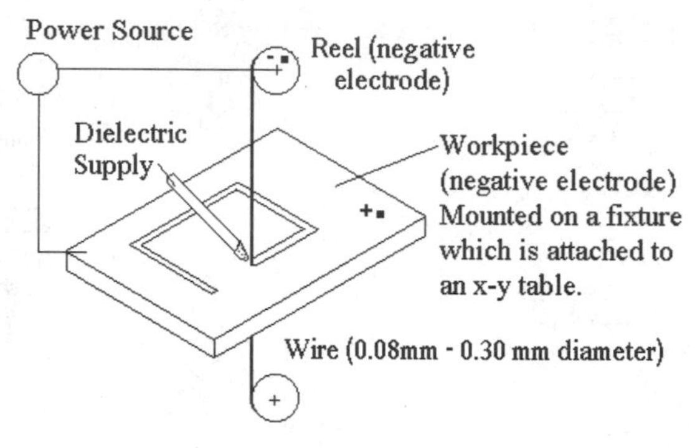

(b) Schematic of traveling wire EDM setup

FIGURE 11.15 *Principal components of electrical discharge machines (EDM).*

workpiece is clamped to the table while the cutting tool, which is attached to a heavy ram, moves horizontally. On the forward stroke, the cutting tool removes metal. After the return stroke of the ram, the worktable is fed to the side in preparation for the next cutting stroke.

Surface and Cylindrical Grinding Machines

In a surface grinding machine the workpiece is fixed to the table that reciprocates longitudinally and is fed laterally. A grinding wheel is fixed to a rotating horizontal spindle and grinds the workpiece as the table reciprocates and is fed. Surface grinding machines are used primarily for improving the tolerance and surface finish of flat surfaces. Cylindrical grinding machines are also used to improve the tolerances and surface finishes of cylindrical surfaces. In this case both the workpiece and the grinding wheel rotate.

Electrical Discharge Machines (EDM)

Electrical discharge machining (EDM) is a process of removing metal by means of an electrical discharge spark. There are two kinds of EDM machines, a "solid" EDM machine (Figure 11.15) and a "wire" EDM machine. In the solid EDM

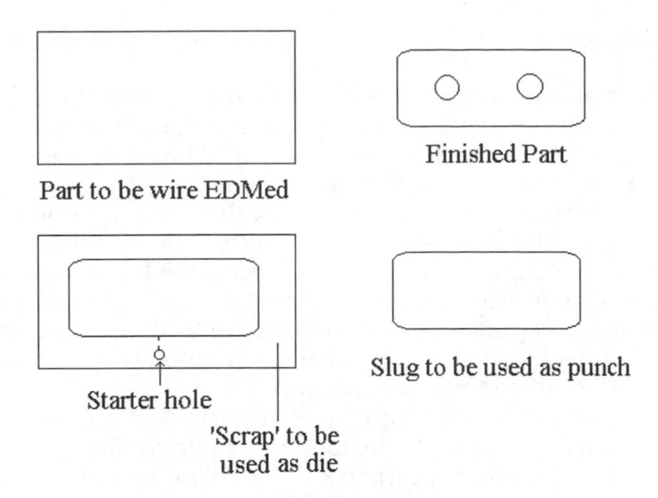

FIGURE 11.16 *Using wire EDM to produce a blanking punch and die.*

machine, a tool (usually graphite, but sometimes copper or brass) that contains the shape to be machined into the workpiece together with the workpiece itself is connected to a dc power source. The workpiece is placed in a tank filled with a dielectric fluid. The tool is fed into the workpiece. When the electrical potential between the two is sufficiently high, a spark is created that removes a small amount of material from the workpiece.

The solid EDM process is used to machine narrow slots, small holes, and complicated shapes. It can be used on any material that conducts electricity, and it is not significantly affected by the material hardness or strength. The EDM process is much slower than the traditional machining process described in Section 11.5 and leaves a "pitted" surface finish that may require further grinding or hand finishing. For these reasons, EDM is avoided when possible.

The wire EDM process is a variation of this process and is often used to produce stamping dies and punches. In this case a computer program is developed to produce the desired shape. Then the workpiece, from which the punch and die are to be fabricated, is prepared with all the necessary holes, including a starter hole for the wire. The thin wire electrode, usually made of brass, is inserted into the starter hole and the punch and die are made.

For example, the setup shown in Figure 11.15b could be used to produce the blanking punch and die for the link shown in Figures 9.41 and 9.42 (see Figure 11.16). For a more detailed discussion concerning the use of EDM to produce stamping dies see Sommer, Carl, 2000.

For a more detailed description of each machining process, as well as other machining processes, see reference *ASM Metals Handbook*, Vol. 16.

11.5.2 Qualitative Guidelines on Design for Machining

In machining, the fact that metal already paid for is removed by use of a sharp cutting tool leads to the following design guidelines:

1. If possible, avoid machining. If the desired geometry can be produced by another process such as casting, molding, stamping, and so on, the cost

will almost certainly be lower (unless, of course, you're making only a few parts).

2. However, if machining cannot be avoided, then specify the most liberal tolerances and surface finishes. Most machining operations are performed in two operations, a roughing operation and a finishing operation. The roughing operation is used to remove large quantities of metal without special regard to tolerances or surface finishes. The finish cut it performed to provide the necessary tolerances and surface finishes. Since surface roughness is directly related to the rate of feed used—large feeds produce high values of surface roughness (rough surfaces), and low feeds produce low values of surface roughness (smooth surfaces)—and low feeds result in longer machining times, surface finishes better than those absolutely necessary for functional purposes should not be specified.
3. For turning operations on a lathe, avoid designs that require sharp internal corners. Corner radii equal to the tool-nose radius of the cutting tool should be specified. Sharp internal corners call for either the use of sharp tools that break more easily or additional operations.
4. For planing and milling operations, avoid sharp internal corners, radiused external corners, and slot widths and shapes other than those available using standard off-the-shelf cutters. Internal corners should be specified so that they are equal to the cutter radius of milling cutters and the tool radius of planing cutters. Since sharp external corners are a natural result of these processes, such corners should be specified.

Although these general machining guidelines apply across the board, a more detailed list of features, which can be readily provided on machined components, can be found in Boothroyd, G., 1975.

11.6 SUMMARY

The purpose of this chapter was to introduce you, as a designer or manufacturing engineer, to some of the other important metal shaping processes. The goal was to give designers a qualitative understanding of these processes to help support their design decisions and their communications with manufacturing engineers. With each process description a set of qualitative DFM guidelines was presented, which are particularly useful in the early stages of design.

REFERENCES

The Aluminum Association. *The Aluminum Extrusion Manual*, 3rd edition, Washington, DC: The Aluminum Association, 1998.

American Society of Metals International. *American Society of Metals Handbook*, Vol. 14. "Forming and Forging." Metals Park, OH: ASM International, 1988.

American Society of Metals International. *American Society of Metals Handbook*, Vol. 16. "Machining." Metals Park, OH: ASM International, 1989.

Boothroyd, G. *Fundamentals of Metal Machining and Machine Tools.* New York: McGraw-Hill, 1975.

Bralla, J. G., editor. *Handbook of Product Design for Manufacturing.* New York: McGraw-Hill, 1986.

Gokler, M. I., Knight, W. A., and Poli, C. "Classification for Systematic Component and Process Design for Forging Operations." Proceedings of the Ninth North American Manufacturing Research Conference, Pennsylvania State University, State Park, PA, May 1981, pp. 158–165.
Kalpakjian, S. Manu*facturing Engineering and Technology*. Reading, MA: Addison-Wesley Publishing, 1989.
Knight, W. A., and Poli, C. "A Systematic Approach to Forging Design." *Machine Design*, January 24, 1985.
Knight, W. A., and Poli, C. "Design for Economical Use of Forging: Indication of General Relative Forging Costs." *Annals of the CIRP*, Vol. 31, no. 1, 1982, pp. 159–163.
Knight, W. A., and Poli, C. "Design for Forging Handbook." Mechanical Engineering Department, University of Massachusetts at Amherst, Amherst, MA, 1984.
Sommer, Carl. *The Wire EDM Handbook*. Houston: Advanced Publishing, 2000.
Wick, C., editor. *Tool and Manufacturing Engineers Handbook*, Vol. 2. "Forming, 9th edition." Dearborn, MI: Society of Manufacturing Engineering, 1984.

QUESTIONS AND PROBLEMS

11.1 Explain what is meant by hot rolling, and list some products that can be made on a rolling mill.

11.2 Figure P11.2 shows a simplified cold roll forming operation for producing an open gutter. Also shown is a potential sequence of rolling operations for producing a semi-closed gutter. With this information as a background, list some products that could be shaped by cold roll forming.

What advantages, if any, does roll forming have over extrusion?

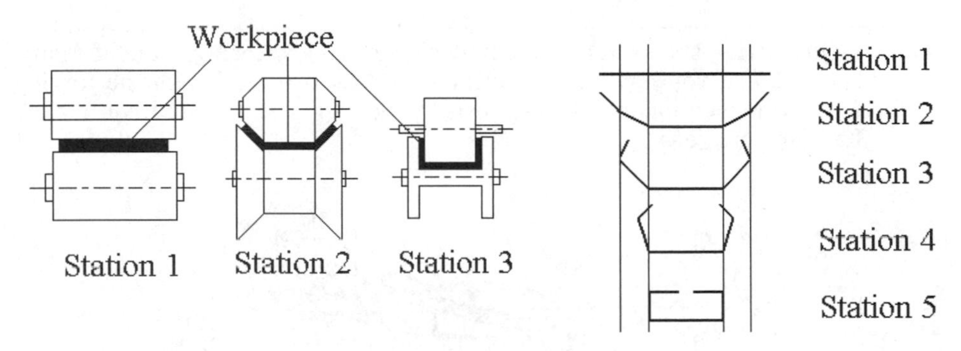

FIGURE P11.2

11.3 Figure P11.3 shows the current design of a stainless steel forging. The spacing between the radial ribs is 22°. Assume that you have been assigned the task of redesigning the forging so as to reduce manufacturing costs. What suggestions would you make? Explain your reasoning in detail.

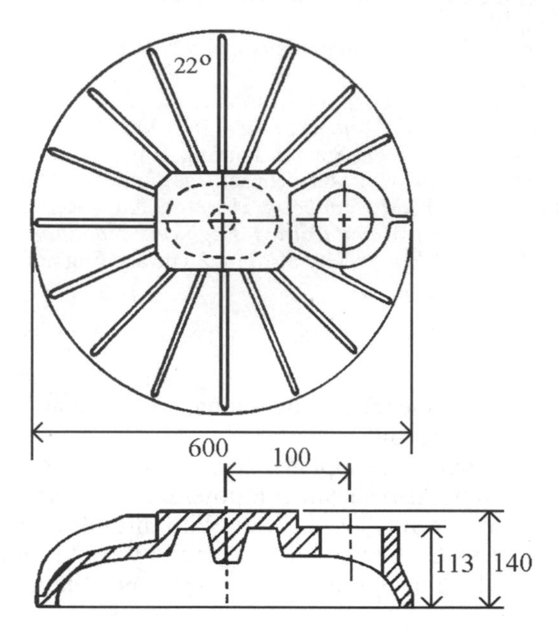

FIGURE P11.3 *(All dimensions in mm, from Knight, W.A. and Poli, C., 1984.)*

11.4 Figure P11.4 shows the current design of a low-carbon steel forging. Assume that you have been assigned the task of redesigning the forging so as to reduce manufacturing costs. What suggestions would you make? Explain your reasoning in detail.

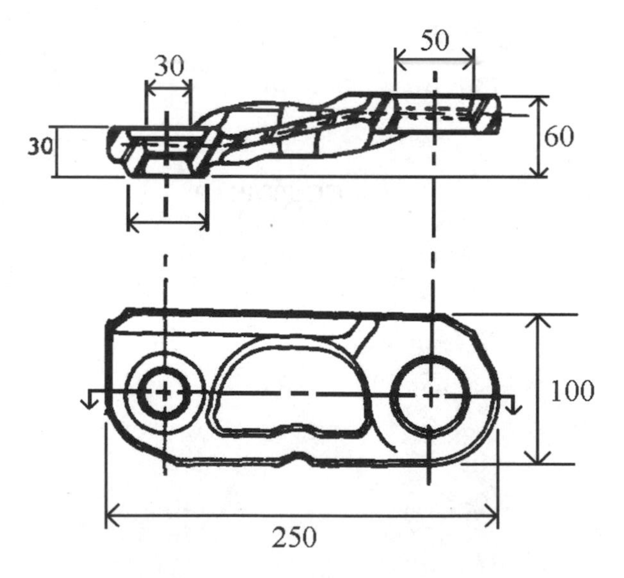

FIGURE P11.4 *(All dimensions in mm, from Knight, W.A. and Poli, C., 1984.)*

11.5 Figure P11.5 shows the current design of an aluminum forging. Assume that you have been assigned the task of redesigning the forging so as to reduce manufacturing costs. What suggestions would you make? Explain your reasoning in detail.

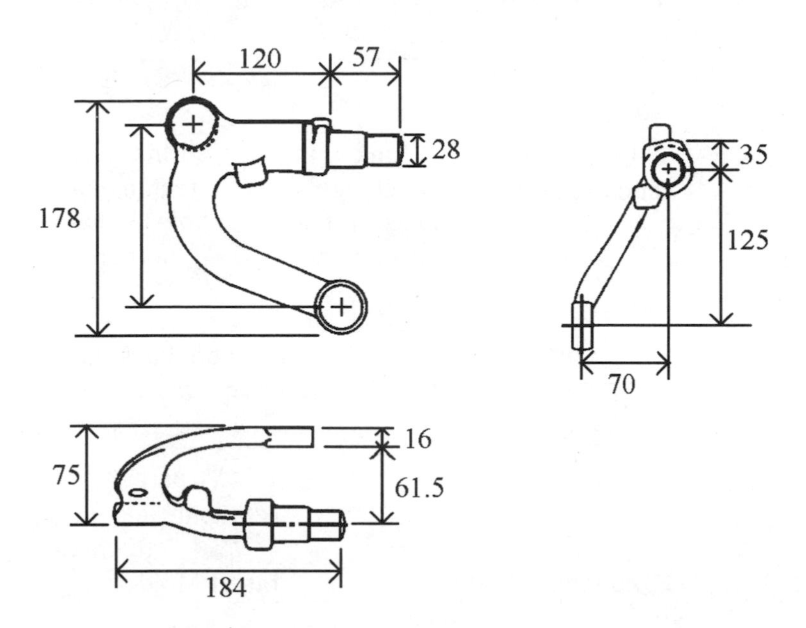

FIGURE P11.5 *(All dimensions in mm, from Knight, W.A. and Poli, C., 1984.)*

11.6 Figure P11.6a shows three versions of an injection-molded link with two holes. Version 1 has circular holes, version 2 has rectangular holes with rounded corners, and version 3 has rectangular holes with sharp corners.

Figure P11.6b shows two alternative mold designs that can be used to produce these links. The first version is that of an integer mold in which the projections required to create the two holes are machined directly into the core. Also shown in Figure P11.6b is a second version that is called an insert mold. In this version two core pins that are inserted into the core replace the projections shown in the first version of the mold.

a) Assuming that the link is to be molded using an integer mold, which of the three hole designs would be least difficult to provide in the mold? Which would be the most difficult to provide?
b) Repeat part (a) under the assumption that an insert mold is to be used to produce the link.

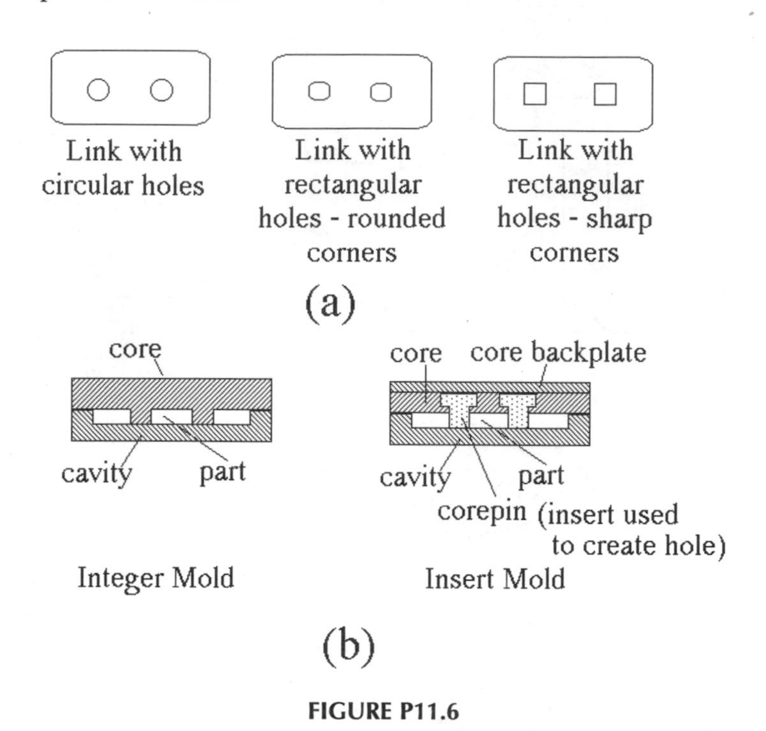

FIGURE P11.6

11.7 Shown in Figure P11.7 are two links with circular holes. In one design the link is provided with sharp external corners, and in the other design the link is provided with rounded external corners. From the point of view of machining the mold required to injection-mold the link, which version is easier (less costly) to produce? Explain.

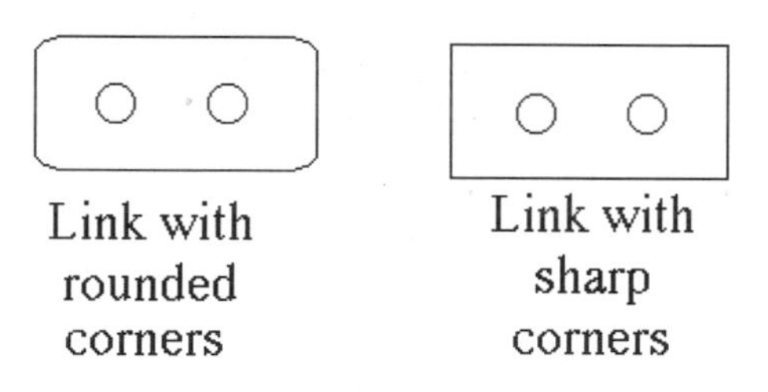

FIGURE P11.7

11.8 Figure P11.8 shows an injection-molded part with two grooves in the side walls. These grooves, as you may recall, are called side shutoffs. Figures P11.8c and P11.8d show two alternative mold designs for producing the part. From the point of view of machining, which mold is less costly to produce? Are there any difficulties, other than machining difficulties, with either of these molds?

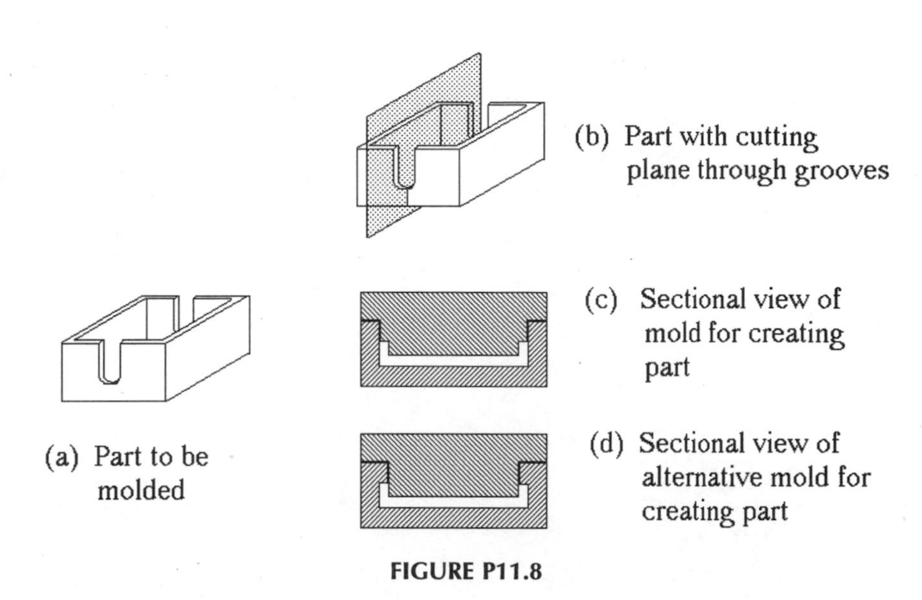

(a) Part to be molded

(b) Part with cutting plane through grooves

(c) Sectional view of mold for creating part

(d) Sectional view of alternative mold for creating part

FIGURE P11.8

11.9 Figure P11.9 shows two alternative molds that can be used to create a deep box-shaped part with a through hole in the base of the box. From a machining point of view, which mold is easier to produce?

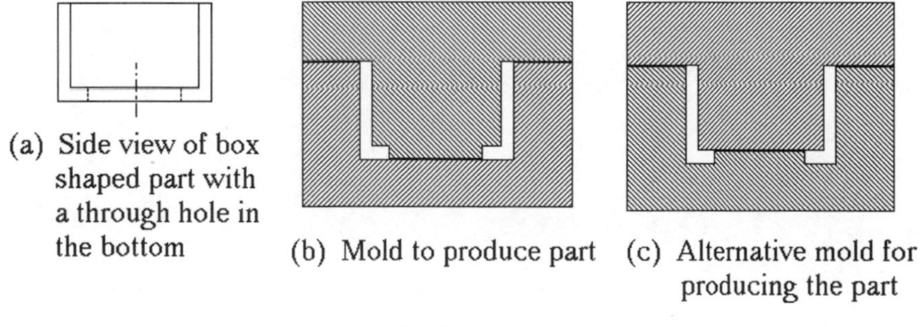

(a) Side view of box shaped part with a through hole in the bottom

(b) Mold to produce part

(c) Alternative mold for producing the part

FIGURE P11.9

Chapter 12

Assembly

12.1 ASSEMBLY PROCESSES

Note: Much of the material on assembly in this book is based on data and other information extracted primarily from the works in the list of references. Readers are urged to consult these original references, especially the pioneering work of G. Boothroyd, for more detailed information and analysis.

The manufacturing process of assembly is generally thought of as consisting of two distinct operations: *handling* followed by *insertion*. Both handling and insertion can be done either manually or automatically.

In the case of manual handling, a human assembly operator stationed at a workbench (Figure 12.1) reaches and grasps a part from a bin, and then transports, orients, and prepositions the part for insertion.

In the case of automatic handling, parts are generally emptied into a parts feeder, such as a vibratory bowl feeder (Figures 12.2 and 12.3), which contains suitable orienting devices (Figure 12.4) so that only correctly oriented parts exit the feeder in preparation for insertion. Feedtracks are then used to transport the correctly oriented parts from the feeder to an automatic workhead. Escapement devices release the parts to the workhead.

In manual insertion, the human assembly operator places or fastens the part(s) together manually. Although power tools may be used, the process is still essentially one of manual insertion under human control.

When automatic insertion is used, automatic workheads, pick-and-place mechanisms, and robots are utilized.

12.2 QUALITATIVE GUIDELINES ON DESIGN FOR ASSEMBLY (DFA)

12.2.1 Reduce the Part Count

As pointed out by Boothroyd in *Product Design for Assembly and Assembly Automation and Product Design,* the two main factors affecting the assembly cost of a product are (1) the number of parts contained in the assembly, and (2) the ease with which the parts can be handled (transported, oriented, and prepositioned) and inserted (placed, fastened, etc.).

It is somewhat obvious that if one product has 50 component parts and if an alternative version of the same product has only 10 parts, then the one with the fewer number of parts will ordinarily cost less to assemble. Thus, the best method available for reducing assembly costs is to reduce the number of parts in the assembly.

Part reduction can be accomplished either by the outright elimination of individual component parts (eliminating screws and washers and using a press or

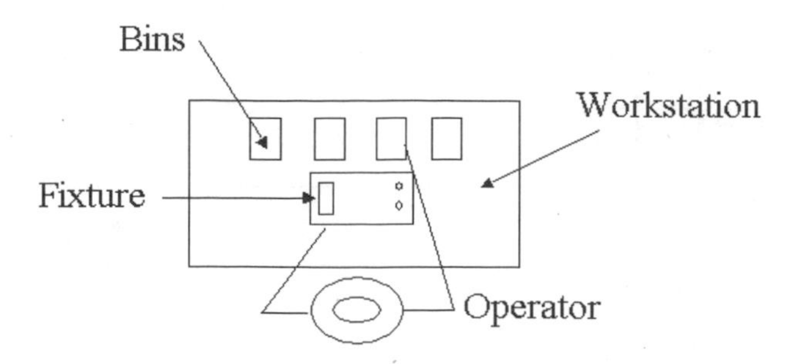

FIGURE 12.1 *A manual assembly operator sitting at a workstation.*

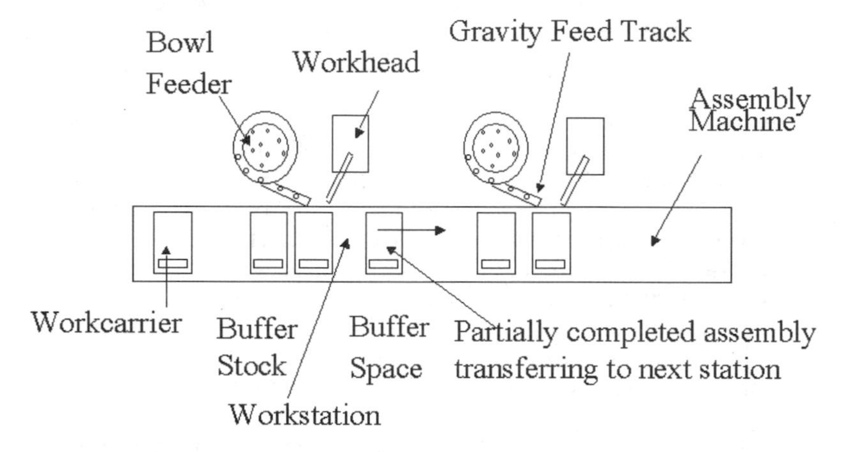

FIGURE 12.2 *Automatic assembly—free transfer system.*

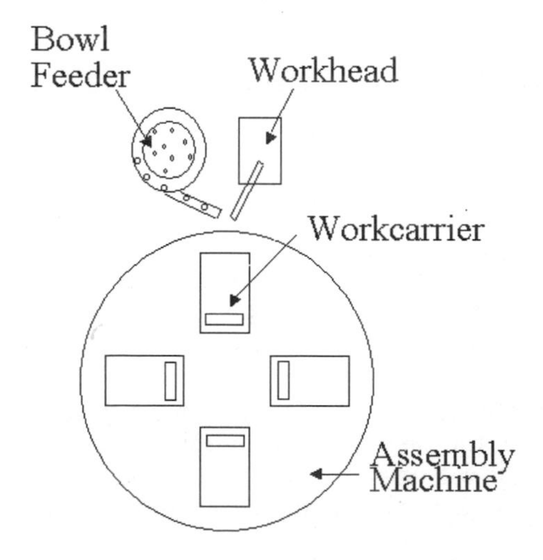

FIGURE 12.3 *A four-station automatic assembly system using a four-station rotary indexing machine.*

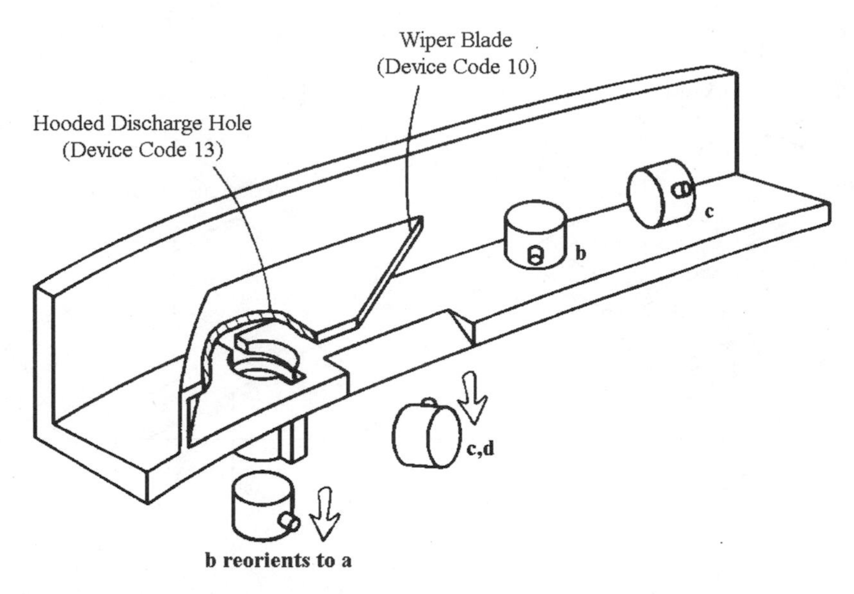

FIGURE 12.4 *Example of tooling for automatic handling. (From "Handbook of Feeding and Orienting Techniques for Small Parts," G. Boothroyd, C. Poli, and L. Murch, Mechanical Engineering Department, University of Massachusetts at Amherst.)*

snap fit to fasten two components, for example) or by combining several component parts into a single, perhaps injection-molded, part. If fasteners cannot be completely eliminated then, Boothroyd suggests to reduce the number or variety of fasteners by incorporating the fastening function by use of another feature. Although screws themselves may not be expensive, handling and inserting them is.

Figure 12.5 shows the back cover subassembly for an electric razor. The original subassembly (Figure 12.5a), consists of nine parts: the back cover, the front cover, two side plates that slide into place on the body and are then held in place by the back cover, four screws to secure the back cover to the body, and a label.

The redesigned back cover subassembly (as done by students) is shown in Figure 12.5b. It consists of two parts, a redesigned back cover, and a screw. The body is redesigned so that the cover can snap into place; a single screw is provided to assist in securing the cover to the body. Although not indicated on the drawing of the redesigned cover, the label has been replaced by lettering molded into the back cover.

It seems obvious that the assembly costs for the second design would be less, but the question remains as to whether or not the overall manufacturing costs have been reduced by this redesign. Later in this chapter we will try to determine the answer to this question.

12.2.2 Reduce the Manual Handling Time

Once a designer has reduced the number of parts contained in an assembly to its "minimum," the remaining parts must be designed so that they are easy to handle and insert. According to Boothroyd in, among other places, *Product Design for Assembly*, a part is easy to handle manually if (1) it is easy to grasp and manipulate with one hand without grasping tools, (2) it is both end-to-end symmetric (as defined below) as well as rotationally symmetric, and (3) its size

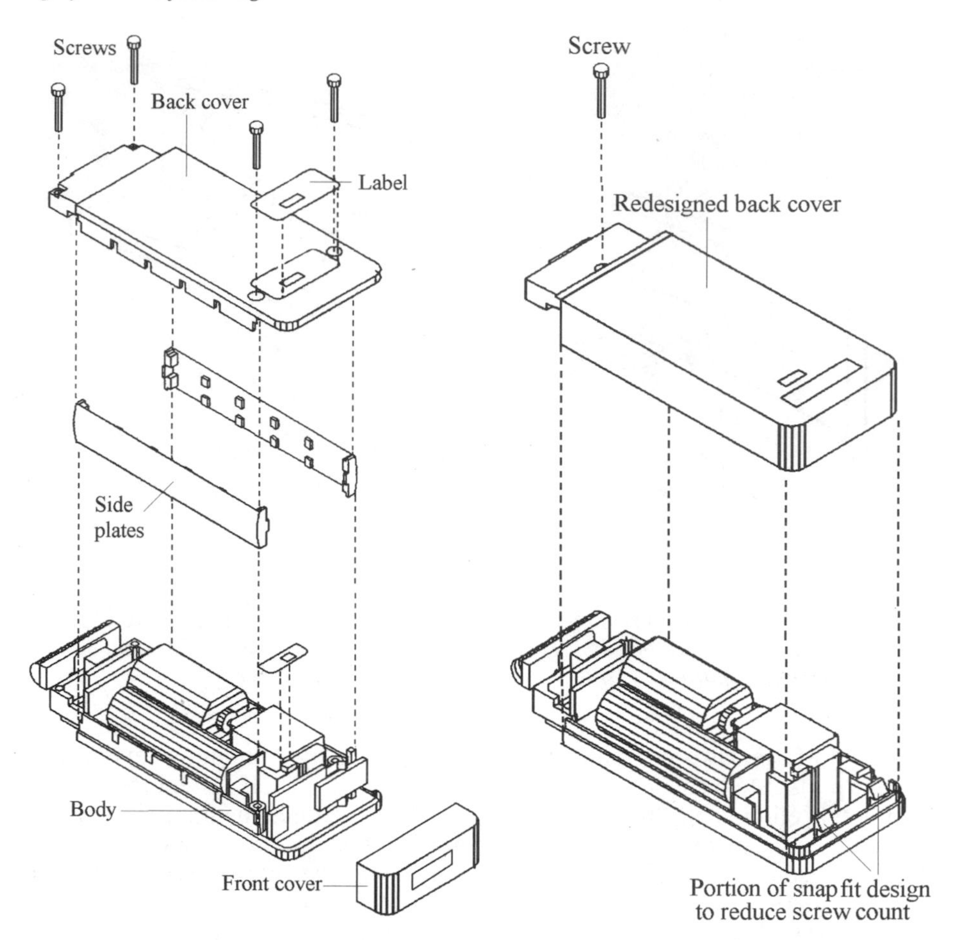

FIGURE 12.5 *The redesign of an electric shaver cover as suggested by students as part of a class project. The drawing shown on the left is that of the original design. The drawing on the right is their suggested redesign.*

and thickness are such that grasping tools or optical magnification are not required.

Tangling and Nesting

Parts that nest or tangle (vending cups, helical springs, etc.; see Figure 12.6) are difficult to grasp singly and manipulate with one hand; they present obvious handling difficulties. In addition, parts that are sticky (a part coated with grease or an adhesive), sharp (razor blade), fragile (glass), slippery (ball bearing coated with light oil), or flexible (belts, gaskets, etc.; see Figure 12.7) are also difficult to grasp and manipulate with one hand, and thus should also be avoided by designers when possible.

Symmetry

To facilitate ease of handling, parts should be designed with symmetry in mind. As pointed out in Boothroyd and Dewhurst, among others, parts that have end-to-end symmetry (that is, parts that do not require end-to-end orientation prior to insertion) require less handling time than parts that are not end-to-end sym-

FIGURE 12.6 *Some examples of parts that tangle.*

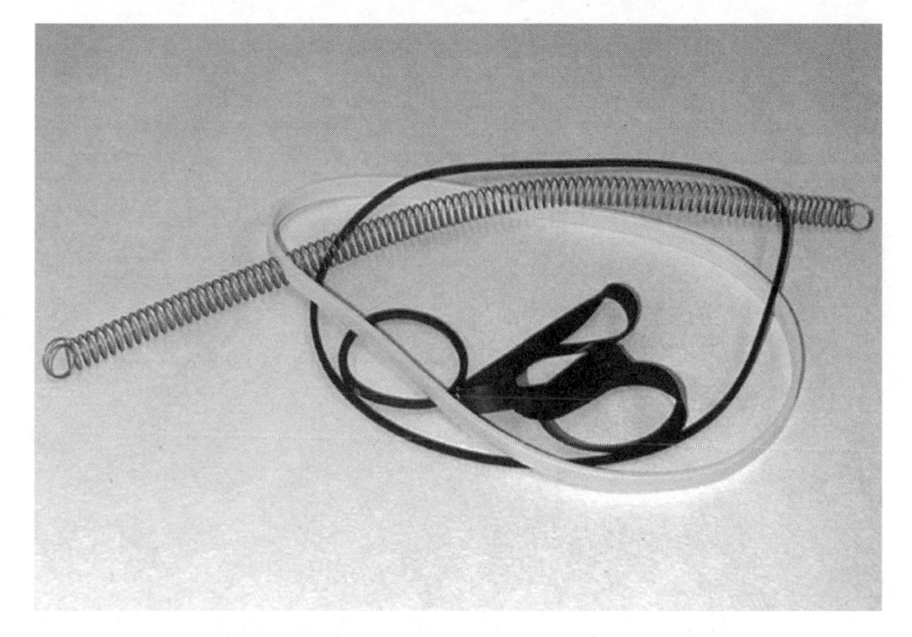

FIGURE 12.7 *An example of flexible parts where two hands are needed to maintain orientation prior to insertion.*

metric. A screw is an example of a part that does not have end-to-end symmetry, but a washer is an example of a part that does have end-to-end symmetry. A screw must be oriented so it can be inserted shank end first, but in the case of a washer either "end" (side) may be inserted first.

Parts that have complete rotational symmetry (i.e., parts that do not require orientation about the axis of insertion such as a screw or washer) take less time to handle than parts that do not have any rotational symmetry—a house key for example. Parts with no symmetry take more time to orient than parts with some symmetry. For example all car keys lack end-to-end symmetry (i.e., only one end of the key can be inserted into the keyhole), but some car keys have 180° rotational symmetry (i.e., either serrated edge can be aligned with the key hole), and others have no rotational symmetry (only one serrated edge exists, and it must be properly aligned with the key hole). Many of us have probably experienced the situation where the key with no symmetry takes more time to properly orient than the key with 180° rotational symmetry. Figures 12.8 and 12.9 show examples of parts with and without symmetry.

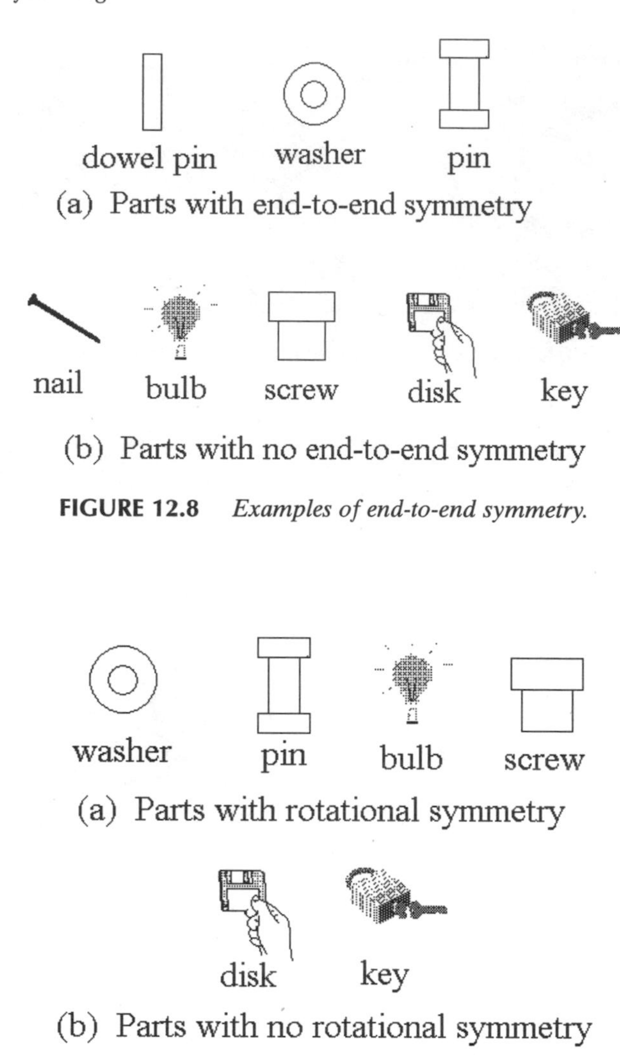

FIGURE 12.8 *Examples of end-to-end symmetry.*

FIGURE 12.9 *Examples of rotational symmetry.*

12.2.3 Facilitate Automatic Handling

Parts can be easily handled automatically if they can be (1) easily fed (e.g., in a bowl feeder), and (2) easily oriented (see Boothroyd and Dewhurst, 1989; Boothroyd, 1992; and Boothroyd, Poli, and Murch, 1982). In general, parts that are difficult to grasp and manipulate manually are also difficult to feed in a feeder. Parts that have end-to-end symmetry as well as rotational symmetry are also more easily and economically oriented in a bowl feeder than parts that do not have any symmetry.

As Boothroyd and Dewhurst (1989) point out, a general rule of thumb is that if the part cannot be made symmetrical, then accentuate its asymmetry. That is, avoid making it almost symmetrical. A complete guide for designing parts for automatic handling can be found in Boothroyd and Dewhurst, 1989; Boothroyd, 1992; and Boothroyd, Poli, and Murch, 1982.

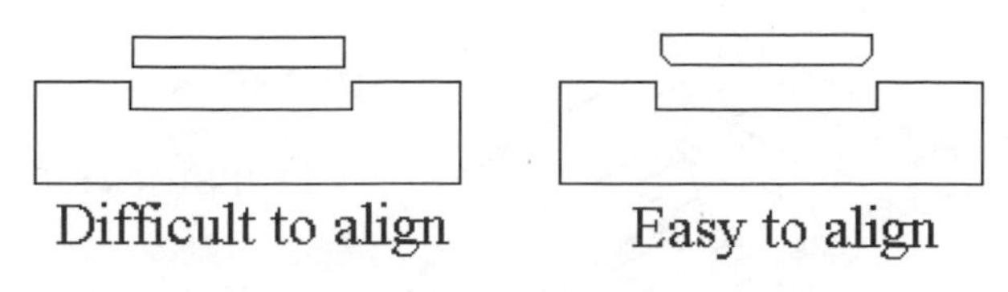

FIGURE 12.10 *Redesign to facilitate alignment.*

12.2.4 Design for Easy Insertion

Based on additional research carried out by Boothroyd (1992; and with Dewhurst, 1989), it has also been shown that insertion costs, whether done manually or automatically, are reduced if parts are designed so that they are easy to align, easy to insert, and self-locating with no need to be held in place before insertion of the next part. Manual insertion costs are also reduced if access and view are not obstructed while attempting to insert one part into the partially completed assembly. In addition, extra operations (reorienting of a partially completed assembly, for example) should be avoided.

A part is easy to align if insertion is facilitated by well-designed chamfers or other features, such as a recess, so that the required accuracy of alignment and positioning is obtained. Figure 12.10 shows an example of a part that was redesigned to facilitate alignment. Figures 12.11 and 12.12 show examples of parts redesigned to provide easy access and to be easily viewed.

Insertion against a large spring force, the resistance encountered with self-tapping screws, and the resistance encountered when using an interference fit (press fit), are all examples of resistance to insertion. The use of small clearances can also result in resistance to insertion. Jamming and wedging that result during insertion can also be considered resistance to insertion. Methods for avoiding jamming in a design are discussed in detail in Boothroyd and Dewhurst (1989) and Boothroyd (1992).

Anyone who has tried to do maintenance or repair work on a car has experienced the situation where parts were designed such that both access and vision were obstructed. Such designs should obviously be avoided.

Figure 12.13 shows two designs. In both designs, the assembly operator places a disk on a plate with the holes aligned. In the original design, shown on the left, the plate is not provided with a recess. Thus, the disk is not self-locating and it must be held such that it maintains its position and orientation relative to the plate prior to insertion of a peg (screw) that will secure the disk to the plate. In the design shown on the right, the plate is provided with a recess making the placement of the disk on the plate self-locating with no need to hold the disk before insertion of the peg.

12.3 TOTAL ASSEMBLY COST

12.3.1 A Rough Rule of Thumb

Precise quantitative methods are available for use by designers and manufacturing engineers to estimate the time or cost to handle and insert parts (Boothroyd and Dewhurst, 1989; Boothroyd, 1992). During the early stages of design, however, much more approximate methods can be helpful in guiding design decisions. One approximate method is based on the assembly advisor shown in Figure 12.14.

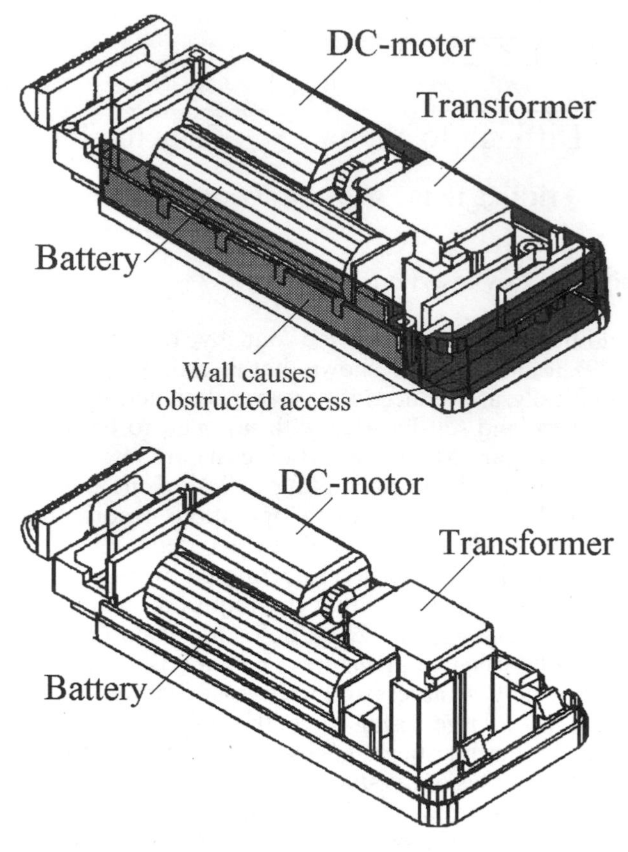

FIGURE 12.11 *An example of obstructed access.*

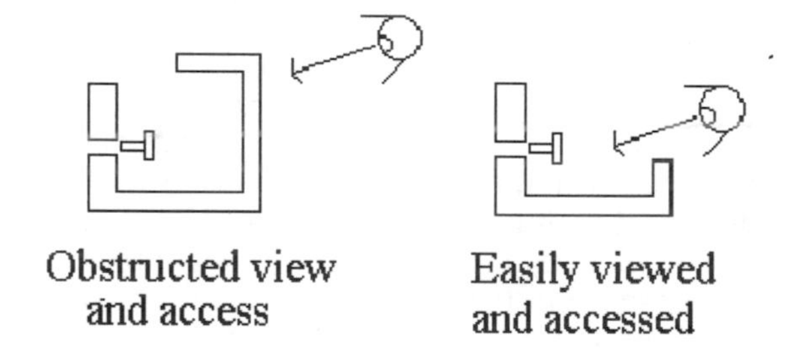

FIGURE 12.12 *An example of obstructed view and access.*

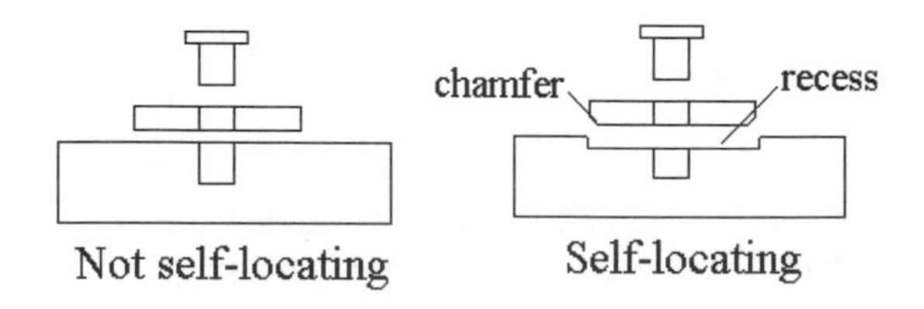

FIGURE 12.13 *Example of redesign in order to make an insertion of the part self-locating.*

INSERTION

HANDLING	Easily Aligned Easily Inserted*	Not Easily Aligned *or* Not Easily Inserted*	Not Easily Aligned *and* Not Easily Inserted*
Easy to Grasp and Manipulate	Good 4.0	POOR 8.0	Costly 13.0
Not Easy to Grasp and Manipulate	Poor 7.5	Costly 11.5	Most Costly 16.5

*Not easily inserted includes difficulties due to obstructed view and/or access, parts jamming and/or difficulties due to lack of chamfers or part geometry,etc. Resistance to insertion due to press and snap fits is not to be included here.

approximate manual assembly time in seconds

FIGURE 12.14 *Assembly advisor.*

In the assembly advisor parts are rated as good designs, fair designs, poor designs, or most costly designs from an ease-of-assembly point of view only. This rough qualitative evaluation depends upon whether or not the part is easy to grasp and manipulate, and on the ease or difficulty of alignment and insertion. The numerical values shown in Figure 12.14 are simply rough approximations of the amount of time in seconds required to handle and insert a part that exhibits the characteristics indicated.

Another rough approximation based simply on part count can also assist in guiding early design. The reasoning is as follows: A "perfectly" designed part from an assembly viewpoint, which is not used as a fastener, will take on average about 3 seconds to handle and insert (Boothroyd, 1992). On the other hand, a difficult-to-assemble part, based on the criteria described above, will take about 11 to 13 seconds to handle and insert. Based on the experience of the author, after analyzing many assemblies and subassemblies, it has been observed that it is possible to obtain a rough estimate of the time required to manually assemble a product by simply assuming that, on average, the handling and insertion time per part is from 7 to 9 seconds. Fasteners will take longer to handle and insert than nonfasteners, and some parts will take longer to handle and insert than others, but if at least 10 parts are contained in the assembly, on average the total assembly time will be 7 to 9 seconds per part or task.

The greater the number of parts in the assembly, the better the above estimate is likely to be. Thus, a reasonably well designed product that contains about 20 parts will be likely to take more or less 180 seconds to assemble manually. A comparable product that has only about 10 parts will be likely to take more or less 90 seconds to assemble. Thus the product that can have its part count reduced in half will likely have about a 50% reduction in assembly costs.

12.3.2 Example 1—Electric Shaver Cover

To illustrate the use of the assembly advisor let us consider the electric shaver shown in Figure 12.5. The original design of the shaver is shown in Figure 12.15 with the handling and insertion characteristics of each part labeled. It is assumed here that the body of the shaver would be inserted into an assembly fixture that is designed so that the body can be easily aligned. Table 12.1 shows a summary of the rough assembly time calculations.

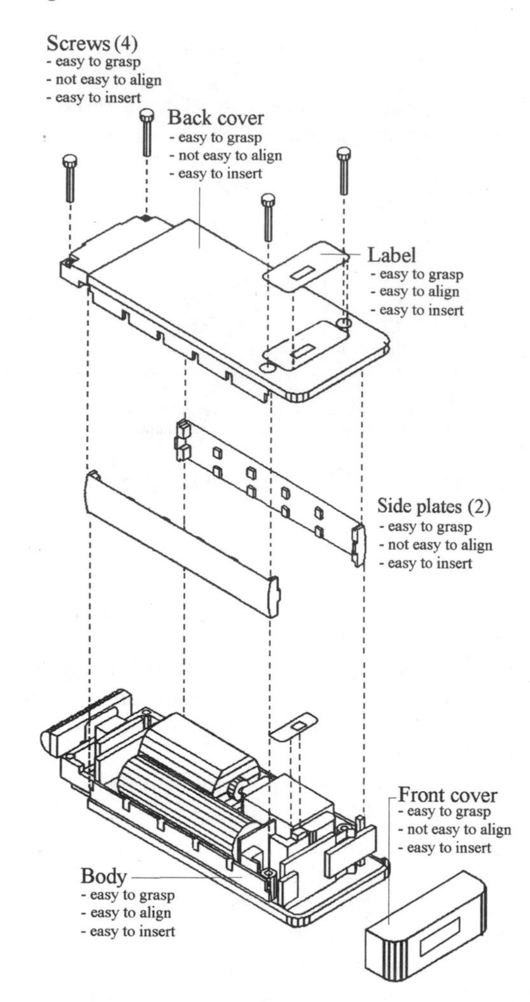

FIGURE 12.15 *Original design of the shaver shown in Figure 12.5 with handling and insertion characteristics of the individual parts labeled.*

Table 12.1 Results of using the assembly advisor to determine a rough approximation to the assembly time for the electric shaver cover shown in Figure 12.15.

Part Name	*Number of*	*Part Design*	*Time per Part*	*Total Time*
Base	1	Good	4	4
Front cover	1	Poor	8	8
Side plates	2	Poor	8	16
Back cover	1	Poor	8	8
Screws	4	Poor	8	32
Label	1	Good	4	4
			Total Assembly Time	72

The redesigned version of this shaver cover is shown in Figure 12.5b and consists of just three parts. It is assumed here that the redesign has been carried out so that all parts, including the screw, are easy to align. Hence, it is easy to show that the assembly advisor estimates that about 12 seconds is required to assemble the cover to the shaver. This redesign results in, roughly speaking, a savings of 83% in assembly time.

For a more precise estimation of manual assembly time the reader should use the Boothroyd methods discussed in detail in his references *Product Design for Assembly* and *Assembly Automation and Product Design*. In addition, *Assembly Automation and Product Design* should be consulted for a detailed discussion of the effects of restricted access and restricted vision on the initial engagement of screws.

12.4 SUMMARY OF DFA GUIDELINES

In summary, in order to reduce assembly costs and to facilitate both handling and insertion, a designer should make every effort to design parts and products such that:

1. The minimum number of parts needed for proper functioning of the product is used. This will probably require the elimination of as many screwing operations as possible and the incorporation of more press and snap fits as a means of fastening. (However, screws are sometimes needed to provide more secure joints. It has also been argued that screws used in efficient robotic assembly can improve the yield—percentage of acceptable products—of an assembly line.) (See Ulrich et al., 1991.)
2. Parts are designed so that they are easy to grasp and manipulate with one hand using no grasping tools (i.e., parts do not nest or tangle, are not sticky, sharp, fragile, slippery, flexible, etc.).
3. Parts are end-to-end and rotationally symmetric as much as possible—or else obviously asymmetric.
4. Parts are designed so that they are easy to align and to insert (i.e., contain chamfers and/or recesses).
5. For manual assembly, both access and vision are not restricted.
6. For automatic assembly, insertion should be in a straight line from above.

Although other criteria exists for the design of parts for ease of assembly, the above are often regarded as the most important.

12.5 REDUCING PART COUNT BY COMBINING SEVERAL PARTS INTO ONE

12.5.1 Introduction and Assumptions

As discussed in the previous section, one of the best methods for reducing total manufacturing costs is to reduce assembly costs. The best way—though not the only way—to reduce assembly costs is to reduce the number of parts to be assembled. This can be done either by the outright elimination of parts (by replacing screws, nuts, and washers with press or snap fits), or by combining two or more individual parts into a single part. In the latter case, the parts involved are most often injection-molded, die-cast, or stamped. When part reduction by combination of parts occurs, the resulting part is usually more complex than the individual parts, and so the question arises as to whether or not the new complex part is in fact less expensive (considering tooling and processing) than the total cost of the individual parts being replaced, including their assembly.

A detailed analysis is presented below that derives relations needed to compare the cost of producing a single, more complex injection-molded or die-cast part with the cost of producing and assembling multiple parts for the same purpose. In the analysis that follows it is assumed that:

1. The wall thickness of the (n) individual parts that are to be combined are approximately equal.
2. The projected area of the single part is no greater than the projected area of the assembled parts that they are replacing. (The two electric shaver covers in Figure 12.5 are examples.) Although on occasion this may not be true, we will see shortly that even if the projected area of the new part is equal to the sum of the projected area of the n parts they are replacing, the conclusions we reach in the next section will remain the same.

12.5.2 Comparative Analysis

To answer the question of whether or not one, perhaps more complex, part is more economical to produce than n, perhaps simpler, parts, we will introduce ΔK, where

$$\Delta K = [\text{Cost to produce n single parts}] + [\text{Cost to assemble the n parts}] - [\text{Cost to produce one functionally equivalent part}] \quad \text{(Equation 12.1)}$$

In terms of symbols:

$$\Delta K = \sum_i K_{ti} + C_{an} - K_{tx} \quad \text{(Equation 12.2)}$$

where

K_{ti} = total cost of producing the ith part,
C_{an} = cost of assembly for the n parts (always a positive quantity)
K_{tx} = total cost of producing the single functionally equivalent part to replace the n parts.

In this section, the subscript x will always be used to denote the single replacement part.

Note that, with this definition, if ΔK is positive, then the single replacement part is the less expensive option.

To begin the analysis, we denote the total cost to produce an injection-molded, die-cast, or stamped part by the following equation:

$$K_t = K_m + \frac{K_d}{N} + K_e \quad \text{(Equation 12.3)}$$

where K_t is the total production cost of the part in units such as dollars or cents, K_m is the material cost, K_e is the processing cost, K_d represents the total cost of the tool, and N represents the production volume or total number of parts to be produced using the tool.

Substituting Equation 12.3 into Equation 12.2 gives

$$\Delta K = \sum_i \left(K_{mi} + \frac{K_{di}}{N} + K_{ei} \right) + C_{an} - \left(K_{mx} + \frac{K_{dx}}{N} + K_{ex} \right) \quad \text{(Equation 12.4)}$$

Comparing Material Costs

In general, the amount of material used to produce the more complex replacement part is about equal to the amount of material required to produce the original parts. Thus, we assume that

$$\sum_i K_{mi} = K_{mx} \quad \text{(Equation 12.5)}$$

and ΔK reduces to

$$\Delta K = \sum_i \left(\frac{K_{di}}{N} + K_{ei} \right) + C_{an} - \left(\frac{K_{dx}}{N} + K_{ex} \right) \quad \text{(Equation 12.6)}$$

Comparing Processing Costs

Now we define

$$\Delta K_e = \sum_i K_{ei} - K_{ex} \quad \text{(Equation 12.7)}$$

That is, ΔK_e is the difference in processing costs between the sum of the individual parts and the single replacement part.

The following argument leads to the conclusion that ΔK_e is a positive quantity:

1. Since the two versions of the design are to be functionally equivalent, the material and wall thickness for both versions will generally be approximately the same. In fact, as we learned for injection-molded and die-cast parts (in Chapters 5 and 7), if good design practice is followed then the wall thickness of the parts should all be constant. Thus, as we can see from Figure 5.2 in Chapter 5, the cycle time for each injection-molded part, including the complex replacement part, should be about the same.
2. In general, each injection-molded part requires its own die and its own injection molding machine. Consequently, the processing cost to produce each part is the product of the machine hourly rate and the cycle time of that part. Since the cycle time for each part is essentially the same, the difference in processing cost between producing n separate parts and the one replacement part is simply the difference between the sum of the n machine hourly rates required to produce the n parts and the machine hourly rate for the single replacement part.
3. In Chapters 5 and 7 we learned that the machine hourly rate is proportional to the machine tonnage required to mold or die cast the part. The machine tonnage in turn is directly proportional to the projected area normal to the direction of mold closure. From Tables 5.5 and 7.5 of Chapters 5 and 7, respectively, it is seen that at least a seven-fold increase in tonnage (i.e., an approximate seven-fold increase in the projected area of the part) is required for the machine hourly rate to double. Since the replacement part is usually about equal in size to the largest of the individual parts it is replacing (recall assumption 2 in Section 12.5.1, Introduction and Assumptions), then the con-

solidation of n parts into a single functionally equivalent part will result in a value of ΔK_e that is greater than zero. (It would be virtually impossible for the projected area of the single replacement part to be such that it will result in a tonnage requirement and hourly rate that would cause ΔK_e to be negative.) Hence, in general,

$$\Delta K_e > 0 \qquad \text{(Equation 12.8)}$$

Thus:

$$\Delta K = \sum_i \frac{K_{di}}{N} + (\Delta K_e + C_{an}) - \frac{K_{dx}}{N} \qquad \text{(Equation 12.9)}$$

where the two terms in parentheses, ΔK_e and C_{an}, are always positive.

Comparing Die Costs

If K_o is the total cost for the reference part, then Equation 12.9 can be written as follows:

$$\frac{\Delta K}{K_o} = \sum_i \frac{K_{di}}{K_o}\frac{1}{N} + \frac{\Delta K_e + C_{an}}{K_o} - \frac{K_{dx}}{K_o}\frac{1}{N} \qquad \text{(Equation 12.10)}$$

As shown earlier, Equation 4.4, for example,

$$\frac{K_{di}}{K_o} = C_{di} = 0.8C_{dci} + 0.2C_{dmi} \qquad \text{(Equation 12.11)}$$

and

$$\frac{K_{dx}}{K_o} = C_{dx} = 0.8C_{dcx} + 0.2C_{dmx} \qquad \text{(Equation 12.12)}$$

Thus,

$$\frac{\Delta K}{K_o} = \sum_i (0.8C_{dci} + 0.2C_{dmi})\frac{1}{N} + \frac{\Delta K_e + C_{an}}{K_o} - (0.8C_{dx} + 0.2C_{dmx})\frac{1}{N} \qquad \text{(Equation 12.13)}$$

Now we define

$$\Delta C_{dm} = \sum C_{dmi} - C_{dmx} \qquad \text{(Equation 12.14)}$$

Since each injection-molded or die-cast part usually requires its own die, and since the replacement part must be approximately equal in size to the largest of the individual parts it is replacing, then the total die material needed to produce n simple parts must be greater than the die material required for the single replacement part. That is,

$$\Delta C_{dm} > 0 \qquad \text{(Equation 12.15)}$$

Total Costs

Substituting the results above into Equation 12.13 for ΔK, we get:

$$\frac{\Delta K}{K_o} = \sum_i \left(0.8C_{dci}\frac{1}{N}\right) + \left[\frac{\Delta K_e + C_{an}}{K_o} + \frac{0.2\Delta C_{dm}}{N}\right] - \frac{0.8C_{dcx}}{N}$$

(Equation 12.16)

where all of the terms in brackets are positive.

If we now define

$$\Delta C_{dc} = \sum_i C_{dci} - C_{dcx} \quad \text{(Equation 12.17)}$$

and if we recall from Chapters 4 and 5 (Figures 4.1 and 5.2) that the minimum value of C_{dc} is 1, then we can conclude that

If $C_{dcx} < n$

then ΔC_{dc} is a positive quantity

and ΔK is positive.

That is, if the relative die construction cost for the replacement part is less than the number of the parts being replaced, then it is certainly less costly to replace these individual parts by a single part.

If C_{dcx} is greater than n, inspection of Equation 12.16 shows that it may still be more economical to combine the parts. However, in this case we would need to first determine the actual relative die construction costs for each individual part using the method described in Chapters 4 (Section 4.6) and 7 (Section 7.2) and then sum these values to determine if Equation 12.16 is positive. If it is positive, then once again it is certainly more economical to use the single, possibly more complex, part.

12.5.3 If Equation 12.16 Is Not Positive

Even if Equation 12.16 is not positive, it may still be more economical to produce the one-piece version since assembly costs and processing costs have not yet been accounted for in detail. In this situation it can be shown (see Poli and Fenoglio, 1991) that if the production volume N satisfies the following expression:

$$N > \frac{0.8K_{do}(C_{dcx} - n)}{K_{eo}t_r\left(\sum_i C_{hri} - C_{hr}\right)} \quad \text{(Equation 12.18)}$$

where

K_{do} = the sum of the tool construction cost (K_{dco}) and tool material cost (K_{dmo}) for the reference washer (values for these are given in Tables 5.1 and 7.6 of Chapters 5 and 7, respectively).

K_{eo} = processing cost for the reference washer (values for these are given in Tables 5.1 and 7.6 of Chapters 5 and 7, respectively).

t_r = the cycle time for the part relative to the cycle time for the reference washer (a method to determine this was discussed in Chapter 5).

C_{hri} = the machine hourly rate for the ith part relative to the machine hourly rate for the reference washer (a method to determine this value was discussed in Chapter 5).

C_{hr} = the machine hourly rate for the replacement part relative to the machine hourly rate for the reference washer (a method to determine this value was discussed in Chapter 5).

then it is more economical to replace the n-part assembly by an equivalent 1-part assembly.

Quite often when combining parts to form a single part, three or more parts are involved. In this case a conservative version of Equation 12.18 is

$$N > \frac{0.8K_{do}(C_{dcx} - n)}{K_{eo}t_rC_{hr}} \qquad \text{(Equation 12.19)}$$

In arriving at Equation 12.16 it was assumed that the replacement part was either injection-molded and/or die cast. It will be left as a student exercise to determine those conditions when it is more economical to replace a multiple part stamping with a single stamped component.

12.5.4 Example 2—Alternative Designs for Electric Razor Cover

Figure 12.5 shows the back cover subassembly for an electric razor. The original subassembly consists of nine parts. The redesigned back cover subassembly consists of two parts: an injection-molded back cover and one screw.

There are two variations of the redesigned cover. One is shown in Sectional View 1 in Figure 12.16 and requires an internal undercut to accommodate the snap fit. The other is shown in Sectional View 2 in Figure 12.16 in which the function is accomplished without need for an internal undercut.

We want to determine whether either version of the new more complex back cover design is less costly to produce than the original back cover subassembly,

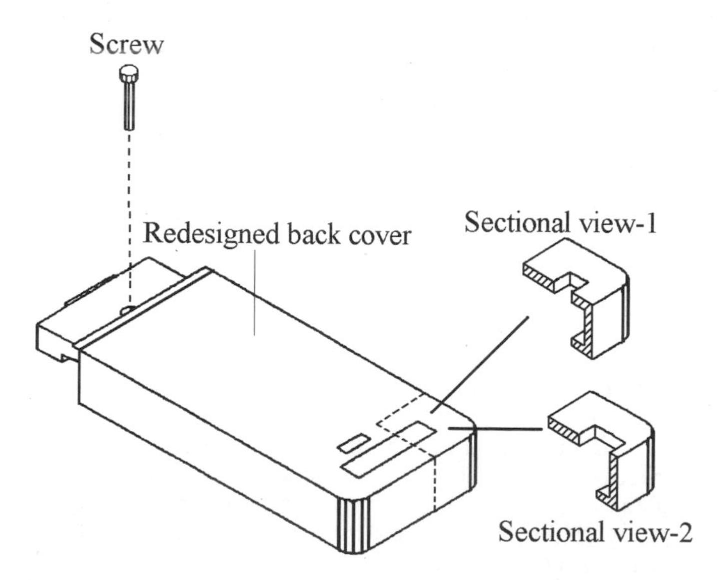

FIGURE 12.16 *Two alternative designs of the revised back cover.*

which consisted of the five molded parts: the back cover itself, two side plates, a front cover, and a label. We ignore the fact for the moment that four screws are needed with the original design versus one for the redesigns.

To compare the costs, we estimate the relative die construction cost for the two designs by using the methodology in Section 4.6 of Chapter 4, "Injection Molding: Relative Tooling Cost." Although the dimensions of the cover are not given in Figure 12.16, the dimensions are such that the part can be considered a "small" part, that is, the largest dimension of the cover is less than 250 mm (10 inches). Using Figure 4.1 and the methodology described in Section 4.6, we can easily show that for the design depicted in Sectional View 1, the part is box-shaped with one internal undercut present, a non-planar dividing surface, and no external undercuts. Thus, C_b is 3.73. From Figure 4.19 we can conclude that the cavity detail is low, hence, from Table 4.1 the subsidiary complexity, C_s, is 1.0. We will also assume commercial tolerances and an SPI 3 surface finish. Hence, from Table 4.2, C_t is also 1. Thus, the relative die construction cost for the redesigned cover, C_{dcx}, is approximately 3.73.

For the design shown in Sectional View 2, no undercuts are present, thus, the value of C_{dcx} is approximately 1.92.

Thus, since $n = 5$, and in both cases the value of C_{dcx} is less than n, then it makes economic sense to replace the four simple parts with this one more complex part. The screws add to the cost savings.

12.5.5 Example 3—$C_{dcx} > n$

Suppose it is determined that the die construction cost, C_{dcx}, for a replacement part in the above example is 6. Then C_{dcx} would be greater than n, the number of parts it is replacing (5 in this case). Hence, we cannot immediately conclude that it is economically viable to replace the four simple parts by this single more complex part.

Let us further assume that by the methods of Chapter 4 the value for C_{dci} for each of the five original parts is such that

$$\sum_i C_{dci} = 5.6 < C_{dcx}$$

which implies that we still cannot conclude that it is more economical to combine these five parts.

To determine the production volume N required to make the single part more economical despite these results, we compute N from:

$$N > \frac{0.8K_{do}(C_{dcx} - n)}{K_{eo}t_r C_{hr}} \qquad \text{(Equation 12.19)}$$

where from Table 5.1 of Chapter 5

$$K_{do} \cong \$7{,}000$$

$$K_{eo} \cong \$0.13$$

and from the original student project report that analyzed the electric razor cover

$$t_r = 2.4$$

$$C_{hr} = 1.21$$

Substituting these values into Equation 12.19 gives the following,

$$N > 0.8(\$7000)(6.0 - 5)/\$0.13(2.4)(1.21) = 14{,}833$$

That is, as long as the production volume is to be greater than about 15,000, then it is more economical to produce the single more complex part than to produce the four simple parts that must then be assembled. (Remember, we still haven't considered the fact that we've eliminated three screws and have fewer parts to assemble.)

Note that the required production is not especially large for a modern mass-produced product, especially one marketed worldwide. Thus one is tempted to conclude that it is almost always better to combine parts when the resulting single part is indeed functionally equivalent and can actually be produced. We know of one case, however, where twenty parts were combined into one awfully complex single part that, it turned out sadly, could not be produced by anyone. That part had to be made in two parts. However, the reduction from twenty to two parts resulted ultimately in great savings in assembly costs. Valuable time was lost, however, in finding out that the single part could not be made.

12.6 SUMMARY

This chapter has presented a very brief introduction to the process of manual assembly. The purpose of this chapter was to introduce you, as a designer or manufacturing engineer, to recognize those features of a part, which affect manual assembly time, hence, cost. Included in the chapter are a set of qualitative guidelines on design for assembly (DFA) and a summary of DFA guidelines.

Also included in this chapter was a detailed analysis that derived the relations needed to compare the cost of producing a single, more complex injection molded or die cast part with the cost of producing and assembling multiple parts for the same purpose.

Since much of the material on assembly in this chapter is based on data and other information extracted primarily from the works in the list references, you are encouraged to consult these original references, especially the pioneering work of G. Boothroyd for more detailed information and analysis.

REFERENCES

Boothroyd, G. *Assembly Automation and Product Design.* New York: Marcel Dekker, 1992.

Boothroyd, G., and Dewhurst, P. *Product Design for Assembly.* Wakefield, RI: Boothroyd-Dewhurst Inc., 1989.

Boothroyd, G., Poli, C., and Murch, L. E. *Automatic Assembly.* New York: Marcel Dekker, 1982.

Boothroyd, G., Poli, C., and Murch, L. E. "Handbook of Feeding and Orienting Techniques for Small Parts." Mechanical Engineering Department, University of Massachusetts at Amherst, Amherst, MA 1978.

Poli, C., and Fenoglio, F. "The Feasibility of Part Reduction in an Assembly." *Concurrent Engineering*, Vol. 1, no. 1, Jan/Feb 1991, pp. 21–28.

Seth, B. "Design for Manual Handling." M. S. Project Report, Mechanical Engineering Department, University of Massachusetts at Amherst, Amherst, MA, 1979.

Seth, B., and Boothroyd, G. "Design for Manufacturability: Design for Ease of Handling," Report No. 9, NSF Grant Apr 77-10197, Jan. 1979.

Ulrich, Kart T., et al. "Including the Value of Time in Design for Manufacturing Decision Making." MIT Sloan School of Management Working Paper #3243-91-MSA, Dec. 1991.

Yoosufani, Z. "Design of Parts for Ease of Handling." M. S. Project Report, Mechanical Engineering Department, University of Massachusetts at Amherst, Amherst, 1978.

Yoosufani, Z., and Boothroyd, G. "Design for Manufacturability: Design of Parts for Ease of Handling." Report No. 2, NSF Grant Apr 77-10197, Sept. 1978.

QUESTIONS AND PROBLEMS

12.1 Figure P12.1 shows a drawing of a type of caster assembly that is commonly found on heavy boxes. Using the assembly advisor, Figure 12.14, evaluate the ease or difficulty of assembling the caster. What suggestions would you make in the design in order to reduce assembly costs? Can you estimate the approximate savings in assembly costs?

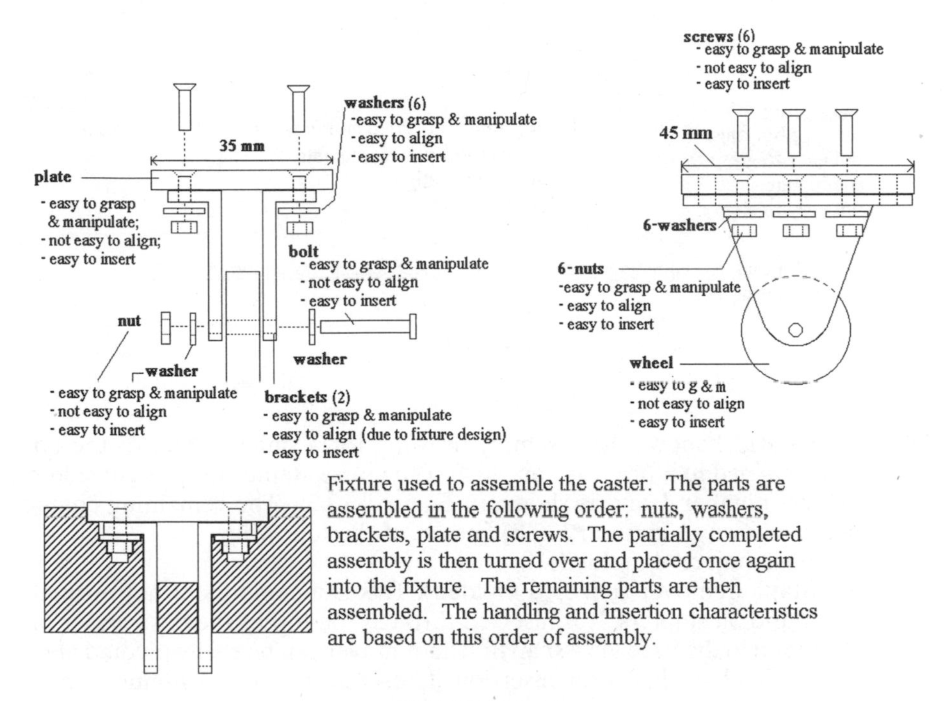

FIGURE P12.1

12.2 Figure P12.2 shows a drawing of a portion of a floppy disk drive that is commonly found in PCs. Using the assembly advisor, Figure 12.14, evaluate the ease or difficulty of assembling the parts shown. What suggestions would you make in the design in order to reduce assembly costs? Can you estimate the approximate savings in assembly costs?

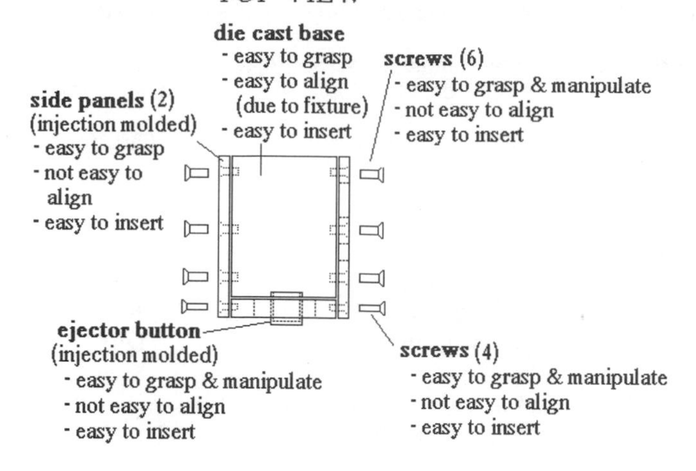

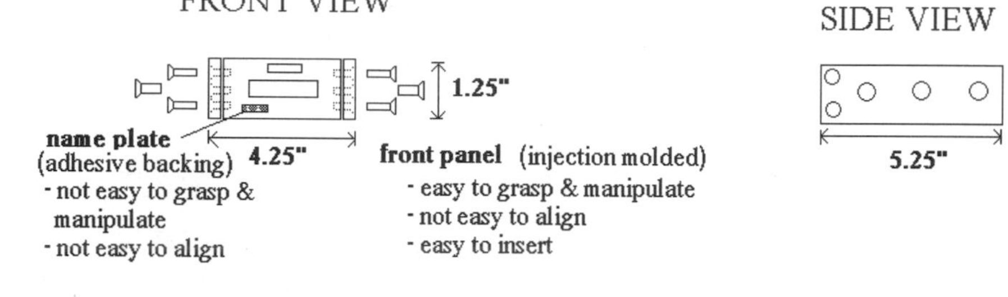

Note: Ejector button is connected to the disk ejector mechanism (not shown) via a coil spring.

FIGURE P12.2

12.3 Figure P12.3 shows the assembly drawing for a small stapler of the type often carried in a briefcase or purse. The entire stapler itself is considered as part number 1 and is shown in Figure P12.3a. The remaining components (shown in Figure P12.3b) consist of the following:

2 – Base (injected molded).

3 – Staple remover (injected molded). This is a component not normally found on staplers. It is inserted through the top of the base and is fastened to the base via a snap fit. The remover can be easily pivoted about one end so that after insertion it rests in a recess contained on the underside of the base (not shown). It can be used to remove staplers from papers that have been stapled together.

4 – Base plate (stamped). Used to fold the staplers as they are used.

5 – Spring plate (stamped). The staple holder compresses this spring as staples are used. It is used to return the staple holder and cover to its normal position after staples are used.

6 – Staple holder subassembly (shown in greater detail in Figures P12.3c and P12.3d).

7 – Top cover.

Using the assembly advisor, estimate the difficulty (or ease) of assembling the stapler. The handling and insertion characteristics of each part and subassembly is indicated in the drawings.

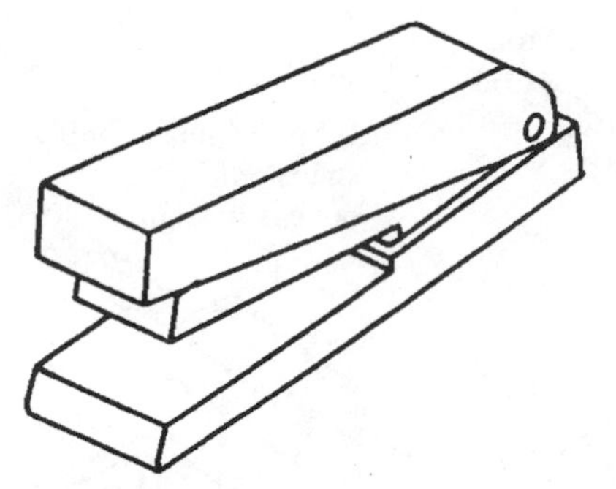

(a) Part 1, The assembled stapler

Note 1: All parts are easy to grasp and manipulate unless otherwise indicated.

Note 2: All parts are easily aligned, viewed, and accessed unless otherwise indicated.

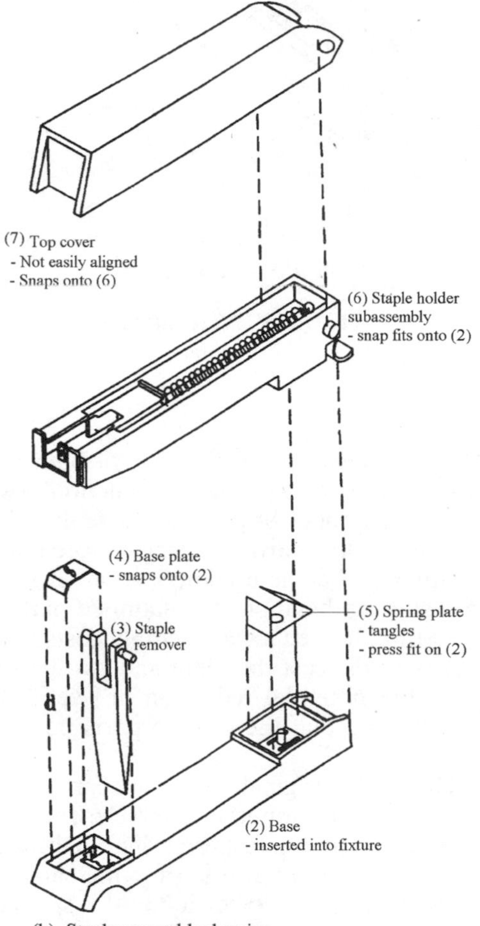

(b) Stapler assembly drawing

FIGURE P12.3 *continued*

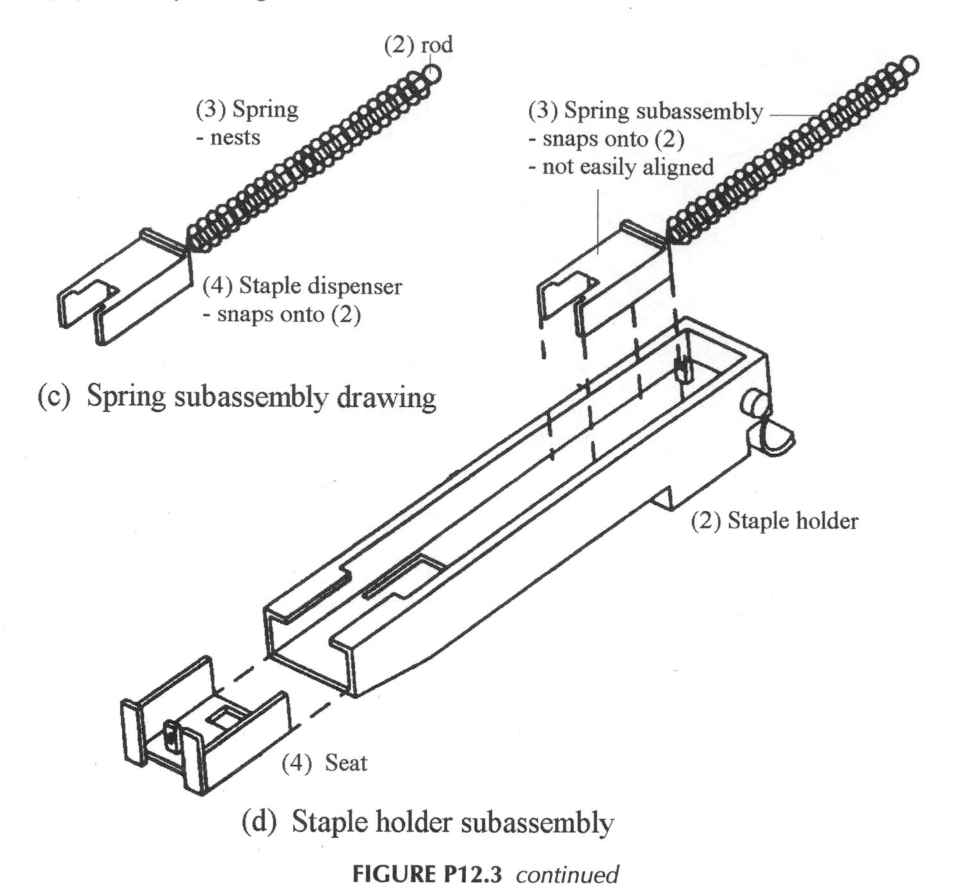

FIGURE P12.3 *continued*

12.4 In Section 12.5 it was shown, for the case of both injection molding and die casting, that if the relative die construction cost for a replacement part is less than the number of parts being replaced, then it is less costly to replace these individual parts by a single more complex part. Following a procedure similar to the one used in Section 12.5, determine whether or not this same result can be applied to stamped parts.

12.5 One possible method for reducing assembly costs for the assembly shown in Figure P12.1 is to die cast the plate and two brackets as a single part. Based solely on the potential reduction in overall manufacturing costs, would you recommend that these three parts be combined into a single component?

12.6 Figure P12.6a shows the original design of a small door (envelope dimensions of 120mm by 65mm) found in a vehicle. The original design consists of five parts including a screw, washer, spring, latch, and door. Figure P12.6b shows an exploded assembly drawing for the same door. Figures P12.6c and P12.6d show the redesigned version of this same door. The redesigned version replaces the screw, washer, spring, and latch by a redesigned latch that incorporates two flexible arms that provide the same function as the original coil spring. The redesigned latch is simply placed on the door as shown in Figure P12.6d and then depressed horizontally so as to snap the

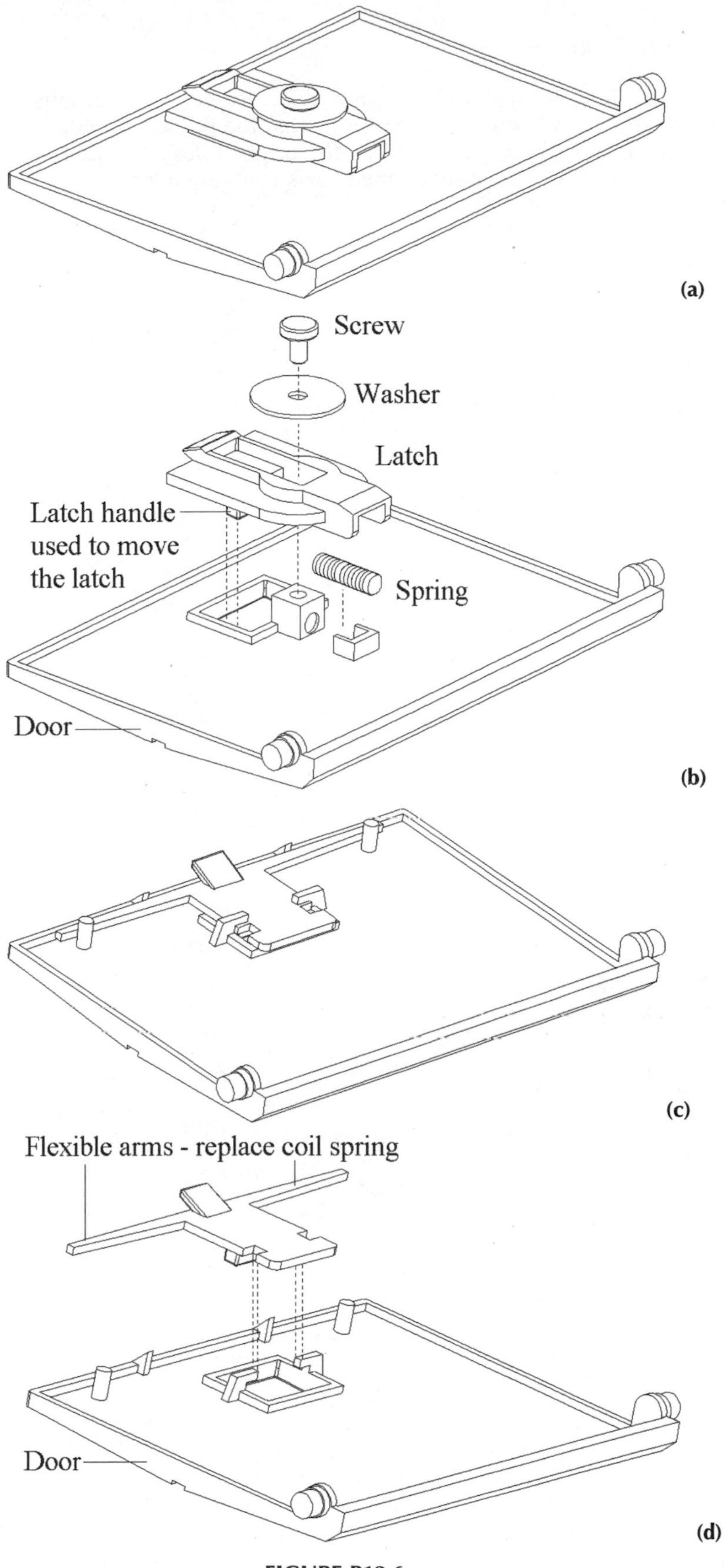
(a)
Screw
Washer
Latch
Latch handle used to move the latch
Spring
Door
(b)
(c)
Flexible arms - replace coil spring
Door
(d)

FIGURE P12.6

latch into the slightly redesigned door. Based on the results obtained in Section 12.5.2, would you conclude that the redesigned latch and door are an economical replacement for the original design? None of the original parts nor any of the redesigned parts contain undercuts.

Chapter 13

Selecting Materials and Processes for Special Purpose Parts

13.1 INTRODUCTION

When it has been determined that a designed object is to be a special purpose part, the task of engineering conceptual design includes

1. Determining the basic material class (e.g., steel, thermoplastic, aluminum, etc.) "that has the properties to provide the necessary service performance . . ." *and*
2. The basic manufacturing process (e.g., injection molding, stamping, extrusion, etc.) to be used to ". . . process the material into a finished part." (Deiter, *A Materials Processing Approach*, 1983.)

In this chapter we describe a methodology for making these important choices at the conceptual engineering design stage. In later chapters, we will discuss how these general choices are made more specific during configuration and parametric design.

Selecting a material class and manufacturing process for a part is a bit like the proverbial chicken-egg argument: Which comes first? That is, should we select a material first and then a process, or vice-versa? Either way, ultimately it is the combination of material and process that must work compatibly during the design, production, and use of a part to meet the requirements of the Engineering Design Specification.

The task of selecting materials and processes is not an easy one. One reason is that there are so many materials and processes to choose from. For each material, there are dozens of material properties that have to be considered. And for each process, there are a variety of process capabilities and limitations to be considered in relation to the design and production requirements of the part. Moreover, the choice is also influenced by other issues having to do with concerns such as safety, cost, availability, codes, disposal, and so on.

To make the problem even harder, the choice of material and process is influenced by a part's size, shape, and geometry, but at the conceptual stage there is little information yet available about either the configuration or the dimensions of the part. For example, stronger materials can lead to thinner-walled parts and, as we learned in earlier chapters, certain processes can better cope with geometrically complex parts. We must nevertheless make basic material-process selections at the conceptual stage before it is feasible to go on to the configuration and parametric stages. We do not always make only a single selection, but the field of possibilities is usually reduced to no more than two or three.

Despite the fact that a part's configuration and parametric design have not yet been determined, there is still information available at the conceptual stage that can be used to guide material and process selection. A great deal of this information is recorded in the Engineering Design Specification (see Section 1.5). For example, the environmental conditions under which the part must perform are specified, and whether the part will be subjected to relatively small or large forces is also known.

In addition to information stated explicitly in the Specification, more can be inferred with reasonable correctness. For example, the approximate size, shape, and degree of geometric complexity are generally understood qualitatively even though the details of configuration and parametric design have not been addressed specifically.

In this chapter, we present a general methodology for selecting one or more material-process combinations for special purpose parts at the conceptual stage. Though this method and some of the data used to support it will also be useful in refining the material-process selection at later stages of the design process, the discussion in this chapter is essentially limited to the conceptual stage. A more detailed approach to material selection can be found in Ashby's *Materials Selection in Mechanical Design.*

Though some data is given, data on all materials and all processes is not included in this book. There are simply too many materials and processes. Thus, some of the most commonly used materials and processes are included here, and it is assumed that readers will have taken courses or have access to information that will provide more complete coverage. The major goal here is to provide a methodology for materials and process selection. As always, designers are strongly urged to consult with materials and manufacturing experts about material-process selections before considering the selection "final."

13.2 TWO APPROACHES—A BRIEF OVERVIEW

13.2.1 Preface

There are two approaches to determining candidate material-process combinations for a part. Designers can use either approach depending on which is most natural to the part being designed. Both approaches end up at the same point. The two approaches are: (1) material-first; or (2) process-first.

In the *material-first approach*, designers begin by selecting a material class—guided by the requirements of the application. Then processes consistent with the selected material are considered and evaluated—guided by production volume and information about the size, shape, and complexity of the part to be made.

In the *process-first approach*, designers begin by selecting the manufacturing process—guided by production volume and information about the size, shape, and complexity of the part to be made. Then materials consistent with the selected process are considered and evaluated—guided by the part's application.

In the next two subsections, an overview of each of these approaches is provided. Then in the remainder of this chapter these approaches are described in greater detail.

13.2.2 Material-First Overview

In the material-first approach, application-related criteria derived from such issues as the environment in which the part will be used, the relative strength

Table 13.1 Preliminary material and process selection: Metals: Cast.

Level I Material	*Level II Material*	*Level III Materials Processes*	*Level IV Material Examples*
Metals	Cast	Steels (carbon, alloy, stainless)	
		Sand Casting	ASTM A27-81
		Investment Casting	ASTM A352-80
		Centrifugal Casting	ASTM A148-80
		Ceramic Mold Casting	ASTM A297
		Forging	
		Aluminum and Magnesium Alloys	
		Sand Casting	
		Investment Casting	Aluminum A380.0
		Centrifugal Casting	Aluminum A413.0
		Ceramic Mold Casting	Aluminum 201.0
		Die Casting	Magnesium AZ91D
		Permanent Mold Casting	Magnesium AZ63A
		Shell Mold Casting	
		Plaster Mold Casting	
		Forging	
		Copper Alloys	
		Sand Casting	
		Investment Casting	C94800
		Centrifugal Casting	C84400
		Ceramic Mold Casting	C80100
		Shell Mold Casting	C81400
		Plaster Mold Casting	
		Forging	
		Zinc Alloys	
		Sand Casting	
		Centrifugal Casting	SAE 903
		Ceramic Mold Casting	SAE 925
		Die Casting	Alloy 7
		Permanent Mold Casting	ILZRO 16
		Plaster Mold Casting	
		Forging	

required, safety or code requirements, and so on, are used to select a set of candidate material classes. These criteria are discussed more completely in Section 13.4. In addition, to assist with material selection, Appendix 13.A contains four tables (Tables 13A.1, 13A.2, 13A.3, and 13A.4) that list a number of material properties and property ranges for broad classes and subclasses of materials. The values are approximate and intended only to assist designers with trial material choices. More exact and complete tables of properties can be found in References. The organization of the tables is as follows:

Table 13A.1 Properties of selected cast metals (alloys of aluminum, magnesium, copper, zinc, and steel).
Tables 13A.2 and 13A.3 Properties of selected wrought metals (alloys of aluminum, magnesium, copper, and steel).
Table 13A.4 Properties of selected plastics (thermoplastics and thermosets).

Tables 13.1 to 13.4 illustrate that with each material class, there is an associated set of feasible processes, as shown in the column labeled Level III for

Table 13.2 Preliminary material and process selection: Metals: Wrought.

Level I Material	*Level II Material*	*Level III Materials Processes*	*Level IV Material Examples*
Metals	Wrought	Steels (carbon, alloy, stainless)	
		Stamping	SAE 1008—hot rolled
		Forging	SAE 1008—cold rolled
		Extrusion	SAE 2330—cold drawn
		Rolling	301 Stainless
		Drawing	410 Stainless
		Aluminum and Magnesium Alloys	Aluminum 2024
		Stamping	Aluminum 2124
		Forging	Aluminum 1100
		Extrusion	Aluminum 6061, 6063
		Rolling	Aluminum 7075
		Drawing	Magnesium AZ61A-F
			Magnesium AZ80A-75
		Copper Alloys	
		Stamping	C23000
		Forging	C37700
		Extrusion	C11000
		Rolling	
		Drawing	
		Zinc Alloys	
		Stamping	ZN-0.08PB
		Forging	ZN-1Cu
		Extrusion	Z300
		Rolling	
		Drawing	

metals and Level II for plastics. For example, if the material class selected is thermoplastics, then Table 13.3 indicates that the feasible processes are injection molding, extrusion, extrusion blow molding, rotational molding, and thermoforming.

Now refer to Tables 13.5 through 13.7. Assuming the part is to be made of a material in a selected class, information about the part size, general shape, and complexity can be used to rule out some processes and point favorably to others. For example, Table 13.7 indicates that injection molding is usually not a good choice if production volumes are less than about 10,000.

In summary, the materials-first approach to materials-process selection maps first from application information to a class of materials by using material data and properties similar to those given in Tables 13A.1, 13A.2, 13A.3, and 13A.4. (More will be said about this in Sections 13.3 and 13.4.) By using Tables 13.1 to 13.4, we are then able to select processes consistent with the materials selected based on application information. Then information about the part is added, and the result mapped into a process type using the guidelines presented in Tables 13.5 through 13.7. Figure 13.1 illustrates the approach in rough schematic form.

13.2.3 Process-First Overview

In the process-first approach, the first step is to select a candidate process type (or types) using information available about production volume and about the

Table 13.3 Preliminary material and process selection: Plastics: Thermoplastics.

Level I Material	*Level II Material Processes*	*Level III Material Examples*	*Level IV Material Examples*
Plastics	Thermoplastics *Injection Molding*	ABS	Magnum 213 (Dow) Cycolac T (GE)
	Extrusion *Blow Molding*	ABS Glass Reinforced	AbsafiL G 11200/30 (Akzo Eng.)
	Rotational Molding *Thermoforming*	ABS Carbon Reinforced	J-1200CF/20 (Akzo Eng.)
		Polystyrene	Polysar 410 (Polysar) PS 318 (Huntsman)
		Polycarbonate Glass Reinforced	R-40FG (Thermofil)
		Polycarbonate Carbon Reinforced	R-40F-5100 (Thermofil)
		Acetal	Delrin 900 (duPont) Celcon M25 (Ceolanese)
		Acetal Glass Reinforced	Thermofil G-40FG (Thermofil)
		Polyamide Nylon 6/6 Nylon 6 Nylon 6 Glass Reinforced Nylon 6/6 C Reinforced	Adell AS-10 (Adell) Zytel 408 (duPont) Ashlene 830 (Ashley) CR1401 (custom Resin) NyLafil G 3/0 (Akzo) Ultramid A3WXH(BASF)
		Polyethylene	Chevron PE1008.5 (Chevron)
		Polypropylene	Excorene pp 122f (Exxon)

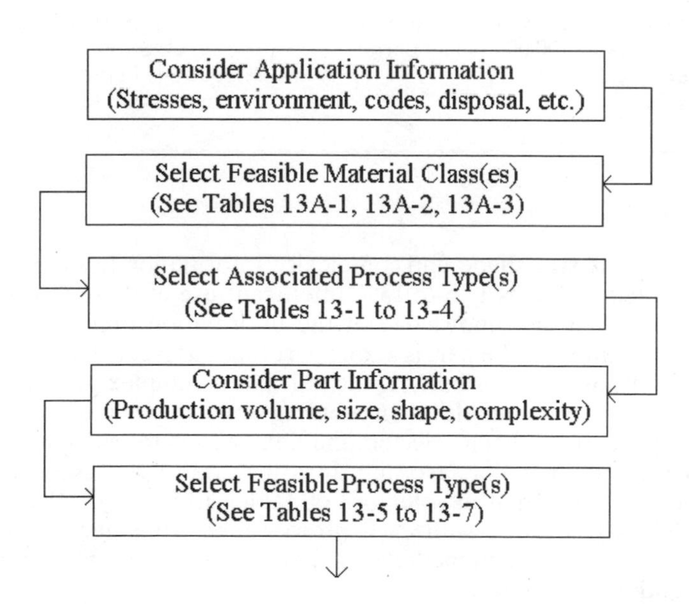

FIGURE 13.1 *A schematic illustration of the materials-first approach.*

Table 13.4 Preliminary material and process selection: Plastics: Thermosets.

Level I Material	*Level II Material Processes*	*Level III Material Examples*	*Level IV Material Examples*
Plastics	Thermosets	Alkyd	Durez 246668 (Occidental)
	Compression Molding	Alkyd Glass Reinforced	Glaskyd 2051B (Am. Cyanmid)
	Transfer Molding	*Epoxy Glass Reinforced*	Rogers 2004 (Rogers) Rogers 1961 (Rogers)
		Polyester Glass Reinforced	Durex 30003 (Occidental)

Table 13.5 Preliminary process and material selection: Wrought Processes.

Wrought Processes	*Materials*	*Production Volume*	*Part Size*	*Shape Capability*
Stamping	Aluminum Steel Copper Brass	Minimum quantity 10,000 to 20,000. For smaller volumes and simple geometries, low-cost steel rule dies can be used in place of progressive dies.	Generally less than 450 mm (18 inches). Larger sizes are done using die lines in lieu of progressive dies.	Moderate complexity is possible; however, moldings and castings are capable of producing more complex shapes.
Forging	Aluminum Magnesium Copper Steel Titanium Superalloys	For medium- and large-sized forgings a production volume of 1,000 to 10,000. For small forgings (under $^1/_2$ lb.) a production volume of 100,000 may be required.	Maximum size is generally less than 800 mm (32 inches).	Moderate shape complexity. No internal undercuts possible. External undercuts limited to simple depressions.
Extrusion	Aluminum Steel Copper Magnesium Zinc	1,000 (larger parts) to 100,000 (small parts)	From $^1/_4$ to 10 or 12 inches (6 to 300 mm) in diameter. Lengths can be very long.	Constant cross-section

part's approximate size, shape, and complexity. Information to support this selection is presented in Tables 13.5 to 13.7. Intelligent use of these tables requires a basic understanding and knowledge of the information presented in previous chapters. For example, if the part is a long part that has a constant cross-sectional area then extrusion is a possibility. If the part has a complex geometric shape (a telephone housing for example) then molding or casting is required.

Once a process has been selected, the next step is to use application information to rule out or point favorably to a material class associated with the selected process. Note that in column 2 of Tables 13.5 to 13.7, there is a set of material classes associated with each process type. Thus, if we decide to use molding to produce a telephone housing then we can select from among a thermoplastic and a thermoset.

Schematically, the process-first approach is shown in Figure 13.2.

Table 13.6 Preliminary process and material selection: Casting Processes.

Casting Processes	*Materials*	*Production Volume*	*Part Size*	*Shape Capability*
Die Casting	Generally aluminum and zinc. Brass and magnesium are also die cast.	Generally greater than 10,000.	Maximum part size usually less than 600 mm (24 inches)	Almost any shape is possible. Internal undercuts should be avoided for practical reasons.
Investment Casting	Steel Stainless Steel Aluminum Magnesium Brass, Bronze Ductile Iron	Generally less than 10,000.	Maximum size generally less than 250 mm (10 inches).	Same as die casting.
Sand Casting	All common metals	Minimum quantity between 1 and 100.	No maximum size. Size limited by carrying capacity of crane.	All shapes are possible. External undercuts are limited due to need to extract pattern from sand.
Centrifugal Casting	Most metals		Large—usually over 100 pounds.	Generally rotationally symmetrical, but nonrotational parts are possible.
Permanent Mold Casting	Aluminum Zinc Magnesium Brass	Minimum quantity about 1,000.	About the same as die casting.	About the same as die casting.
Plaster Mold Casting	Mainly aluminum and copper.	Minimum quantity about 10.	Generally limited to parts weighing less than 100 pounds.	Undercuts are difficult to provide.
Ceramic Mold Casting	All common metals.	Minimum quantity about 10.	Generally limited to parts weighing less than 100 pounds.	Undercuts are difficult to provide.

13.3 A HIERARCHICAL ORGANIZATION OF MATERIAL ALTERNATIVES—TABLES 13.1 TO 13.4

13.3.1 Introduction

At the conceptual stage, though it may occasionally happen, it is unlikely to be necessary to go so far as to propose specific materials such as 1030 hot rolled steel or 6063-T5 aluminum or Zytel 408 (a nylon 6/6 polyamide thermoplastic made by duPont). The reason we don't generally need such specificity yet is that the evaluation of the physical concept seldom depends on it. The finer discriminations can therefore generally be postponed—consistent with least commitment—until the configuration or parametric stage.

With this in mind, Tables 13.1 to 13.4 have been prepared to support the selection process by organizing materials into a hierarchy of classes. The tables organize the most commonly used materials in four levels, from very broad (e.g., metals or plastics) to very specific (e.g., Lexan 101). Here are two examples that illustrate the meaning of the four levels:

Table 13.7 Preliminary process and material selection: Plastic Processes.

Plastic Processes	*Materials*	*Production Volume*	*Part Size*	*Shape Capability*
Injection Molding	Thermoplastics (Unfilled, reinforced)	Generally greater than 10,000. However, less expensive tooling can be used for smaller production volumes. Seldom used for volumes less than 1,000.	Maximum part size usually less than 600 mm (24 inches).	Almost any shape is possible, including internal and external undercuts.
Compression Molding Transfer Molding	Thermoplastics (Unfilled, reinforced)	Same as injection molding.	Same as injection molding.	Same as injection molding.
Extrusion	Thermoplastics (Unfilled, reinforced)		Most common extruders have maximum diameters of about 200 mm (8 inches). Some are available up to 12 inches, or more.	Constant cross-sectional.
Blow Molding	Thermoplastics		Generally between 1 ounce and 1 gallon. Maximum size about 55 gallons.	Hollow thin-walled parts. Minor undercuts okay.
Rotational Molding	Thermoplastics Thermosets (some)	Low (compared to injection molding)	Limited by size of molding machine. Usually must fit within 5-foot diameter sphere.	Hollow thin-walled parts.
Thermoforming	Thermoplastic sheets and films		Usually 1 foot to 6 feet. However, some as large as 10 feet by 30 feet.	Simple flat- and boxed-shaped parts. Holes and openings cannot be formed; secondary operations required for these.

	Example A	Example B
Level I	Metal	Plastic
Level II	Wrought	Thermoplastic
Level III	Aluminum	ABS
Level IV	Aluminum Alloy 6061	ABS Magnum 213 (Dow)

13.3.2 Level I of the Hierarchy

At the highest level (I) of the class hierarchy in Tables 13.1 to 13.4, the material classes are

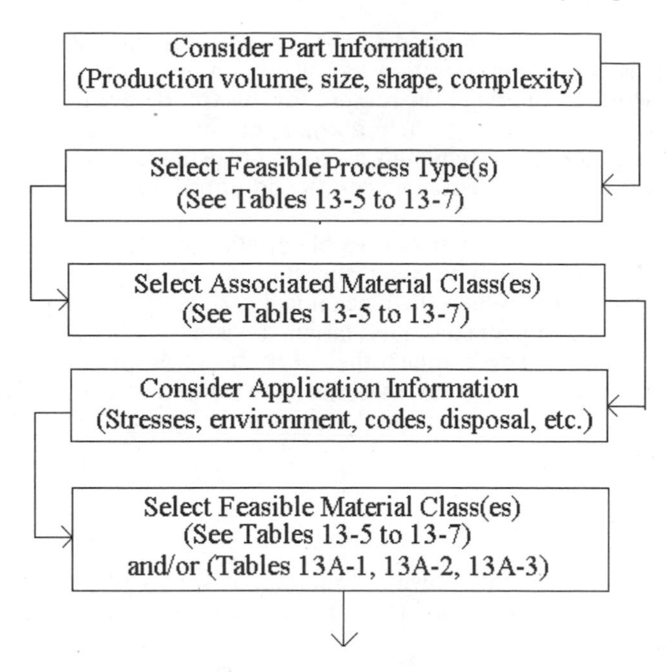

FIGURE 13.2 *A schematic illustration of the process-first approach.*

metals,
plastics,
ceramics,
wood, and
concrete

Only metals and plastics are considered below this top level in this book.

Metals vis-à-vis Plastics

Tables 13A.1, 13A.2, 13A.3, and 13A.4 list some of the most commonly used metals and plastics along with their more important mechanical and physical properties. It is evident from this data that, in general, plastics tend to be less dense (hence, lighter in weight than a comparable metal part), less costly, better able to resist corrosion, and better insulators than metals. In addition, as we learned in Chapters 3 to 5, plastics can be processed into almost any conceivable shape, with any desired surface finish (from mirror to textured)—and often require no finishing.

Metals, however, tend to have better mechanical properties. The elastic modulus of steel, for instance, is almost 60 times higher than that of an unreinforced rigid plastic, and the moduli of aluminum, zinc, and copper alloys are about 20 times higher than unreinforced plastic. (The moduli for reinforced plastics are about 3 times that of unreinforced plastics.) Also, steels have higher temperature capabilities, better thermal and electrical conductivity, and are capable of being processed by a very fast processing method: stamping. On the other hand, carbon steels usually need costly protective procedures such as pickling, galvanizing, priming, and painting to prevent rust.

Metals and plastics, especially reinforced plastics, often compete head to head as the material of choice. In recent years, plastic has been used increasingly to replace or substitute for metal. Some examples are automobile bumpers, automobile fuel tanks, lawn tractor hoods, wheel covers, instrument panels, and electric staple guns. Some consideration has also been given to the use of a plastic (long-fiber nylon) automobile. The reasons for this trend have to do with a variety of factors, including weight (to help automakers meet the U. S. Government's average fuel economy requirements), corrosion resistance, insulating effects, consolidating parts (to reduce assembly costs), and improving ergonomics (i.e., user comfort).

Because of the wide variety of possible applications and evaluation criteria, it is not possible to state any very specific rules about when plastic or metal will be the preferred choice. When the mechanical properties (strength at operating temperatures, creep, impact resistance, fatigue resistance, etc.) of plastics are adequate for the task, however, their other very desirable properties will usually make them the choice. However, the total cost (material cost, tooling cost, and processing cost) of a plastic part is not necessarily less than that of a functionally equivalent metal part.

13.3.3 Level II of the Hierarchy

At Level II in the hierarchy shown in column 2 of Tables 13.1 to 13.4, metals are classified as either cast or wrought, and plastics are subdivided into thermoplastics and thermosets.

Metals: Cast vis-à-vis Wrought

Two of the most important manufacturing considerations in selecting a material for a design application are (1) how easily can the material be formed into a finished part; and (2) how might the material's mechanical and physical properties change during the forming of the part?

In forging, for example, as we learned in Chapter 11, "Other Metal Shaping Processes," a "thin" cup-shaped part is formed by squeezing (or hammering) a metal billet between two die halves that are attached to a hammer or press. Actually, the metal in a forged part will first go through a couple of intermediate stages: an ingot is cast, then worked (i.e., wrought) by rolling or extruding the ingot into a billet, and finally it is forged.

In casting, as you recall from Chapter 6, "Metal Casting Processes," this cup-like part is produced by pouring or injecting molten metal into a die. The solidification process used to produce castings results in parts whose metal structure is usually composed of large grains, small voids, and small pockets of impurities called inclusions.

Working the metal alloy from an as-cast ingot to a wrought state—as in forging—is accompanied by a marked improvement in its mechanical properties. The associated plastic deformation refines the cast structure of the metal, closes the shrink voids that occur, and breaks up the inclusions. Since the porosity and inclusions that occur in castings are often the source of fractures, their removal in wrought metal is responsible for considerable increased ductility and strength in alloys of the same composition. For nonferrous metals this increase in ductility can be seen by comparing the Elongation Percent for cast and wrought metals in Tables 13A.1, 13A.2, and 13A.3.

The increase in tensile and yield strengths of wrought metals over those of a cast metal of the same composition is not easily seen by comparing the values shown in Tables 13A.1 and 13A.2. The difficulty comes in finding cast and wrought alloys of the same composition. It is certainly possible, however, to find a cast alloy that exhibits greater ductility and strength than a wrought alloy. Tables 13A.1, 13A.2, and 13A.3 show, for example, that a high-carbon steel casting can have superior mechanical properties to those of a low-carbon steel sheet. However, for the same composition, wrought metals do have superior mechanical properties.

It is interesting that cast and wrought carbon steel can be ordered from suppliers simply by specifying not the composition but the desired mechanical properties alone. Producers can adjust the composition of the steel to meet specified mechanical properties. Thus, given that it is possible to select a cast metal that has mechanical properties equal to a wrought metal, the question becomes—what should guide designers in the choice between cast and wrought?

Of course, we should always consider a casting process (hence cast metals) when it will be more economical. In addition, however, we should consider a cast metal when (1) parts have complex shapes and/or hollow sections; (2) large parts are being designed (sand castings can produce parts of almost any size); or when (3) overall manufacturing costs can be reduced by combining several simple parts into a single more complex part and thereby reduce assembly costs.

Conversely, we should select a wrought metal (and consequently a wrought process) when the geometry of the part is simple, and relatively fast and/or inexpensive processes such as stamping or extrusion can be used to produce the geometry.

Plastics: Thermoplastics and Thermosets

Table 13A.3 contains a representative list of some of the more commonly used thermoplastics and thermosets. It is evident from the data contained in this table that, generally, thermosets have a much higher flexure modulus. Thus, thermosets have greater rigidity than thermoplastics. In addition, some thermosets, alkyd for example, have good electrical properties (including arc resistance), low water absorption, and retention of electrical properties at elevated temperatures. Thus, alkyds are most often used for automotive distributor caps, coil caps, and circuit breakers. Other thermosets are used for electrical connectors and switch gears, and circuit boards.

Thermoplastics, however, are faster and easier to process. The cycle time is strongly dependent on wall thickness, but most parts made of thermoplastic materials have cycle times between 15 and 60 seconds. The cycle time for thermosets parts, however, ranges between 30 seconds and 5 minutes because the material cures relatively slowly during the molding process. As implied in Chapter 3, "Polymer Processing," the tooling costs for injection molding, compression molding, and transfer molding are comparable; thus the lower processing costs for injection-molded parts generally makes thermoplastics the plastic of choice when it is capable of satisfying functionality requirements.

13.3.4 Level III of the Hierarchy

At Level III, metals are subdivided into the various broad categories such as steel (carbon, alloy, and stainless), aluminum alloys, and copper alloys. Plastics at Level III are subdivided into the various classes of thermoplastics and thermosets such as polycarbonates and polyesters. Usually Level III is about the right one for specifying and evaluating conceptual design alternatives for most special purpose parts.

Metals

Tables 13A.1 to 13A.3 contain a representative sample of some of the most commonly used ferrous and nonferrous metals and metal alloys. A study of these tables shows that, in general, the nonferrous alloys (aluminum, magnesium, copper, and zinc) are much more resistant to corrosion, lighter in weight (less dense), and are better thermal and electrical conductors than steel. In addition, the nonferrous alloys are usually also easier (and less costly) to fabricate since they have lower tensile and yield strengths than steel. They are also usually easier (and less costly) to cast since they have a lower melting temperature than steel. Indeed, because of steel's high melting temperature, some casting processes (e.g., such as die casting) are not suitable for steel.

As a class, however, Tables 13A.1 to 13A.3 show that ferrous metals have higher tensile and yield strengths, and a higher modulus, than nonferrous alloys. Tables 13A.2 and 13A.3 show, for example, that:

	1015 Hot Rolled Steel	*1100 H18 Aluminum*
Modulus, E (psi)	30×10^6	10×10^6
Tensile Strength (psi)	50×10^3	24×10^3
Yield Strength (psi)	27.5×10^3	22×10^3

Thus, similar parts made of steel are stiffer than those made of aluminum or any of the other nonferrous alloys. Consequently, for similar parts made of aluminum and steel and subjected to the same loading, the steel component ($E = 30 \times 10^6$ psi) will deflect only about one-third that of the aluminum component ($E = 10 \times 10^6$ psi). Thus, aluminum components must be designed in configurations that provide additional stiffness. Alternatively, aluminum, magnesium, copper, and zinc alloys can be strengthened and (strain) hardened by heat treatment, cold rolling, or drawing.

Low carbon steels (less than 0.20% carbon) are used primarily in the production of bars, sheets, strips, and drawn tubes. Low carbon steels possess good formability (the ductility, as seen by it's percent elongation, is high) and are adaptable to low-cost production techniques such as stamping. For this reason, low carbon steels are primarily used in the production of parts for inclusion in many consumer products.

Increasing the carbon content of steel increases its strength and hardness but decreases its ductility and formability, thereby increasing the cost to process it. Mild steel (carbon content between 0.15% and 0.30%) is used in the production of structural sections, forgings, and plates. High carbon steel (greater than .80% carbon) has low ductility, but its hardness and wear resistance are high. It is thus used in the production of hammers, dies, tool bits, etc.

Alloy steels are steels that contain small amounts of elements like nickel, chromium, manganese, etc., in percentages that are greater than normal (in carbon steel). The alloying elements are added to improve strength or hardness or resistance to corrosion.

Plastic

At Level III, plastics are divided into various broad family classes such as ABS, acetal, nylon, etc. Table 13A.4 contains a representative sampling of these plastics in both their neat (i.e., unmodified by the addition of fillers or powders) state, as well as in a composite material state. In the composite state, fine fibers (0.01 mm diameter) of additives (e.g., glass fibers) are chopped to very short lengths and mixed with the resin in order to improve its mechanical properties.

Table 13A.4 shows that in every case the strength and modulus of the resin is increased by the addition of glass fibers. See, for example, how the strength and modulus of nylon 6 with 30% glass compares with that of nylon 6 without glass.

Metal powders and fibers are also mixed with plastics to improve their electrical and thermal conductivities, and to provide EMI (electromagnetic interference) shielding.

In general, the term composite includes plastics that come with additives, fillers, and reinforcing agents like glass fibers, carbon, or graphite. A composite can also have additives of metal, wood, foam, or other material layers, in addition to the resin/fiber combination. For the purposes of this chapter, composites are restricted to resins that contain reinforcing additives to improve the mechanical and thermal properties of the resins.

A study of Table 13A.4 shows that polycarbonate has the highest impact resistance of those listed. In fact, polycarbonate has the highest impact resistance of any rigid, transparent plastic. For this reason, and its relative ease of producibility, it is used in the production of many consumer products that might get dropped or struck. Examples are: computer parts and peripherals, business machine housings, power tool housings, vacuum cleaner parts, sports helmets, windshields, lights, boat propellers, etc.

Table 13A.4 shows that Nylon™ (Trademark of duPont) is resistant to oils and greases and most common solvents. It has high strength (for a plastic) and a high modulus, and it has a relatively high maximum service temperature. The coefficient of friction of Nylon is also low and for this reason it is frequently used in the production of bearings, bushings, gears, and cams.

Acetals are strong, stiff, and highly resistant to abrasions and chemicals. Like Nylon they have a low coefficient of friction when placed in contact with metals. They are used in the production of brackets, gears, bearings, cams, housings, and plumbing.

13.3.5 Level IV of the Hierarchy

At Level IV of the hierarchy in Table 13.1 to 13.4, we have listed only a small number of specific examples of the material classes in Level III. Examples of metals are listed according to specific alloy grade (e.g., ASTM A27 6535). The mechanical and physical properties of a representative sampling of metals are given in Tables 13A.1 to 13A.3. Although Level III is about the right level for evaluating conceptual design alternatives, the selection of a specific material based on specific mechanical and physical properties must be done at Level IV. The selection at Level IV can usually be delayed until the parametric stage of design, though in some cases it must be made sooner.

For plastics, the physical and mechanical properties given in Table 13A.4 are based on the generic version of a given plastic. For this reason, a range of values is generally given for each class of resins (such as ABS or Nylon 6). The existence of a range of values is partially due to the difference in properties between competing commercial brands.

As noted above, Tables 13A.1 to 13A.4 provide just a small sampling of materials and material properties. The American Society of Metals references must be consulted for a more complete listing, as well as the *Modern Plastics Encyclopedia* and *Manufacturing Handbook and Buyers' Guide*, for more complete and accurate material properties, and for additional assistance in the selection of material and process combinations for various applications.

13.4 APPLICATION ISSUES IN SELECTING MATERIALS

How and where a part is to be used—that is, the conditions of stress and the environment it will encounter when in use—obviously have a great influence on the choice of a suitable material for the part. In this section, we list and discuss briefly a number of the most common application issues. There may well be others that will be relevant in certain applications, so designers must be thorough in considering other possible issues in addition to those listed. (Remember, what you forget to consider is the biggest danger!)

13.4.1 Forces and Loads Applied

Magnitude

To a certain extent, of course, the internal stresses experienced by the materials of parts can be reduced by competent configuration design. However, it is still generally true that the larger the external forces, the greater the need for stronger materials. Thus, the magnitude of external forces is a factor relevant to the required tensile and compressive yield strength of materials. Though steels are generally the strongest, other metal alloys and plastic composites can also be made very strong.

Creep

If steady loads are applied over a long period of time (i.e., months or years), and if the temperature of the part is elevated, then creep (the slow continuous deformation of material with time) must be considered a possibility.

At room temperature, the deformation or strain for most metals (excluding lead) is a function only of the stress. As the temperature increases, materials subjected to loads that cause no plastic (permanent) deformation at room temperature may now deform plastically (creep) depending upon the combination of stress, temperature, and time. Since most metals (and ceramics) have high melting temperatures (over 2700°F for steel, about 1100°F for aluminum), they start to creep only at temperatures well above room temperatures. A general rule of thumb is that metals begin to creep when the operating temperature is greater than 0.3 to 0.4 times the melting temperature (see Ashby and Jones, 1980).

Creep is often a more important concern with plastics. In fact, many plastics creep at room temperature. Plastics have no well-defined melting point; thus, for plastics the important temperature to consider is known as the glass transition temperature, T_g. Roughly speaking, below the glass transition temperature a plastic is in a glassy, brittle state; above T_g, it is in a rubbery state. Well below T_g, plastics do not creep. For example, the glass transition temperature for Nylon 6 is about 50°C (122°F), but that for polycarbonates is about 150°C (302°F). For epoxies, T_g is usually greater than 100°C (212°F).

When designing with plastics that support loads, therefore, temperature is an especially important consideration. Unfortunately, creep data for specific plastics is still difficult to find in the open literature, and so must be sought from the resin manufacturer.

Impact Loads

Under certain conditions, even ductile materials (as defined by the uniaxial tensile test) will have a tendency to behave in a brittle manner. This is particularly true when parts are subjected to impact or sudden dynamic loads. Thus, if a part in use will be subjected to impact loads, then the impact strength of the material to be used is a relevant design evaluation issue.

Impact strengths are measured by either a Charpy test or an IZOD test (Kalpakjian, 1992). Although it is difficult to relate the results of these tests directly to design requirements, in general, materials that have high-impact resistance also have high strength, high ductility, and high toughness. These impact strengths are perhaps more useful in comparing various types of grades of metals or plastics within the same material family, and not as useful in comparing one metal with another metal or one plastic with another plastic.

Cyclic Loads and Fatigue

Many products and components—the door handle on a car, the latch mechanism on a briefcase, a tennis racquet, a paper clip, and so on, are subjected to repeated fluctuating loads. Despite the fact that the loads on these parts and products may fall well below the tensile strength of the material, and even below the yield strength of the material, the door handle falls off, the latch comes apart, the tennis racquet breaks, and the paper clip fails. Such failures are usually due to fatigue. In fact, fatigue failures are responsible for the majority of the failures in mechanical components.

If stresses in a material fluctuate, then the concept of endurance limit becomes relevant. When the average internal stress is zero, it has been found that there will be no fatigue failure if the fluctuating load amplitude is kept below a critical value called the endurance limit or endurance stress.

Tables 13A.1, 13A.2, and 13A.3 show values for the endurance limit for some metals and metal alloys. It is clear that in general steel is better than the non-ferrous metals. In fact, aluminum alloys do not have an endurance limit.

Although plastics also exhibit fatigue, for many plastics there is no well-defined fatigue strength (the stress level at which the test specimen will sustain N cycles prior to failure) or endurance limit. Thus, as in the case of metals, large numbers of cyclic stress reversals can be expected to cause failure in plastic parts even if the applied stress is low. Also, in the case of plastics, the mode of deformation and the strain rate have a much more profound effect on the results than they do for metals. In addition, high-frequency cyclic loading can cause the plastic to warm up and soften, thus further reducing the load-carrying capacity of the part.

13.4.2 Deformation Requirements

If the degree of a part's deformations as a result of the applied loads is crucial, then the resistance of material to deformation will be a factor. The relevant material property is the Modulus of Elasticity, or E. Low values of E will result in a part that, when loaded, will have large deflections (as compared to the same part/load configuration of a part made of a material with a large value of E). At times this may be desirable—for example in the design of springs and beams that are to act as springs. The value of E also affects the natural frequency of a part or assembly. For example, a beam or spring with a low value of E (i.e., low stiffness) has a lower natural frequency than one with a high modulus.

As seen from the values for the modulus of a representative sampling of metals, metal alloys, and plastics in Tables 13A.1 to 13A.3, the values of E can be summarized as follows:

Engineering Thermoplastics (no additives or fillers)	E = 0.3 – 0.5 × 106 psi (2000–3500 Mpa)
Engineering Thermoplastics (glass reinforced)	E = 1.3 – 2.0 × 106 psi (9000–14,000 MPa)
Nonferrous Metal Alloys	E = 10 – 20 × 106 psi (68,950–13,800 MPa)
Ferrous Metals and Metal Alloys	E = 30 × 106 psi (206,850 MPa)

13.4.3 Other Application Factors

In addition to the force and deformation requirements discussed above, there are many other factors related to application that may have to be considered. Among them are the following.

In-Service Temperature

Is it low? Is it elevated? Will the material be capable of functioning at these temperatures? Metals can be used at higher temperatures than plastics. Thermosets can be used at higher service temperatures than thermoplastics. The American Society of Metals references, as well as the *Modern Plastics Encyclopedia* and the *Manufacturing Handbook and Buyers' Guide* and other references, can assist us in determining the effect of temperatures on materials.

Exposure to Ultraviolet (Sunlight)

Will the optical properties of the material, especially plastic materials, change after exposure to sunlight? Once again the American Society of Metals references, as well as the *Modern Plastics Encyclopedia* and *Manufacturing Handbook and Buyers' Guide* and other references, can assist us in determining the effect of sunlight and other elements on materials.

Exposure to Moisture (Fresh Water, Sea Water)

Will the exposure be intermittent or continuous? Will the part or product be immersed in water? Again, the American Society of Metals references, as well as those of the *Modern Plastics Encyclopedia* and *Manufacturing Handbook and Buyers' Guide*, contain information that can assist us in determining which metals and metal alloys can best resist corrosion and moisture, and the effect of moisture on materials. For example, according to the *Metals Handbook* (Vol. 2), "Aluminum alloys of the 1xxx, 3xxx, 5xxx, and 6xxx series are resistant to corrosion by many natural waters." The *Handbook* goes on to state that service experience with these same wrought aluminum alloys in marine applications demonstrates their good resistance to and long life in sea water.

Chemicals (Acids, Alkalies, Etc.)

Will the part or product be subjected to chemicals? Will the materials we select be able to resist corrosion by these chemicals?

Weather

Some materials are better able to deal with atmospheric conditions than others. Ferrous alloys need protection since they corrode when exposed to air. Most aluminums have excellent resistance to atmospheric corrosion, and in many outdoor applications they do not require a protective coating. Stainless steels are suitable for exposure to rural and industrial atmospheres. Plastics, of course, do not typically corrode when exposed to atmospheric conditions. Again, the American Society of Metals references, as well as the *Modern Plastics Encyclopedia* and the *Manufacturing Handbook and Buyers' Guide*, are an excellent source of information concerning the effect of atmospheric conditions on materials.

Insulating and Conducting Requirements (Electrical and Thermal; Conductivity and Emissivity)

These can include both electrical and thermal transfers. In thermal cases, note that heat may be transferred by conduction, convection, and radiation.

Transparency and Color Requirements: Safety and Legal

These issues may include flammability, Food and Drug Administration regulations, Underwriters Laboratory standards, National Sanitation Foundation requirements, and other codes.

The list above is just a brief reminder of things that one should consider. There may be—in fact there probably are—others. Remember, don't forget to consider everything! The thing that you forget to consider will come back to haunt you.

13.5 COST

Once one or more materials have been selected then, as explained above in Section 13.2, an indirect selection of alternative processes also occurs. For example, if we decide that stainless steel must be used to produce a part of moderate geometric complexity (thin walls, external undercuts, several ribs or bosses, etc.), then we have ruled out, as discussed in Chapter 6, the possibility of using die casting. If external undercuts are present, and if these undercuts are due to projections, then forging is also eliminated as a possible means of forming the part. (If the projections are due to circular holes, then forging is still a possibility.) Thus, from Tables 13.1 and 13.2 it is seen that we are essentially left with choosing from among sand casting, investment casting, and ceramic mold casting. If, on the other hand, an aluminum alloy could have been selected, then several other casting processes would have been available to choose from, including die casting.

Although the selection of one material or alloy may be based in part on material cost, it is the total cost of a part that should be considered when making the final selection. And the total cost of the part is a function not only of material cost but also of tooling cost and processing cost. One may not always be free to select the material process combination that results in the lowest overall part cost, but elimination of that material process combination should only be done with good reason.

The American Society of Metals references, as well as those of the *Modern Plastics Encyclopedia* and *Manufacturing Handbook and Buyers' Guide*, contain a comprehensive listing of materials along with a description of designations, mechanical properties, thermal properties, composition, and typical uses, and the processes best suited for the various metal alloys and grades of plastics. Those references should be used in conjunction with Tables 13.1 to 13.7 and Tables 13A.1 to 13A.4.

The material prices listed in Tables 13A.1 to 13A.4 are incomplete. The prices of plastics, both thermoplastics and thermosets, are easy to obtain. A pricing update appears each month in the trade journal *Plastics Technology*. Metal prices are more difficult to obtain. A partial and incomplete list of base metal prices can be found in each issue of *American Metal Market*. The actual cost of a metal depends, however, on metallurgical requirements, dimensions and shape (bar, plate, channel, etc.), tolerances, surface treatment, thermal treatment, and quantity. The actual price of a metal can only be obtained by obtaining a quotation from a vendor.

13.6 PART INFORMATION IN SELECTING PROCESSES—TABLES 13.5 TO 13.7

The number of possible manufacturing process alternatives, though not as huge as the number of material class alternatives, is nevertheless substantial. Tables 13.5 to 13.7 contain a list of the more commonly used processes together with the associated material classes and data about practical production volumes, part sizes, shapes, and approximate complexity.

The first group of processes listed in Table 13.7 are melting and molding processes that can be used with both thermoplastics and thermosets. This group of processes offers probably the greatest flexibility in geometric shape complexity. Because of the faster processing times achievable when using thermoplastics, injection molding is generally the process of choice for "engineering type" plastic parts.

The group of processes listed in Table 13.6 are casting processes. Casting also offers a great deal of flexibility in shape complexity but, because of the difficulty of producing internal undercuts, they offer somewhat less shape capabilities than molding. However, as we have learned in Chapter 6, "Metal Casting Processes," undercuts, especially internal undercuts, should be avoided as much as possible. Some casting processes can be used to cast almost any metal (sand casting, for example) while others, such as die casting, can be used only to cast a few metals. Because die casting has a faster cycle time than other casting processes, and because it is more automated than other processes, it is the casting process of choice when aluminum, zinc, or magnesium will satisfy the necessary material requirements.

The group of processes listed in Table 13.5 are the wrought processes of stamping, forging, and extrusion. These processes offer less flexibility in our choice of shapes; however, amongst all manufacturing processes, stamping is the fastest. Based on material cost and the mechanical properties of metals, low-carbon cold rolled steel is generally the material choice for stampings.

For each process listed in Tables 13.5 to 13.7 a representative list of the materials that can be formed by the process, along with minimum production volumes, part sizes, and a rough guideline as to the geometric shape capabilities of each process, is given. Thus, based on considerations of size, production volume, and shape, a preliminary process selection (and, indirectly, a preliminary material selection) can be made from the list contained in Tables 13.5 to 13.7.

In using Tables 13.5 to 13.7, designers involved in the design of parts for consumer products should keep in mind that from a strictly economic point of view, a product is not likely to be successful unless it has a minimum production volume of 10,000 to 20,000. The most economic processes to use at these production volumes are injection molding, die casting, and stamping.

Although forging can be used at production volumes as low as 5,000, it is usually more costly. Forgings are typically used only when a part is to be subjected to high shock and fatigue, often as parts of machines.

Thus, from a practical point of view, we will in a large majority of designs be faced with selecting from among a cold rolled steel stamping, an aluminum die casting, and an engineering thermoplastic molding.

Part size can also eliminate processes from consideration. For example, due to machine size limitations, most die castings, plastic moldings, and stampings are less than 450 mm to 600 mm in length. Most forgings are less than 800 mm in length. Parts larger than this are often fabricated by joining or fastening together standard shapes (sheets, tubes, beams, etc.) to form the necessary geometry.

13.7 EXAMPLES

13.7.1 Example 1—Process-First Approach

Figure 13.3a shows a drawing of a generic type of small handheld stapler of the sort we might typically carry around in a briefcase or purse. The maximum dimen-

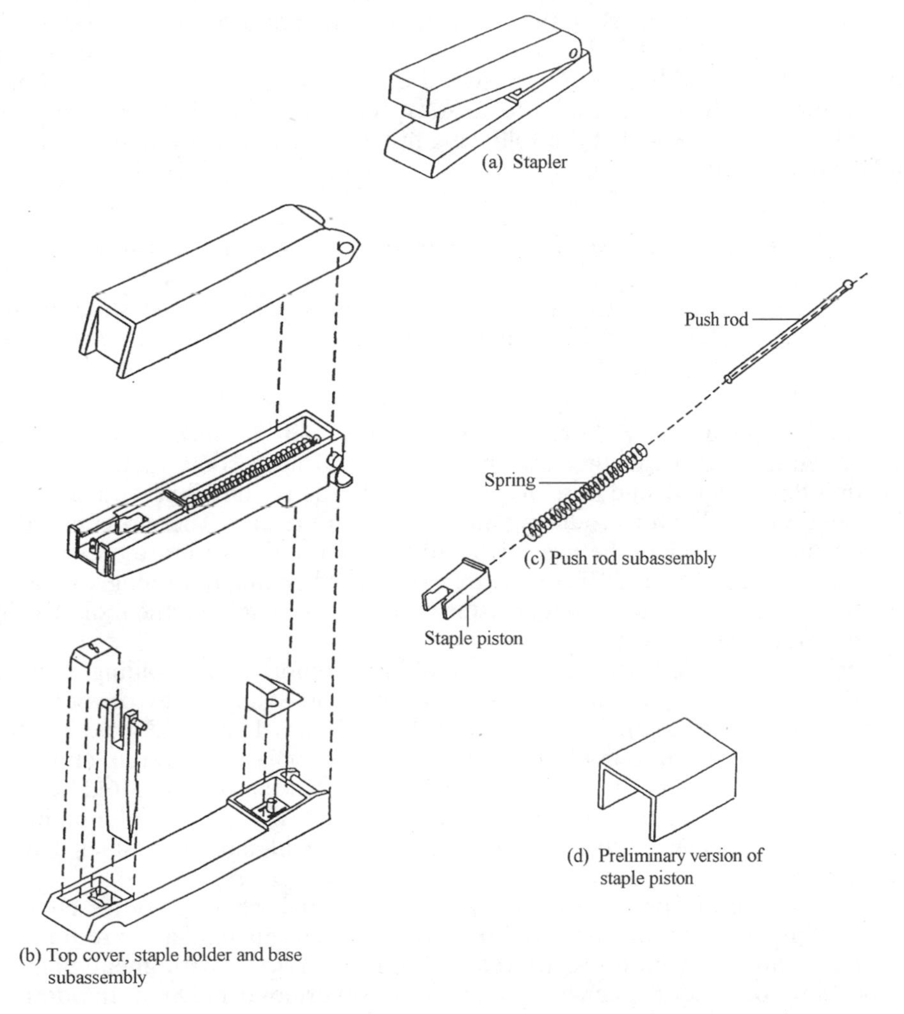

FIGURE 13.3 *Stapler for Example 1—as drawn by students as part of a DFM project.*

sion of the stapler is less than 4 inches (100mm). Figure 13.3b shows the top cover, base, and staple holder subassemblies of one brand of stapler on the market. For the purposes of this example, we focus on a portion of the staple holder subassembly, namely, the push rod subassembly shown in Figure 13.3c. Various versions of this subassembly are found in all staplers, since its function is to push the staples forward as they are used.

Let us assume that we are involved in the design (or redesign) of a small stapler that is to compete with the one discussed above. Let us also assume that we are at the early stage of the design and that the materials and processes for the various components have not yet been selected. The exact geometrical shapes of the components have not been determined. For the purposes of this example we will concentrate on the preliminary selection of a process and materials combination to produce an alternative design for the staple piston shown in Figure 13.3c.

A study of various staplers will show that a wide variety of staple piston geometries exist. In general, however, at the early design stages they are all basically a simple U-shaped part similar to the one shown in Figure 13.3d. The exact geometrical configuration of this part, as well as the other components in the stapler, will depend in part on the processes and materials chosen to produce the various components. Thus, let's determine the alternative material/process combinations that could be used to produce a U-shaped part for this application.

From a study of Tables 13.5 to 13.7, as well as our knowledge of the previous chapters, we know that the following processes are all capable of producing a U-shaped part:

injection molding, compression molding, transfer molding, extrusion—all of which are plastic forming processes, and

die casting, investment casting, sand casting, permanent mold casting, plaster mold casting, ceramic mold, stamping, forging, and extrusion—all of which are metal forming processes.

Let's assume that we have estimated that we will produce about 20,000 of these staplers. Let us further assume that we will need some method of connecting the push rod, and possibly the spring, to the piston. Thus we may possibly need the facility for producing an undercut in the part. Therefore, based on production volume and the possible need to produce one or more undercuts in the part, Tables 13.5 to 13.7 indicate that we should eliminate forging, extrusion, investment casting, sand casting, plaster mold casting, and ceramic mold casting from further consideration.

We are now left with injection molding, compression molding, transfer molding, die casting, permanent mold casting, and stamping as possible processes for producing our U-shaped part. From Tables 13.5 to 13.7 we see that using injection molding implies the use of a thermoplastic material, and using compression molding or transfer molding implies that thermosets could be used. Die casting and permanent mold casting permit the use of aluminum or zinc. Stamping allows us to consider the use of wrought aluminum, steel, copper, or brass.

Staplers of this type are typically used in normal school or office environments, thus they are not subjected to corrosive environments and are not subjected to high temperatures. Therefore, from a service environment point of view, there does not appear to be a need for a thermoset polymer. In addition, since from our knowledge of Chapter 3, "Polymer Processing," we are aware that the use of thermosets requires considerably longer processing times than thermoplastics, we eliminate the use of thermosets from further consideration. From Tables 13.1 to 13.4 (or Tables 13.5 to 13.7) we see that this in turn eliminates from further consideration the use of compression molding and transfer molding.

Since we are not concerned about corrosion, Tables 13A.1 to 13A.3 indicate that we can also eliminate from consideration the need to use more expensive materials such as stainless steel, brass, and copper. We are basically left, therefore, with selecting either a thermoplastic (and hence the use of injection molding), or aluminum or steel. The use of aluminum implies the use of either die casting or stamping, and the use of steel implies using stamping.

In summary, then, by the use of Tables 13.5 to 13.7 and our knowledge of previous chapters, we arrive at the following possible process/material combinations for producing the staple piston, namely,

injection molding/thermoplastic
die casting or permanent mold casting/cast aluminum
stamping/wrought aluminum or wrought steel

To reduce the above list of material/process combinations still further, other factors must be taken into consideration. Among these factors is, of course, cost. Later, in Section 13.8, we will learn that for parts whose wall thickness is less than 2 mm (0.08 inches), injection molding is generally less costly than die casting. This then allows us to reduce the process selection to one of either injection molding or stamping. At this early stage in the design process it would be difficult to decide which of these two processes would result in a less costly part. However, if cost is the overriding issue, then the technique discussed in Section 13.9 could be used to help decide between injection molding and stamping.

13.7.2 Example 2—Material-First Approach

Let us approach the materials/process selection for this same stapler part using a materials-first approach.

Again we assume that there is no need to consider the use of materials that are particularly good at high temperatures or in a corrosive environment. Thus, as explained in the previous example, Tables 13A.1, 13A.2, 13A.3, and 13A.4 allow us to eliminate from consideration the need to use materials such as copper, brass, stainless steel, and thermosets. This leaves us basically with thermoplastics, aluminum (cast and wrought), and steel (cast and wrought).

From Tables 13.1 to 13.4, and our knowledge of Chapter 3, "Polymer Processing," we see that using thermoplastics implies that injection molding and extrusion are capable of providing the necessary U-shape to the part. We also see that the use of aluminum implies that sand casting, investment casting, ceramic mold casting, forging, extrusion, and stamping can be used. Using aluminum also allows us to add for consideration the use of die casting and permanent mold casting to our list of possible processes.

Once again, we assume a production volume in the vicinity of 20,000 and the possible need to produce one or more undercuts in order to assemble the push rod and spring to the piston. Hence, based on the information provided in Table 13.5 to 13.7, we eliminate from further consideration the processes of extrusion, forging, plaster mold casting, and ceramic mold casting.

Thus, as in the previous example, we arrive at the same conclusion, namely, that at this stage of the design the following material/process combinations are possible for producing a U-shaped part:

thermoplastics/injection molding
cast aluminum/die casting or permanent mold casting
wrought aluminum/stamping
wrought steel/stamping

13.7.3 Example 3—Material/Process Selection for a Fishing Reel Crank Handle

Figure 13.4 shows an assembly drawing of a fairly common type of fishing reel. One of the subassemblies that makes up this reel is the crank handle subassembly, which consists of a crank, two knobs, a nut, and two rivets for connecting the knobs to the crank. For the purposes of this example we will concentrate on the crank itself. It will be assumed that some 25,000 of these reels are to be produced and that the reel is intended primarily for use in streams and lakes where fish are relatively small.

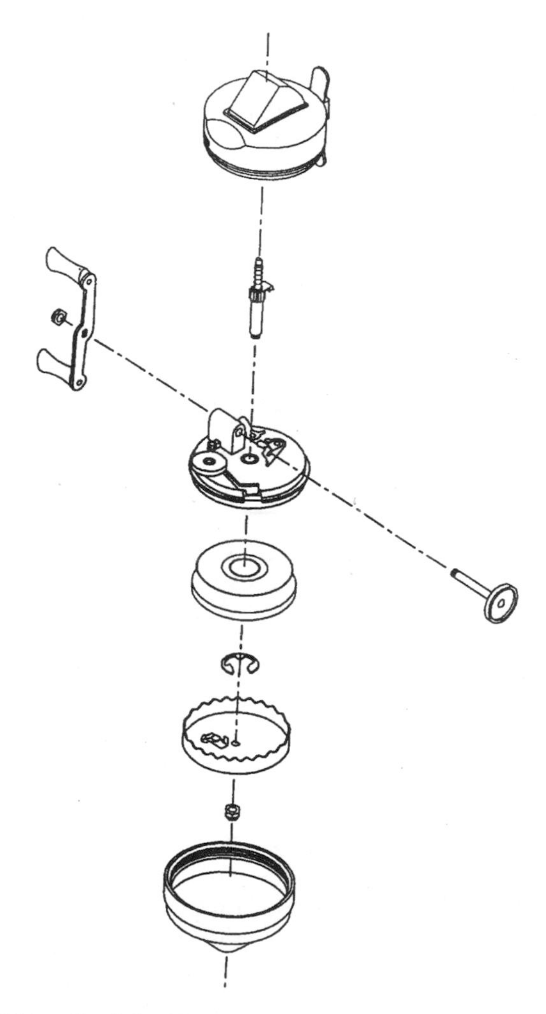

FIGURE 13.4 *Fishing Reel for Example 1—as drawn by students as part of a DFM project.*

The forces to which the crank will be subjected are not large (certainly under 20 pounds). The material should be one that will not rust. For these reasons, it appears that the part could be made from either a plastic or a metal that does not easily rust.

In spite of the fact that injection molding may be the least costly method for producing the crank, let us assume that for marketing reasons we reject its use because of the perception by many fishermen that metal is better.

Although the reel is not primarily intended for use in salt-water fishing, we need to consider the possibility that it may at times be used for such a purpose. Thus, the metal to be used should be one that would resist salt-water corrosion. From Tables 13A.1 to 13A.3 we conclude that aluminum or stainless steel, cast or wrought, could be used. From Tables 13.5 to 13.7, and our knowledge of Chapter 6, "Metal Casting Processes," we see that any one of a number of casting processes could be used to produce the part from either aluminum or stainless steel. In addition, we know that stamping could also be used to produce the same geometry.

From our earlier discussions of stamping and the various casting processes it is easy to reason that for a production volume of 25,000, reasonably high production rates (with their lower processing costs) may be desirable. Thus, from the point of view of higher production rates, die casting and stamping are the processes of choice. Thus, it is concluded that the following material/process combinations could be used to produce the crank, namely

aluminum/die casting
aluminum/stamping
stainless steel/stamping

As discussed earlier, other factors must be considered in order to reduce this list still further. Among these factors would be the selling cost of the reel and the material to be used for the remaining components. For example, all component parts should be capable of resisting corrosion to about the same level. Thus, to make some components of aluminum and some of stainless steel would probably not make sense. From Table 13A.3 we can also see that stainless steel is considerably more costly than aluminum. Thus, if cost is a major factor (and it usually is), then aluminum and stamping are probably the material and process of choice.

13.7.4 Example 4—Material/Process Selection for a Contact Lens Case

As our final example, let's consider we are involved in the design of a small storage case for an individual's contact lenses. These cases have two separate cylindrical compartments, one for each of two lenses. There is a cap for each of the compartments, which screw onto the main case and are tightened by hand, to prevent leakage of the neutral saline solution used for soaking the lenses during sterilization. The sterilization temperature of the case is controlled at 220°F.

Let's begin the initial materials/processes selection using a process-first approach. We know that the case must be designed as a box-shaped enclosure with two separate compartments. Thus, geometry must be either fabricated by assembling individual components or it must be produced as a one-part assembly.

As we learned in Chapter 12, "Assembly," to minimize assembly costs, we should design the case with a minimum number of parts. Again, from our knowledge of previous chapters we know that the only processes capable of producing the main body of a lens case with two cup-shaped enclosures (minus the two covers) as a single part are casting, injection molding, transfer molding, and compressions molding.

From Tables 13.5 to 13.7 we see that compression molding and transfer molding, processes that are slower than injection molding, imply the use of a plastic thermoset. The temperature and corrosion environments in this case are not severe enough to warrant the need for a thermoset; thus, we reject the use of compression molding and transfer molding. We are left with deciding from among injection molding a thermoplastic material or casting a metal.

Because of the need to resist corrosion by saline solution, Tables 13A.1, 13A.2, and 13A.3 indicate that aluminum or stainless steel would be the metal of choice. If this case is to sell for something in the order of a couple of dollars, then with a comparison of the material costs (shown in Tables 13A.1 to 13A.4) for aluminum, stainless steel, and engineering thermoplastics, we can eliminate stainless steel from further consideration. Thus, we once again arrive at the choice between an aluminum die casting or an injection-molded thermoplastic.

Once again, based on information provided in Section 13.8, it is easy to conclude that, based on price, injection molding is the process of choice. Consequently, a thermoplastic is the material of choice.

The choice of which particular thermoplastic to use must be based on additional considerations such as:

The impact strength required. In this case, normal impact strength is needed. That is, normal handling, occasional dropping, but no sharp or heavy blows.
The material will come into contact with moisture/steam at 220°F.
With the caps screwed on for eight to ten hours at 220°F, creep must be considered a factor.
The flexural modulus is somewhat important to keep the threads from deforming, but again there are no large external loads to consider.
Must have FDA approval that is the same as food additives.

The information and data provided in Table 13A.4 of this chapter is insufficient to make a selection of a specific thermoplastic to satisfy all of these requirements. Table 13A.4 is simply a sampling of a few thermoplastics with only some of their properties. To select the specific material to satisfy all of the functional requirements, we must refer to a reference such as the American Society of Metals' *Engineered Materials Handbook*, Vol. 2.

13.8 INJECTION MOLDING VERSUS DIE CASTING

13.8.1 Introduction

The examples discussed in the previous section indicated that we are often faced with choosing between (1) a thermoplastic injection molded part, or (2) a geometrically similar die-cast aluminum part. The purpose of this section is to provide additional information, based solely on cost, to assist in the selection between two such parts. It is assumed here that the size, shape, wall thickness, and dimensions of subsidiary features (such as holes) are identical.

In the analysis that follows the subscript "a" will refer to an aluminum die casting, and the subscript "p" will refer to a plastic injection molding.

13.8.2 Relative Cost Model

We define ΔK_{ap} as follows:

ΔK_{ap} = [Cost to produce a part as an aluminum die casting] –
[Cost to produce the same part as a themoplastic injection molding]
(Equation 13.1)

or

$$\Delta K_{ap} = K_{ta} - K_{tp} = \left(K_{ma} + \frac{K_{da}}{N} + K_{ea}\right) - \left(K_{mp} + \frac{K_{dp}}{N} + K_{ep}\right)$$
(Equation 13.2)

where

K_{ta} = the total cost to produce the part as an aluminum die casting
K_{tp} = the total cost to produce the part as a thermoplastic injection molding
K_{ma}, K_{mp} = material cost of the part
K_{da}, K_{dp} = tool or die cost required to produce the part
N = production volume
K_{ea}, K_{ep} = processing cost for the part

If ΔK_{ap} is greater than zero, then it is more economical to produce the part as an injection-molded part.

From Chapters 4 and 7, we can see from a study of the classification systems for basic tool complexity for injection molding and die casting (Figures 4.1 and 7.4), respectively, that the tooling costs for moldings and die castings are essentially the same. Thus, Equation 13.2 reduces to:

$$\Delta K_{ap} = K_{ta} - K_{tp} = (K_{ma} + K_{ea}) - (K_{mp} + K_{ep}) \quad \text{(Equation 13.3)}$$

Now define

$$\Delta K_m = (K_{ma} - K_{mp}) \quad \text{(Equation 13.4)}$$

Hence, Equation 13.3 becomes

$$\Delta K_{ap} = \Delta K_m + (K_{ea} - K_{ep}) \quad \text{(Equation 13.5)}$$

Unfortunately, metal prices published in the open literature are sparse, and not as reliable as plastic prices. In general, because metal prices depend on a multitude of factors, true metal prices can only be obtained by direct quotation from vendors when the metals are needed. In spite of this, however, from the data provided in Tables 13A.1 to 13A.4, it can be seen that the cost per unit volume for an unreinforced engineering thermoplastic is less than that of aluminum. It can also be seen that the costs of reinforced plastics are in the same price range as aluminum. Thus, we can at least say that ΔK_m is generally a positive quantity.

Now Equation 13.5 shows that if the processing cost for injection molding is less than the processing cost for die casting—that is, if

$$K_{ep} < K_{ea} \quad \text{(Equation 13.6)}$$

then ΔK_{ap} is certainly positive and it is less costly to injection mold than to die cast.

We now introduce the two reference parts introduced earlier, one for injection molding and one for die casting. We will use these reference parts to compare processing times (see Tables 5.1 and 7.6) and, hence, processing costs for alternative competing designs. Let K_{eao} and K_{epo} represent the processing cost for the aluminum die-cast reference part and the injection-molded reference part, respectively.

If, as before (Equations 5.2 and 7.10)

$$C_{ep} = \frac{K_{ep}}{K_{epo}} \qquad \text{(Equation 13.7)}$$

and

$$C_{ea} = \frac{K_{ea}}{K_{eao}} \qquad \text{(Equation 13.8)}$$

where C_{ep} and C_{ea} represent the cost of a part relative to the reference part, thus, from Equation 13.6, if

$$C_{ep}K_{epo} < C_{ea}K_{eao}$$

or if

$$\frac{C_{ep}}{C_{ea}} < \frac{K_{eao}}{K_{epo}} = \frac{(C_{ho}t_o)_{Al}}{(C_{ho}t_o)_{plastic}} \qquad \text{(Equation 13.9)}$$

then ΔK_{ap} is positive and it is less costly to injection mold than to die cast.

Based on data provided in Chapter 5 and 7, Tables 5.1 and 7.6, which are reproduced below as Tables 13.8 and 13.9, it can be shown that

$$\frac{(C_{ho}t_o)_{Al}}{(C_{ho}t_o)_{plastic}} = \frac{62.57(\$/hr)(17.23s)}{27.53(\$/hr)(16s)} = 2.5$$

Thus, Equation 13.9 reduces to the following condition:

When

$$\frac{C_{ep}}{C_{ea}} < \frac{(C_{ho}t_o)_{Al}}{(C_{ho}t_o)_{plastic}} = 2.5 \qquad \text{(Equation 13.10)}$$

then it becomes more economical to mold than to die cast.

That is, when the relative processing cost for the injection-molded version of the design is less that 2.5 times the relative processing cost for the die-casting version of the design, then it is more economical to injection mold.

Table 13.8 Relevant data for the injection-molded reference part.

Part Material	*Polystyrene*
Material cost, K_{po}	1.46×10^{-4} cents/mm^3
Part volume, V_o	1244 mm^3
Cycle time, t_o	16 s
Mold machine hourly rate, C_{ho}	\$27.53

Table 13.9 Relevant data for the die-cast reference part.

Part Material	*Aluminum*
Material cost, K_{po}	0.0006 cents/mm^3
Part volume, V_o	1885 mm^3
Cycle time, t_o	17.23 s
Mold machine hourly rate, C_{ho}	\$62.57

Table 13.10 Machine tonnage and relative hourly rate based on data published in *Plastic Technology* or obtained from die-casting vendors (June 1989).

Machine Tonnage	*Injection Molding Relative Hourly Rate, C_{hrp}*	*Die Casting Relative Hourly Rate, C_{hra}*
<100	1.00	1.00
100–299	1.19	1.11
300–499	1.44	1.25
500–699	1.83	1.47
700–999	2.87	2.02
>999	2.93	2.11

In Chapters 5 and 7 (Sections 5.15 and 7.10) we showed that machine hourly rates and, consequently, relative machine hourly rates C_{hrp} and C_{hra} are functions of the machine tonnage required to mold or die cast the part. The machine tonnage, in turn, is a function of the projected area of the part normal to the direction of die closure. In both injection molding and die casting, the tonnage required is approximately 3 tons per square inch. Thus, comparable parts require comparable sized machines. The relative hourly rates for the two machines, however, are not equal. Table 13.10 shows how C_{hrp} and C_{hra} vary with machine tonnage. It can be shown that $0.70 \le \frac{C_{hra}}{C_{hrp}} \le 1.0$. Thus:

when

$$\frac{t_{rp}}{t_{ra}} < 0.7(2.50) = 1.75 \qquad \text{(Equation 13.11)}$$

then it is more economical to injection mold.

13.8.3 Results

In Chapter 5, "Injection Molding: Total Relative Part Cost," we introduced the concept of partitionable and non-partitionable parts. Very briefly, partitionable parts are those parts that can be easily and completely divided (except for add-ons like bosses, ribs, etc.) into a series of simple plates (see Figure 5.12, for example). Non-partitionable parts are those parts whose complex geometries make it difficult to partition into a series of simple plates. Figure 5.9 of Chapter 5 shows an example of a non-partitionable part.

Partitionable parts have exterior plates and interior plates. Simply stated, the exterior plates of a part are its peripheral side walls, whereas interior plates are nonperipheral side walls and the base of the part. Methods for partitioning parts and classifying plates are discussed in detail in Chapter 5. For now we concentrate only on the results relevant to comparing costs.

Using the data contained in Chapter 5, "Injection Molding: Total Relative Part Cost," and Chapter 7, "Die Casting: Total Relative Part Cost," along with the data provided in Table 13.10, the plots shown in Figures 13.5 through 13.9 were obtained. In obtaining these results, it is assumed that the plate controlling the cycle time is always an external plate, and that tolerances are easy to hold. Since internal walls do not play a major role in determining the cycle time for injection-molded parts but do often control the cycle time for die castings, the results depicted in Figures 13.5 to 13.9 are, from the point of view of injection molding, conservative. In addition, the effect of part surface quality or surface finish is ignored. This is done since it is difficult to compare moldings and die castings based on part surface finish requirements. Also, since engineering thermoplastics cannot generally be used for parts whose wall thickness is greater than 5 mm, the results shown in these figures are restricted to wall thickness less than 5 mm. In general, when the wall thickness is greater than 5 mm, one would probably use either foamed materials or metals.

We see from an examination of Figures 13.5 to 13.9, that for parts whose cycle time is controlled by relatively thin plates (wall thickness less than 2 mm), it is always less costly to mold the part. As the size of the part increases (i.e., as C_{hrp}/C_{hra} increases) and the wall thickness increases, then the certainty that injection molding is less costly becomes less.

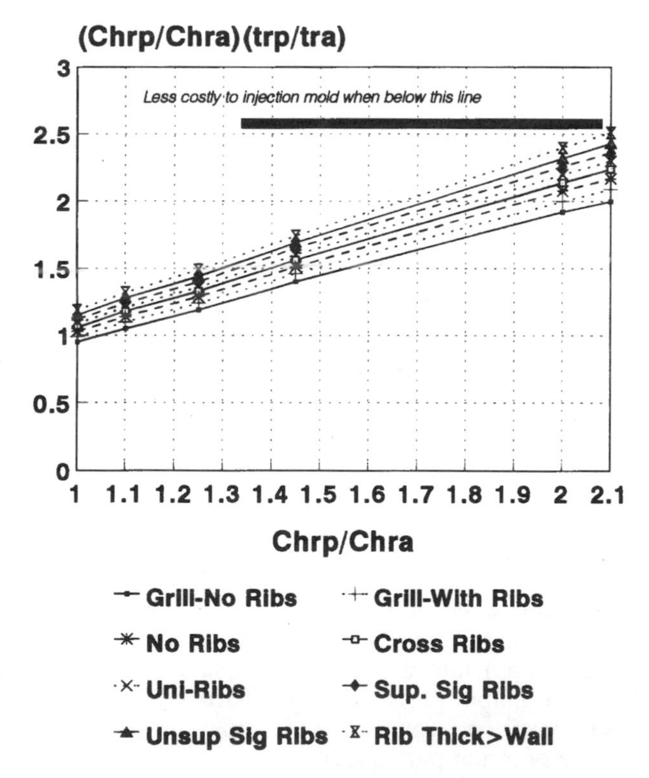

FIGURE 13.5 *Injection molding versus die casting, partitionable parts: 1 mm < wall thickness < 2.01 mm.*

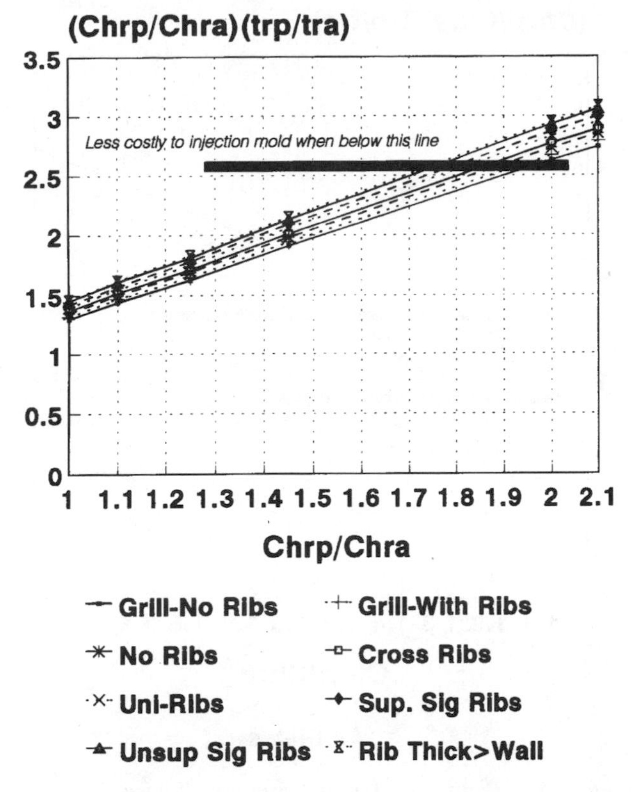

FIGURE 13.6 *Partitionable parts: 2 mm < wall thickness < 3.01 mm.*

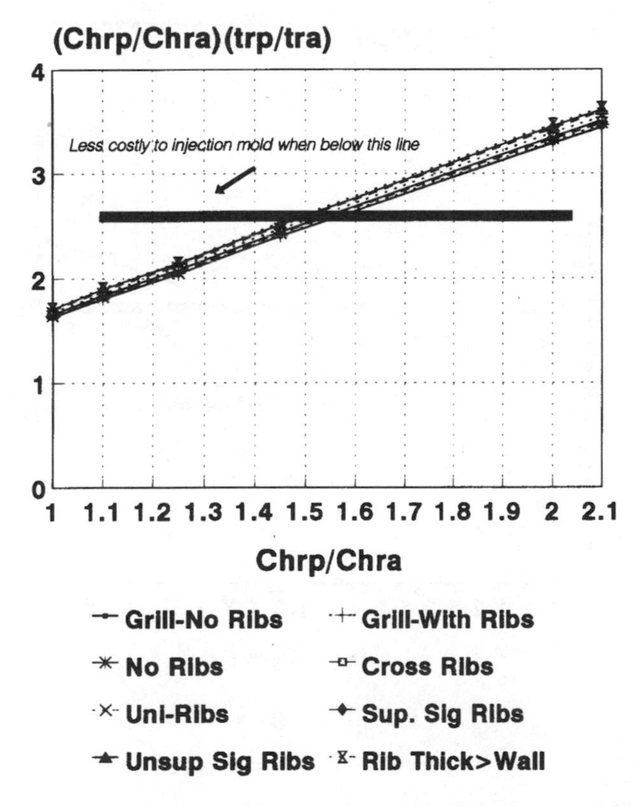

FIGURE 13.7 *Partitionable parts: 3 mm < wall thickness < 4.01 mm.*

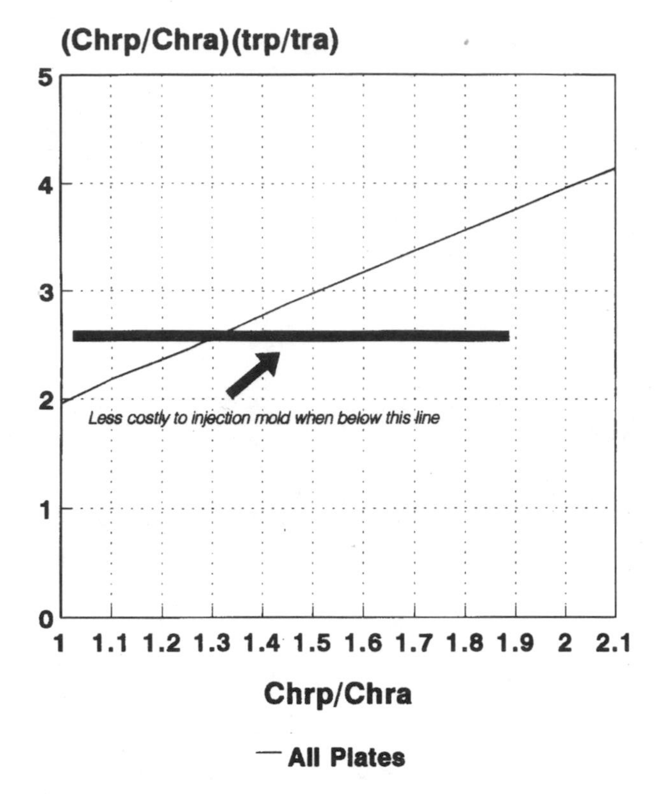

FIGURE 13.8 *Partitionable parts: 4 mm < wall thickness < 5.01 mm.*

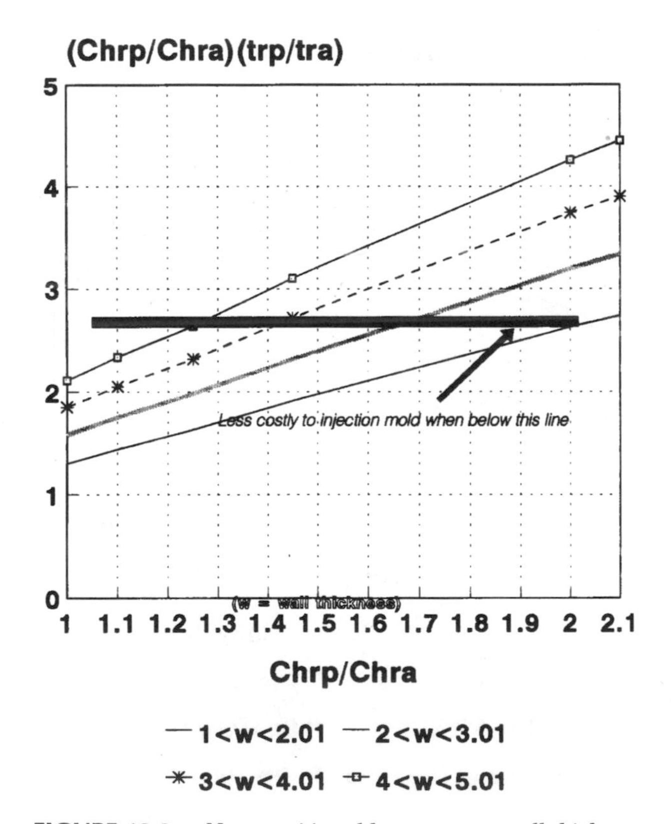

FIGURE 13.9 *Non-partitionable parts; w = wall thickness.*

For values of $(C_{hrp}/C_{hra})(t_{rp}/t_{ra}) > 2.5$ it is still not necessarily true that an injection-molded part is more costly. In this situation, Equation 13.5 must be used in place of Equation 13.10 or Equation 13.11 in order to determine which of the two processes is more economical. In this case, the exact cost of the material must be known and, for the die-cast version, the nature (internal or external) of the plate controlling the cycle time must be determined.

13.8.4 Example—Injection Mold or Die Cast a Fishing Reel Crank Handle?

Consider the crank shown in Figure 13.4. In section 13.7.3 the possibility of injection molding this crank was rejected for marketing reasons. We assume here that we wish to reexamine this issue, but strictly from the point of view of manufacturing cost. It is assumed that functionally a crank with a wall thickness of 3.5 mm will suffice.

The crank is a slender part without ribs. Thus, from Figures 5.2 and 7.11 we find that for a wall thickness of 3.5 mm, $t_{rp} = 1.70$ and $t_{ra} = 1.48$.

A ballpark figure for the projected area of the crank normal to the direction of mold closure is 1875 mm^2 or 3 in^2. Thus—recalling that about 3 tons per square inch is required to mold or cast a part—the tonnage required is about

$$F = 3Ap = 3(3) = 9\,\text{tons}$$

Hence, from Table 13.10 we see that the relative machine hourly rates for both versions of the design are 1.0. Hence,

$$C_{hrp}/C_{hra} = 1,\ (C_{hrp}/C_{hra})(t_{rp}/t_{ra}) = 1.15$$

and from Equations 13.11 or 13.10 we conclude that it is more economical to mold this part from plastic than to die cast it from aluminum.

13.9 INJECTION MOLDING VERSUS STAMPING

13.9.1 Introduction

In the previous section we determined, for the case of two geometrically identical parts, when it would be more economical to mold than to die cast. The purpose of this section is to discuss, for two "functionally" equivalent geometries, when it is more economical to injection mold a part and when it is be more economical to stamp a part that has a functionally equivalent geometry. We use the term functionally equivalent here because it is seldom possible to stamp a part that has exactly the same geometry as an injection-molded part. The differences in the physics of the two processes make certain geometries more difficult to produce in stamping than in injection molding. For example, it is simple to form an injection-molded box-shaped part with each of the walls connected to the adjoining wall. In stamping, while a box-shaped part could be drawn to the same geometry, it would be easier to wipe form the walls as discussed in Chapter 8, "Sheet-Metal Forming." In this situation the adjoining walls would not be connected to each other; they would, however, be connected to the base. Thus, when a box-shaped part whose walls are not connected will satisfy the functional requirements, then we can think of these two parts as being "functionally" equivalent.

The database provided in previous chapters dealing with injection molding, die casting, and stamping are sufficiently precise to allow a cost comparison between alternative designs within a given manufacturing process. Because of the similarities between injection molding and die casting they are also sufficiently precise to allow a cost comparison between two geometrically similar parts, one produced by injection molding and the other produced by die casting. Unfortunately, they are not sufficiently precise to permit an exact comparison of alternative designs between two different processes. This is due in part to the fact that, in using the group technology approach of Chapters 4, 5, and 7, each cell of the matrices contains groups of parts that are similar but not identical. The relative tooling cost data provided is for a representative part that would fall within that cell. The absolute tooling costs for all parts that would fall within that cell, however, are slightly different.

Although there are limitations to using the data base provided in previous chapters to determine whether or not a part is cheaper to mold or to stamp, it nevertheless does provide us with a quick first approximation based solely on cost. Hu and Poli, in their 1997 articles, provide a more precise technique for determining whether or not a part is cheaper to mold or to stamp.

In the analysis that follows the subscript "s" will be used to indicate stamping, and the subscript "p" will indicate injection molding.

13.9.2 Relative Cost Model

In Chapter 5 we found that the total cost of an injection-molded part, K_{tp}, was given by the following equation, namely,

$$K_{tp} = K_{op}C_{rp} = (C_{mp}f_{mp} + (C_{dp}/N)f_{dp} + C_{ep}f_{ep})K_{op} \qquad \text{(Equation 13.1)}$$

where

$$f_{mp} = K_{mop}/K_{op}$$
$$f_{dp} = K_{dop}/K_{op}$$
$$f_{ep} = K_{eop}/K_{op}$$

represent the ratio of the material cost (K_{mop}), tooling cost (K_{dop}), and processing cost (K_{eop}) of the reference part to the total manufacturing cost of the reference part (K_{op}), and

$$C_{mp} = K_{mp}/K_{mop}$$
$$C_{dp} = K_{dp}/K_{dop}$$
$$C_{ep} = K_{ep}/K_{eop}$$

are, respectively, the material cost (K_{mp}), tooling cost (K_{dp}), and equipment operating cost (K_{ep}) of an injection-molded part relative to the material cost, tooling cost, and equipment operating cost of the injection-molded reference part. The production volume of the plastic part is denoted by N.

If we assume, as in Chapter 5, that the production volume of the reference part is 7970, then the cost of the reference part is $1 and the total cost of an injection-molded part is given by Equation 5.12, namely,

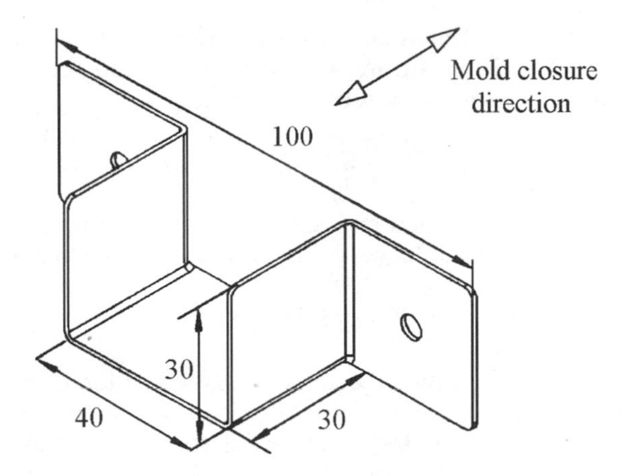

FIGURE 13.10 *A bracket producible by either injection molding or stamping. Wall thickness is 3 mm for the molded version and 1 mm for the stamped version.* $L_{ul} = 100\,mm$, $L_{ub} = 60\,mm$.

$$K_{tp} = C_{rp} = 0.00182C_{mp} + (6980/N)C_{dp} + 0.1224C_{ep} \quad \text{(Equation 5.12)}$$

Similarly, for a stamped part the total cost of the part is given by Equation 10.12, namely,

$$K_{ts} = C_{rs} = 0.0215C_{ms} + (6561/N)C_{ds} + 0.00177C_{es} \quad \text{(Equation 10.12)}$$

where once again the equation applies when the cost of the reference stamping is $1.

The processing cost of a stamped part, compared to the processing cost of an injection-molded part, is negligible. Hence, Equation 10.12 reduces to the following:

$$K_{ts} = C_r = 0.0215C_{ms} + (6561/N)C_{ds} \quad \text{(Equation 13.2)}$$

If the difference between the cost to produce the part as a thermoplastic injection molding and the cost to produce the part as a stamping is greater than zero, that is if $(K_p - K_s) > 0$, then it is less costly to produce the part as a stamping.

Example 5

Figure 13.10, taken from Hu and Poli's "To Injection Mold, to Stamp, or to Assembly? Part I: A DFM Cost Perspective," shows the drawing for a mounting bracket producible by either injection molding or stamping. The injection-molded version is made of nylon 6 and has a wall thickness of 3 mm. The stamped version is made of aluminum 3003 and has a wall thickness of 1 mm.

Injection-Molded Bracket

The injection-molded version of the bracket is a small, box-shaped part with no undercuts. In addition, the peripheral height from a flat dividing plane is not constant. Hence, from Figure 4.1 the basic tool complexity part code is F20 and $C_b = 1.92$.

Cavity detail is low, commercial tolerances apply, and the part is assumed to have an SPI finish of 3. Thus, C_s and C_t are both 1.0 and the relative tool construction cost for the plastic version of the part, C_{dc}, is

$$C_{dc} = C_b C_s C_t = 1.92$$

Following the procedure outlined in Section 4.8 of Chapter 4, we can show that the value of C_{dm} is about 1.2. Hence, from Equation 4.4 we have

$$C_{dp} = 0.8C_{dc} + 0.2C_{dm} = 0.8(1.28) + 0.2(1.2) = 1.26$$

With regard to cycle time, the bracket is considered non-slender with all plates identical to each other. The plates contain no projections, and they are not grilled or slotted. Thus, from Figure 5.2 we have a classification code of N22 and a basic relative time, t_b, equal to 2.67. There are no threads or inserts, thus, from Table 5.3, $t_e = 0$. We will assume low surface requirements and easy to hold tolerances, thus, from Table 5.4, $t_p = 1$. Hence, from Equation 5.8, we have

$$t_r = (2.67 + 0)(1.0) = 2.67$$

From Equation 5.10, the tonnage required to mold this part is

$$F_p = 0.005A_p = 0.005(100)(30) = 15\,\text{tons}$$

Thus, from Table 5.5, the machine hourly rate, $C_{hr} = 1.00$, and the relative processing cost, C_{ep}, is

$$C_{ep} = t_r C_{hr} = 2.67$$

Finally, the relative material cost, C_{mp}, is given by

$$C_{mp} = (V/V_o)C_{mr} = [(14{,}400\,\text{mm}^3)/(1244\,\text{mm}^3)](2.79) = 32.3$$

where the value of C_{mr} for Nylon 6 was obtained from Table 5.2.

Thus, the total cost to produce this bracket as an injection-molded bracket is found, from Equation 5.12, to be given by

$$\begin{aligned} K_{tp} = C_{rp} &= 0.00182C_{mp} + (6980/N)C_{dp} + 0.1224C_{ep} \\ &= 0.00182(32.3) + (6980)(1.26)/N + 0.1224(2.67) \quad \text{(Equation 5.12)} \end{aligned}$$

or

$$K_{tp} = 0.385 + 8794/N \qquad \text{(Equation 13.3)}$$

Stamped Bracket

Using Tables 9.1 and 9.2 we find that the total number of active stations required to stamp the bracket shown in Figure 13.10 is 6. From Equation 9.8 we find that the design hours required for the tool needed to stamp the part is

$$t_d = 18.33N_a - 3.33 = 107\text{ hours}$$

Table 13.11 Build hours required for the stamped version of Example 5.

Station	*Operation*	t_{basic}	t_i
1	Pierce 2 pilot holes	30	45
2	Pierce 2 standard holes	30	45
3	Notch	40	40
4	Wipe form—1st bend	40	40
5	Wipe form—2nd bend	40	40
6	Blank out part	40	40
	t_b		250

and using the data base provided in Table 9.3 we find that the build hours for the tool are 250 (see Table 13.11). Hence, the total die construction time, t_{dc}, is 357 hours, and the relative die construction cost is

$$C_{dc} = 250/168 = 2.13$$

It was assumed here that both F_t and F_{dm} equal 1.0.

We will assume here that the strip layout of the part is such that L_{ub} = 100 mm and L_{ubn} = 60 mm. Hence from Table 9.8 we see that no idle stations are required for shearing and forming, and that 2 idle stations are required for wipe forming. Following the algorithm depicted in Table 9.8 we can show that A = 925 mm, B = 185 mm, and S_{ds} = 171,125 mm^2. Hence, using Equation 9.5 we conclude that C_{dm} = 3.27. Thus,

$$C_{ds} = 0.8(2.13) + 0.2(3.27) = 2.36$$

The relative material cost, C_{ms}, is found to be

$$C_{ms} = [(4800)/(3750)](1.79) = 2.29$$

where the relative material price for aluminum, C_{mr} = 1.79, was obtained from Table 10.2.

Thus, the total cost to produce this as a stamped bracket is found, from Equation 13. 2, to be given by the following expression

$$K_{ts} = C_r = 0.0215C_{ms} + (6561/N)C_{ds}$$
$$= 0.0215(2.29) + (6561)(2.36)/N = 0.0492 + 15484/N \quad \text{(Equation 13.4)}$$

If we subtract Equation 13.4 from Equation 13.3, we obtain the difference in cost, ΔK_{ps}, to produce the part as a stamping and as an injection molded part, namely,

$$\Delta K_{ps} = 0.336 - 6690/N \quad \text{(Equation 13.5)}$$

If $\Delta K_{ps} > 0$ then stamping the part is less costly. That is, if the production volume, N, is greater than about 20,000 then the stamped version of the part is more economical to produce.

13.10 SUMMARY

In this chapter, we presented a general methodology for selecting one or more material-process combinations for special purpose parts at the conceptual stage. Two approaches were presented: (1) material-first or (2) process-first.

In the material-first approach, designers begin by selecting a material class guided by the requirements of the application. Then processes consistent with the selected material are considered and evaluated, guided by production volume and information about the size, shape, and complexity of the part to be made.

In the process-first approach, designers begin by selecting the process guided by production volume and information about the size, shape, and complexity of the part to be made. Then materials consistent with the selected process and the part's application are considered.

Designers can use either approach depending on which is most convenient or natural to the specific part being designed. Both approaches end up at the same point. As always, designers are strongly urged to consult with materials and manufacturing experts about material-process selections before considering the selection "final."

It is not always possible to reduce the choice of material/process combination to a single alternative. It is thus often necessary to move on to configuration design with two or three possibilities. This will usually necessitate designing configurations for each of the material-process combinations, since different materials and processes may well lead to different optimal configurations. Of course, there will be common features, but configuration design must be done with a material-process combination (at least to Level III) in mind.

REFERENCES

Ashby, M. F. *Materials Selection in Mechanical Design*. New York: Pergamon Press, 1992.

Ashby, M. F., and Jones, D. R. H. *Engineering Materials 1: An Introduction to Their Properties and Application.* Oxford: Pergamon Press, 1980.

American Society of Metals International. *Engineered Materials Handbook*, Vol. 2. "Engineering Plastics." Metals Park, OH: ASM International, 1988.

———. *American Society of Metals Handbook*, 9th edition, Vol. 1. "Properties and Selection: Iron and Steels." Metals Park, OH: ASM International, 1978.

———. *American Society of Metals Handbook*, 9th edition, Vol. 2. "Properties and Selection: Nonferrous Alloys and Pure Metals." Metals Park, OH: ASM International, 1979.

———. *American Society of Metals Handbook*, 9th edition, Vol. 3. "Properties and Selection: Stainless Steels, Tool Materials, and Special-Purpose Metals." Metals Park, OH: ASM International, 1980.

Deiter, George. *A Materials and Processing Approach.* New York: McGraw-Hill, 1983.

Hu, W., and Poli, C. "To Injection Mold, to Stamp, or to Assembly? Part I: A DFM Cost Perspective." Proceeding of the 1997 ASME Design Engineering Technical Conference, September 14–17, 1997, Sacramento, CA.

———. "To Injection Mold, to Stamp, or to Assembly? Part I: A Time to Market Perspective." Proceeding of the 1997 ASME Design Engineering Technical Conference, September 14–17, 1997, Sacramento, CA.

Kalpakjian, Serope. *Manufacturing Engineering and Technology*, 2nd ed. Reading, MA: Addison-Wesley, 1992.

Manufacturing Handbook and Buyers' Guide, 1991–1992. New York: Plastics Technology, 1991.

Modern Plastics Encyclopedia 1991. Hightstown, NJ: Modern Plastics, October 1990.

QUESTIONS AND PROBLEMS

13.1 Assume that the caster shown in Figure P13.1 is at the early configurational design stage and that neither the part material nor the processes to be used to produce the component parts has been decided upon. Assume that the caster is to be used on dollies utilized in moving "heavy" boxes being unloaded from ships docked in various harbors around the world. For an estimated production volume of 50,000,

Which material/process combinations would you consider as possibilities for producing the brackets?

Which material/process combinations would you consider as possibilities for producing the plate?

Note: In deciding on alternative material/process combinations for the parts, you should allow the geometry of the parts to vary slightly from that shown in Figure P13.1 in order to take advantage of alternative processes.

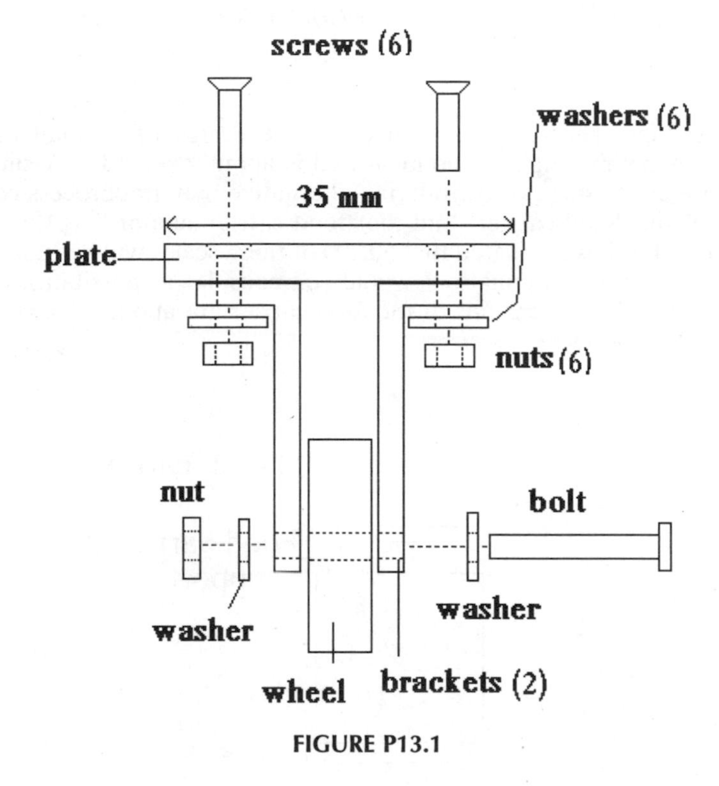

FIGURE P13.1

13.2 In an effort to reduce assembly costs, a suggestion has been made to combine the two brackets and the plate shown in Figure P13.1 into a single part. Which material/process combinations would you consider as possibilities for producing the redesigned part?

Note: In deciding on alternative material/process combinations for the part, you should allow the geometry of the part to vary slightly from that shown in Figure P13.1 in order to take advantage of alternative processes.

13.3 Figure P13.3 shows the proposed design for the spinner head being utilized as part of the spinning reel shown in Figure 13.4. It is estimated that 20,000 reels will be produced. What material/process combinations would you consider as possibilities for the part? Note: The diameter of the head is about 60 mm and the length is about 15 mm.

FIGURE P13.3

13.4 Assume that you are involved in the design of a small kitchen scale (maximum weight to be measured is about 2 pounds). Assume that you are at the stage of considering alternative material/process combinations for the "food carrier" and the "food carrier support" as shown in Figure P13.4. If it is estimated that 50,000 of these scales will be built, what material/process combinations would you consider as possibilities for the two parts? The dimensions of the food carrier are about 4 inches by 2 inches by 0.5 inches.

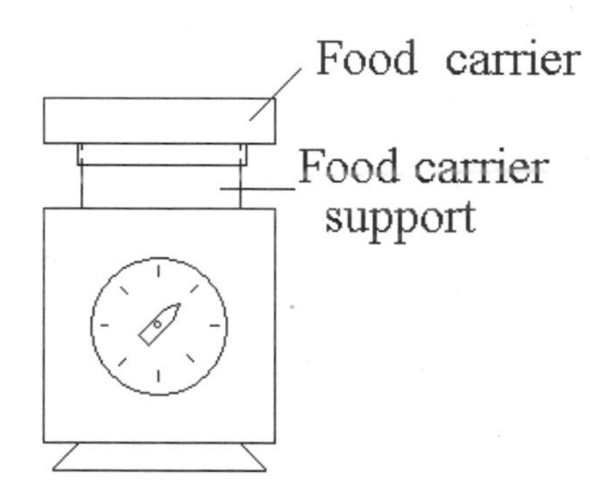

FIGURE P13.4

13.5 Repeat Problem 13.4 for the case where the food carrier and food carrier support are produced as a single part.

13.6 Tables 13A.1, 13A.2, 13A.3, and 13A.4 give a representative sample of materials with some of their mechanical and physical properties. In spite of the fact that a product would be subjected to a corrosive environment, can you list applications where you might select a high-density corrosive material over a low-density "noncorrosive" material? Can you suggest applications when a low-density material would be a better choice than a high density material?

13.7 One possible method for reducing assembly costs for the assembly shown in Figure P13.1 is to cast or mold the plate and two brackets as a single part. Make a sketch, to scale, of what this single part will look like. Based solely on cost considerations, would you die cast the part or injection mold it?

13.8 In Figure P13.8 is the original design of a die casting analyzed in Chapter 7 (Example 7.1 in Section 7.3)? Assume that as part of a job interview you have been asked to comment on whether or not it would be less costly to produce this as a polycarbonate part. Based solely on cost, what would you recommend? Assume that by pure coincidence you had a copy of this book with you.

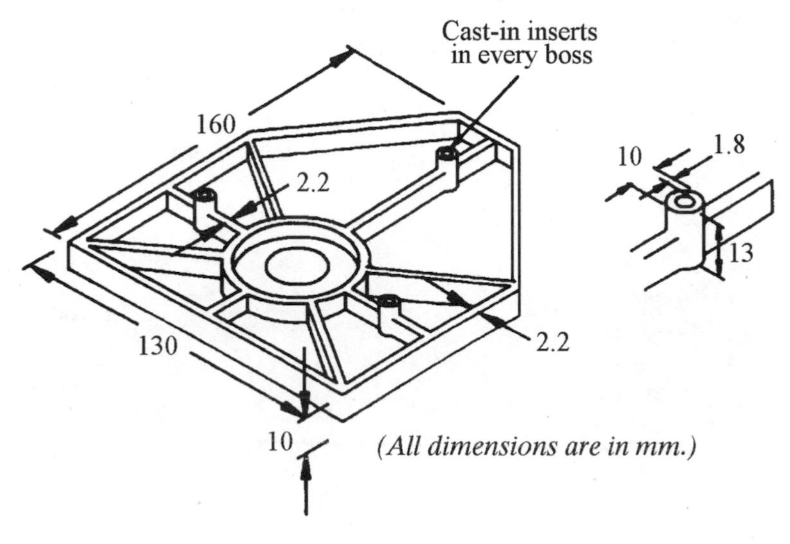

FIGURE P13.8

APPENDIX 13.A

Some Properties of Selected Materials

Table 13A.1 Properties of selected cast alloys.

Material	Casting Processes Used In	Corrosion Resistance	Cost (cents/in^3)	Tensile Strength (kpsi)	Yield Strength (kpsi)	Shear Strength (kpsi)	Modulus (Mpsi)	Elongation Percent	Hardness Brinell	Endurance Limit (kpsi)	Density (lb/in^3)	Thermal Conductivity, 68 deg F (BTU-ft/hrft^{2}F)	Coef. of Thermal Expansion 68-212 deg. F (10^{-6}in/in/F)	Electrical Conductivity percent IACS	Typical Uses
Aluminum Alloys															
A380.0	Die Casting	Good	7.8	47	23	28	10.3	4	80		0.097	58	11.7	25	Auto parts, motor frames, housings, vac. Cleaners. Most widely used of all die casting alloys.
A413.0				35	16	25	10.3	3.5	80		0.096	70	11.4	31	Misc. parts. Used where resistance to corrosion needed.
201	Sand, permanent, and investment	T7 temper recommended for opt resistance to stress corrosion cracking.		65	55		10.3	9	110		0.101		10.7	30	Structural members, aerospace housings, truck and trailer castings.
Magnesium Alloys															
AZ91D	Die Casting	Good		34		20		3	63	14	0.065	31	14	13	
AZ63A	Sand, permanent Mold	resistance to atmosphere, attacked by salt water unless finished	10.8	29	14	18	6.5	6	50	12	0.065	39	14	15	
Copper Alloys															
C80100				25	9		17	40	44	9	0.323	226	9.4	100	
C81400	Centrifugal, continuous, investment, plaster, and sand.	Corrosion, and oxidation resistant		52	40		16	17		15	0.319	182	10	82	Electrical parts that meet RWMAClass II standards. Low press. valves and fittings. General hardware supplies.
C84400				38	14		12	14	35	55	0.314	49	11	16.4	
Zinc Alloys															
SAE903	Die Casting	Excellent		41		31	12.4	10	82	6.9	0.238	65.3	15.2	27.5	Auto parts, household utensiles, office equipment, toys.
Steel															
Low Carbon (<.2%)	Sand, investment, continuous, ceramic, and centrifugal.	Rust easily. Corrosion rate increased if salt is present		45-65	25-45		30	23-30	115-140		0.283				
Med Carbon				65-95	35-50		30	23-30	115-140	19-22	0.283	8-8.3			
High Carbon (.5-1.0%)				100-125	50-60		30	23-30	115-140		0.283				
Alloy				68-205	38-170		30	32	137-401	32-88	0.283				
Stainless	Sand, ceramic & centrifugal.	High resistance.		69-120	36-100		25-29	2-60	140-390		0.272-0.294	6.4-10.3			

(1 psi = 6895 Pa; 1 kpsi = 6.895 Mpa; Example 10 kpsi (6.895) = 69 Mpa)

Table 13A.2 Properties of selected wrought alloys.

Material	Processes Used In	Corrosion Resistance	Cost (cents/in^3)	Tensile Strength (kpsi)	Yield Strength (kpsi)	Shear Strength (kpsi)	Modulus (Mpsi)	Elongation Percent	Hardness Brinell	Endurance Limit (kpsi)	Density (lb/in^3)	Thermal Conductivity, 68 deg F (BTU-ft/hrft^{2}F)	Coef. of Thermal Expansion 68-212 deg. F (10^{-6}in/in/F)	Electrical Conductivity percent IACS	Typical Uses
Aluminum Alloys															
1100 (soft) Annealed		High to rural, industrial, and marine atmosphere, and good to others.		13	5	11	10	35-45	23	5	0.098		13	59	Sheet metal parts, drawn or spun parts.
1100-H14 (half-hard)				18	17	13	10	9-20	32	7	0.098	128	13.1		
1100-H18 (Hard)				24	22	18	10	5-15	44	9	0.098		13.1	57	
2014 Annealed	Stamping, extrusion, and forging	High to rural, poor to sea water and others		27	14	38	10.6		45	13	0.101		12.8	50	Heavy duty forgings, plate and extrusions for aircracft fittings, wheel and major structural components; truck frame and truck components.
2014-T4				62	42	42	10.6	20	105	20	0.101	111	12.8	34	
2014-T6				70	60	12	10.6		135	18	0.101		12.8	40	
6061 Annealed		High to good	6.9	18	8	22	10	25-340	30	9	0.098	104	13	47	Trucks; places where high str. and corrosion res. Needed.
7075 Annealed		Good		33	15		10.4	17	60		0.101		13.1		Structural components where high str. and corr. resistance needed.
Magnesium Alloys															
AZ61A-F	Extrusion, forging	Good to atmosphere. Attacked by sea water.		38-46	21-33	20	6.5	7-17	55	20	0.065	34	14	12.5	General purpose forgings and extrusions.
Copper Alloys															
C23000 Annealed	Stamping, extrusion	Fair	50	39	10	31	17	48	F56		0.316	92	10.4	20	Elec. conduits, sockets; fastners, pipes, costume jewelry.
C23000 Hard				70	57	42	17	5	B77						
C37700 Annealed	Forging	Good to excellent		52	20	35	15	45	F78		0.305	69	11.5	27	Forgings of all kind.

(1 psi = 6895 Pa; 1 kpsi = 6.895 MPa; Example 10 kpsi (6.895) = 69 MPa)

Table 13A.3 Properties of selected wrought alloys.

Material	Processes Used In	Corrosion Resistance	Cost (cents/in^3)	Tensile Strength (kpsi)	Yield Strength (kpsi)	Shear Strength (kpsi)	Modulus (Mpsi)	Elongation Percent	Hardness Brinell	Endurance Limit (kpsi)	Density (lb/in^3)	Thermal Conductivity, 68 deg F (BTU-ft/hrft^{2}F)	Coef. of Thermal Expansion 68-212 deg. F (10^{-6}in/in/F)	Electrical Conductivity percent IACS	Typical Uses
Carbon Steel															
Rods, Bars, Forgings															
1015 (low carbon) hot rolled				50	27.5			27.5	101						
1015 (low carbon) cold drawn				56	47			47	111						
1045 (med carbon) hot rolled	Forging, extrusion	Poor		82	45		30	16	163		0.283	27	8.1	18	
1045 (med carbon) cold drawn				91	77			12	179						
1095 (high carbon) hot rolled				120	66			10	248						
1095 (high carbon) cold drawn				99	76			10	197						
Sheets															
Commercial				52	38			30	55						
Drawing Quality	Stamping	Poor		44	27			42	42		0.283	27	8.1	18	
Alloy Steel															
Rods, Bars, Forgings															
1340	Forging, stamping	Poor		100-282	76-235			9-25	235-578						
Sheets															
ASTM A606 Cold Rolled	Stamping	Improved resistance.		65	45			22							
Stainless Steel															
302 Annealed	Stamping	Excellent including food products.		90	40			50	Rb85						
304 Cold Rolled	Stamping, forging		42.6	110	75		28	60	240		0.29	9.4	9.6	73	General purpose.
316 Annealed	Stamping	Best of all standard stainless.	57.7	84	42			50	Rb79				8.9	74	Parts exposed to severe corr. media and stressed parts subject to hi temp.

(1 psi = 6895 Pa; 1 kpsi = 6.895 MPa; Example 10 kpsi (6.895) = 69 MPa)

Table 13A.4 Properties of selected plastics.

Material	Processes Used In	Chemical Resistance	Cost (cents/in^3)	Tensile Strength (kpsi)	Flexural Strength (kpsi)	Tensile Modulus (kpsi)	Flexural Modulus (kpsi)	Hardness Rockwell	Impact Izod-Notch (ft-lb/in)	Density (lb/in^3)	Thermal Conductivity, 68 deg F (BTU-ft/hrft^{2}F)	Coef. of Thermal Expansion 68-212 deg. F (10^{-6}in/in/F)	Typical Uses
Thermoplastic													
ABS (Medium Impact)	Injection molding and extrusion.	Hi to acqueous acids, alkilis and salt	3.4	6.3-8.0	9.9-11.8	340-400	350-400	80	2.4-4.0	0.038	0.96-2.16	3.2-4.8	Appliance parts; toys; office, lawn and garden equipment.
Acetal	Injection molding, extrusion, blow molding, and rotational molding	Excellent to most. Poor for strongacids and alkalis.	6.4	10	14.1	520	410-450	80	1.5	0.052	1.56	4.5	Appliance parts, gears, bushings, auto and plumbing parts
Acetal 20% glass			8.4	8.5	16.5	1300	800	110	0.8	0.056		2.0-4.5	
Nylon 6		Resists weak acids, alcohol and common solvents.	5.9	5.5-13	10.0-11.6	200-500			0.8-3.0	0.039	1.2	1.6-8.3	Bearings, gears, bushings, rod, tubing.
Nylon 6 30% glass			7.7	22-26	26-34	1000-1450			2.3-3.0	0.05	1.2-1.7	1.2-3.0	General purpose parts requiring stiffness.
Nylon 6/6		Attacked by strong concentrations of mineral acids.	6.5	11.8		385-475	410		1.0	0.041	1.7	1.7	Bearings, gears, bushings, rod, tubing.
Nylon 6/6 30% glass			9.8		26-35	1400-2000	1300		2.2	0.05	1.5	1.5	
Polycarbonate	Injection molding and extrusion.	Resists weak acids and alkilis, oils and grease.	6.7	8.5-9.0	12-14.2	325-340	310-350	63	12-18	0.04	1.35-1.41	3.75	Electrical parts, portable tool housings, lenses, sports goods, impellers, and auto parts.
Polycarbonate 40% glass	Blow molding and thermo-forming.		10.4	23	27	1680	1400	50	2.5	0.055	1.53	0.93	
Thermosets													
Alkyd	Compression and transfer molding	Resistant to weal acids. Unattacked by organic liquids (alcohol, fatty acids and hydrocarbons).	5.5	7-8	19-20	1950	2500	70-75	2.2	0.079	4.2-7.2	1-3	Encapsulation of resistors, coils and small electronic parts, switches, relays, connectors, sockets, circuit breakers, parts for transformers, motor controllers, and auto ignition systems.
Alkyd and glass				5-9	12-17	2250	2500	70-80	8-12	0.073	2.4-3.6	1-3	
Epoxy and glass		Highly resistant to water and bases.		8-11	19-22		1500-2500	75-80	0.4-0.5	0.069	1.2-6	1-2	Electrical molding such as condensers, resistors, coils, etc.

(1 psi = 6895 Pa; 1 kpsi = 6.895 MPa; Example 10 kpsi (6.895) = 69 MPa)

Chapter 14

Communications

14.1 INTRODUCTION

It probably seems strange that in a book devoted to design for manufacturing we include a chapter on communications. However, it is difficult to overemphasize how important effective communications are to the efficient achievement of quality designs. The communication abilities of the individuals and groups involved are critical, as is the communication system in the organization. As one manager of a complex design activity said: "If I could have one wish granted, it would be to make our internal communications fewer in number and far more effective."

Communication is also extremely important to personal success. An engineer who can communicate well in person, in written reports, through sketches and diagrams, and in oral presentations—and who can also listen effectively—is far more likely to be given greater responsibility and advancement. There is no substitute for technical competence, but without the ability to communicate well, even the most competent engineer will have difficulty advancing in business organizations. An engineer simply cannot fully realize the benefits of technical ability unless he or she is also able to communicate effectively with others.

For example, a very successful former student of mine confided recently that he attributes his rapid rise primarily to his ability to communicate effectively. In addition, this engineer believes that his ability to communicate effectively was developed by practicing the ideas in this chapter while working on various engineering projects during his undergraduate years.

In this chapter, we will discuss three types of communication: written, oral, and graphical. We point out here some of the major principles of good communication, and provide some practical tips and guidelines that will help young engineers get their communication efforts off to a better start. But I also strongly urge lots and lots of practice coupled with lots and lots of effective criticism.

Because so many design and manufacturing courses involve project work in which students are required to submit written reports and drawings, and to present the results of their work orally, this last chapter is often the first chapter dealt with in the course.

14.2 WRITTEN COMMUNICATIONS

14.2.1 Preparation for Writing

There are several types of written communications, including notes, memos, letters, and several types of reports. We will discuss the specific types in later sections; first, we describe some general guidelines that apply to all of them.

Before You Write.

Question whether writing is the best way to communicate. That is, ask yourself whether this message for this audience is best delivered in writing, or in person, over the telephone, or in an oral presentation where more interaction and discussion are possible.

You may want to write so there will be a record, or so that your whole message can be heard without interruption. If the message is highly technical, writing provides the receivers of your communication with an opportunity to study it and to refer to it later. Some messages need both a written form (e.g., for a record or for a permanent reference) and a personal or oral presentation that permits discussion. In this case, however, both the written and the oral messages must be complete in themselves.

Explicitly state for yourself the purpose of the communication. What is it exactly that you want to accomplish? Is there something you want the receiver(s) to do? Is there information you want them to have, or that they need to know?

Explicitly describe the intended receiver(s) for yourself. Who are they? Why are they interested in or needing to get the message? What symbols (both words and graphics) do they understand and not understand?

Often, there are several intended receivers for messages in a company. Each must then be described, and the communication constructed and organized accordingly. In a written report, for example, this may call for a brief abstract for top managers, a main section (or body) for other engineers involved in the project, and appendices with all the details for colleagues most involved.

List and organize the critical information that is to be communicated. This need not be a formal outline (though it can be), but you need to decide at an early stage, before you begin to write, exactly what are the essentials of your message. You must be brief here. For example, consider what such a list might have looked like as we prepared for writing this chapter in this book:

1. Principles and Practical Guidelines Useful to Students and Young Engineers Relative to

   Written communications
   Oral communications
   Graphical communication

Then for each of these major sections, another list, more detailed, is prepared before it is written. For example, for this section on written communications:

A. Preparations for Writing

   Know the purpose,
   Know the readers,
   Know the message.

Select the written format that is most appropriate to the audience, your goals, and the message. After you have decided to write, you have your choice of many different writing formats: a handwritten note, memo, letter, electronic mail, or one of several types of reports. We will subsequently discuss two types of reports: the Research Report and the Business Technical Report. There are also notes and records of various kinds that you keep for yourself. You could even write a book.

14.2.2 The Expected Standard for Spelling and Grammar

Suppose you got a technical report to review, and it began as follows:

> Introduction
>
> *A major problem that exists with the production of injection molded plastic parts is acheiving the demensions and shape of the targeted design. Due to the complex numbers of interactive shrinkages that typically are developed from the molding of the part it is virtually impossible for even the most skilled designer to take an emperical approach in predicting what their net effect will be on the final molding. Therefore it is expected that the part cannot be fully evaluated demensionally or mechanically until a mold is actually built and the part produced.*

Considering the misspellings—achieving, dimension (twice), empirical—how would you feel about these authors, and about their technical capability? How do you feel about the wordiness? For example, how about those "complex numbers of interactive shrinkages"!

We assume that readers of this book are able to write proper sentences, and know about such writing basics as the proper structure for sentences and paragraphs. We realize, however, that these basics unfortunately present problems for some engineering students, so we emphasize again that writing is extremely important to successful engineering practice in any organization. If basic writing ability is a problem for you, then you should immediately and aggressively take whatever steps are needed to correct it.

In all business and technical communications, spelling and grammar must be *perfect*. Yes, *perfect*—because that is the expected professional standard. Anything less will hurt you and the effectiveness of your communications significantly.

The reason that perfect is the standard is that if there are spelling or grammatical errors, the people who receive the message will—consciously or unconsciously—infer that you have made other errors as well. That is, they will assume that since you did not care enough to get the spelling and grammar right, then you either (a) don't care much about them, or (b) didn't take care to get the content right either, or (c) both. Usually they assume (c). Many people will stop reading altogether when they begin to encounter more than a rare spelling or grammar error. Of course, an occasional typographical error may get through. No doubt you have found some in this book. A few such mistakes are inevitable, but the standard to strive for in spelling and grammar is perfection.

If you are not confident of your spelling and grammar, then let someone you know is capable read your messages before they are sent. If you are using a word processor, you can use the spell-checker first to save yourself some embarrassment. But don't assume that spelling and grammar are the same thing; spelling can be perfect, but grammar may still be bad.

14.2.3 The Importance of Conciseness

In the design of a part or product the tried and true saying is Keep It Simple, Stupid, or KISS. KISS also applies to business and professional communications, though in a slightly modified form. Here it means: Keep It Short, and then Shorten it some more. In written communication, conciseness is a primary virtue. But "short" is not easy to attain. There is a story about the person who received a long, long letter that concludes: "Sorry I didn't have time to write you a short letter."

How concise? Well, usually a written communication should be much more concise than you, the writer, would like it to be. It is probably even more concise than you think at first it is possible to be. But any message can be made any length, and the shorter the better. For example, forced to reduce this section to one short sentence, we could say: Keep written communications short.

The Take-Aways

As a help to achieving conciseness, you may want to try the following exercise. Suppose that an intended reader of one of your communications has just finished reading it, and is then asked by another person passing by in the hall, "What did he or she have to say?" Now what, exactly, in just a few brief sentences, would you want your reader to reply in answer to that question?

Your answer to this question is called the *take-away*. That is, the take-away is that brief essential central message you want your readers to take away with them after getting your communication. Knowing what your desired "take-aways" are can often help you shorten a communication to include just the take-aways plus whatever is absolutely necessary to support them. The take-aways are what you should emphasize in your message, and hence knowing what they are tells you what can be eliminated.

Reduce the Noise

Another technique for achieving conciseness is to keep the noise level down. By noise we mean the frequent communication of extraneous or irrelevant information. It is true that you don't always know exactly what all reader(s) will consider useful for their purposes, and you don't want to leave out something useful. But too much noise causes people to tune out, thus missing what you want to communicate to them. Stick to just the take-aways.

Eliminate Wordiness and Clutter

One of the best ways to achieve concise writing is to eliminate unnecessary words. You can also change fancy words and jargon to plain English. Look again at the third subsection above headed with "The Take-Aways." Here is a rewrite that eliminates some unnecessary words:

> The Take-Aways. To achieve conciseness, try the following exercise. Suppose a person who has just read your communication is asked "What did it say?" The short quick answer you hope for is called the take-away.

We have reduced the word count from 71 to 37, without significantly changing the content. This kind of word cutting can almost always be done at least once or twice after a first draft has been written. For example, we can do it again to the above paragraph:

> *Take-aways. Try this: Ask someone who has just read your communication, "What did it say?" The short quick answer you hope for is the take-away.*

Now we have only 27 words, about one-third the original, but essentially the same content. Of course, you may not want the terseness that extreme word reduction can sometimes create. But too short is seldom the problem. The problem with most writing is excessive length due to unnecessary words; rewriting to eliminate them will improve the result significantly.

14.2.4 The Writing Process

The writing process is, like design, done by guided iteration; that is, by writing, evaluating, and rewriting until an acceptable result is obtained. At least 90% of writing is rewriting.

The best process is this: Write a complete first draft, no matter how rough; then rewrite, rewrite, and rewrite until you get the result you want. Do not try to get any one part perfect except during a near-final complete rewrite.

The first complete, rough draft may be little more than an outline, or just the main ideas jotted down crudely in the order you want them. Then you begin to expand and rewrite and rewrite and rewrite until you have the desired quality. Good writing has generally been rewritten and shortened ten to twenty times or more.

The process recommended here is not only the best way to get a quality result, it is the fastest way to get a quality result. If you are in a hurry, the absolute worst thing to do is to try to get the first paragraph or first section nearly perfect before going on to the next paragraph or section. Invariably when you do this you will find later that you must come back to earlier sections to make major changes or additions. Then some of the time you spent making the first part perfect will have been wasted. The final quality and effectiveness of your writing is determined by the last draft, not by the first draft.

14.2.5 Two Pieces of Practical Advice

Always keep a copy of all written communications, even informal notes and E-mail messages.

Never, ever, write anything that you don't want the whole world to see. Sooner or later, the whole world may see it. This is especially true if it is something you would rather the whole world did not see.

We now turn our attention to the specifics of the more common types of written communication required of young design engineers.

14.2.6 Types of Written Communications

Personal Notes and Records

Design engineers must keep a notebook for patent and liability reasons. However, a notebook is essential for other reasons as well. Memory sometimes fails, and when you are busy, memory often fails. Thus a personal daily log of activities and important communications is a personal and professional necessity.

Hardly anyone likes to keep notes; most of us are not used to doing it, and it seems a distraction from pressing work that needs to be done right now. It is therefore important for personal note-keeping to include the important items and omit the unimportant ones. In most cases, it will be about right if you devote 10 or 15 minutes a day (on average) to personal notes, and supplement the notes with documents (or references to them) received from others, such as memos, letters, literature, and other communications. As noted above, you will also keep copies of all of your own written communications to others.

With your personal notes, you are both sender and receiver. Even so, it is still helpful to think about the purposes of the notes. It is most likely that you will keep them concise to save yourself time, but watch out for too much conciseness here. Your memory a week or a month or a year from now won't be as good as you think. What you fail to include might cost you a lot of time or trouble later.

Informal Handwritten and Electronic-Mail Messages

Informal handwritten or electronic-mail messages are often the most convenient and appropriate medium in day-to-day working relationships. Messages can be prepared and sent quickly. You can carry on some discussions via E-mail.

In contrast to longer, more formal media, E-mail messages generally get read and responded to immediately—unless you send too many of them or they are too long. As always, keep the general noise down if you want to be heard when it is important.

Memos and Letters

Business memos and letters are generally used to express opinions, to make requests, to provide notification of one kind or another, and to provide information that is not highly technical. Memos and letters may provide summaries of the results of technical work, or recommendations based on such results, but usually in this case they will refer to a longer report that contains the complete technical information. When letters or memos accompany technical reports, they are called "covering" or transmittal memos or letters.

Memos are generally used for communications inside firms, and letters for outside communications. However, there are no hard rules about this.

Any business memo or letter longer than one page is highly undesirable. Probably there ought to be law against it. Certainly two pages is the absolute maximum, and that should be done only rarely. If the memo or letter is longer than one page, then the first and last paragraphs must provide the desired take-away information. The reason is that the rest, especially the second page (except perhaps for the last paragraph), is not very likely to get read very carefully—if it is read at all.

Here is an example memo:

Date: March 5, 1993

To: T. S. Jones, Vice-President, Advanced Manufacturing Engineering

From: J. K. Smith, Manager, Manufacturing Engineering

Subject: Systems Manufacturing Testing

We have had double-digit growth in productivity throughout the recent period of rapid technological advancement. This growth has continued despite manufacturing system test methods that have not changed for fifteen years. However, as we enter new markets, and our competition increases in the present ones, our development and production processes should now be automated to maximize productivity. Since testing follows a product through its entire life cycle, there is much to be gained from automation of the existing test methods. Moreover, some tests can be eliminated.

Preliminary trials conducted on our LS-2 indicate that an automated test system can significantly reduce costs in all our plants by

- *Decreasing the time of highly skilled people doing testing,*
- *Reducing capital and maintenance costs, and*
- *Eliminating redundancy*

Additional trials with the automated testing system will be conducted during the next quarter, and an updated report issued at the end of the quarter.

Two Types of Technical Reports

A technical report describes the results of activities such as research, development, analysis, design, experimentation, and related topics including specifications, marketing studies, and the like. In this book, we describe two kinds of technical reports:

1. Research Reports
2. Business Technical Reports

Though we will discuss these forms separately as essentially pure forms, there are often circumstances when a combination of these two formats is the best solution. This is discussed later in Section 14.5.

If a technical report includes specific recommendations for decisions or actions, or if it includes results or conclusions that will be used in the short term as a basis for making or supporting decisions or actions affecting specific current activities or plans, then we call it a *Business Technical Report.*

If a technical report contains general information, basic data, or research results useful in the longer term—but not immediately and directly relevant to specific current operations—then we call it a *Research Report.* Research reports do not include recommendations for decisions or actions, except perhaps about continuing research.

The reason for making a distinction between Business Technical Reports and Research Reports is that, because of their different purposes, the two types are best written in different formats. We describe two recommended formats below.

There are special customs in some companies about the form of internal technical reports, and if so, readers should follow those customs unless there is general agreement in the company on desired changes. If there are no customs, or possibly as a model for change, we recommend the formats that follow in the next two subsections.

To help make the following discussion more efficient, we will use these definitions:

Data: Experimental data, input values used, survey results, results of previous studies, material, machine or product parameters, etc.;
Analyses: Analytical and computational procedures employed, including simulations;
Results: Generalizations of data, and results produced by analyses;
Conclusions: Conclusions of a technical or business nature based on an accumulation of results;
Recommendations: Recommendations for decisions or actions based on the results or conclusions.

Research Reports follow the usual path taught in science courses: first the goal of the project is described; next the experiment is described; then the data, analyses, and results; and finally the conclusions and any recommendations.

Business Technical Reports use almost the reverse order; that is, they begin with recommendations, and follow with conclusions, results, analyses, and data. This order is used because different people in a business organization read reports for different reasons. Higher managers may want to know only what the recommendations are, and these managers should not have to read through a lot of preliminary detail to find them. If they want to know the reasoning and the details, they can read on, but it should be their option.

We now describe the two types of technical reports separately and in more detail.

14.3 RESEARCH REPORTS

14.3.1 Overview

Most Research Reports are usually written for internal company use only. However, they are sometimes also prepared for external publication in a trade magazine or research journal. Most companies require that such reports be cleared by the firm's management or legal departments before they are submitted for publication.

Though there are certainly variations possible depending on the content, Research Reports are generally structured more or less as follows:

- Title
- Abstract
- Introduction
- Literature Review
- Main Body
  - Hypothesis and Method of Attack
  - Experiments, Analyses, or Simulations Performed
  - Results
- Discussion of Results
- Summary and Conclusion
- Appendices (if any)

When a research paper is to be submitted for possible publication in a journal, the journal's guidelines regarding format, length, style, and content should be obtained and studied before the report is written. Though there are many similarities among journals on these issues, there are also differences, and it is always a good idea to submit papers to journal editors in the form in which they are wanted.

Though it is certainly a matter of personal choice, many researchers feel that it is better to avoid use of the first person (that is, I or we) in Research Reports. This necessitates use of the passive voice (e.g., "six tests were conducted" instead of "we conducted six tests") which can produce a rather pedantic style that sometimes gets boring. Whatever choice is made about issues of voice and tense, however, must be maintained consistently throughout the report. Changing back and forth causes readers a great deal of distress.

14.3.2 Titles

Whether a report is a Business Technical Report or a Research Report, the title is important. This discussion of titles applies to all reports.

A title should indicate the content of the report clearly and concisely. The title provides information needed by a reader to decide whether or not to bother reading the abstract. The title should be accurate; that is, it should not mislead readers into thinking that the report is something that it is not. Nor should a title leave readers guessing about what the report is specifically about, which happens when a title is too short and hence too general.

For example, consider this title of a Research Report:

Assembly Analysis of Can Opener

It provides too little information. There are many types of can openers (e.g., electric or manual); which type is being analyzed? Or are all can openers being analyzed? Or maybe it is a study of can openers in general? Is it manual or automatic assembly of the can opener that was studied? A better title would tell the reader more accurately and completely what the report is about. For example:

Analysis and Comparison of the Manual Assembly Times for Three Electric Can Openers

Here is another short title that is too general:

Comparisons of Robust Design Processes

A more helpful title would be:

A Comparative Study of Three Robust Design Processes: Optimization, Taguchi methods, and Guided Iteration Using Dominic

Titles that provide good descriptions of the content of a report can get too long. There are no hard and fast rules about length, but ten to fifteen words is generally not too long. Twenty words usually is too long.

14.3.3 Abstracts

In a Research Report, the purpose of the abstract is to enable people to determine whether or not they want to read the rest of the paper. Thus, the Abstract for a Research Report usually contains descriptive as well as summary information. That is, it contains a brief description of what was done as well as a brief summary of the results obtained. When writing an Abstract for your own paper, be sure to write it not as if it is a review of someone else's work but as descriptive summary of your work.

Here is a good abstract for a Research Report:

A design for manufacturing (DFM) analysis of a TOT-50 Swingline stapler is described. The results show that incorporating six simple redesign suggestions can reduce piece part costs by about 25% and manual assembly costs by about 35%.

Notice that the first sentence in this Abstract is descriptive—it says what was done. The second sentence is a summary of the results. This Abstract is brief, it describes the project well, and it contains the most important results.

For comparison, the following is an abstract (taken from a student group report) that is too long and too filled with information that is not important in an abstract (i.e., noise). The writing is far below professional standards in other ways, too:

The objectives which we used to guide us through the analysis of our part were simple. We analyzed the current design of the part using techniques learned in class. Each com-

ponent of the part was evaluated for relative die construction cost, relative processing cost, relative material cost, and assembly time. We looked at the analysis and came up with two redesign suggestion.

The first suggestion was to incorporate the cover and the body into one stamped piece. This eliminates a part which reduces the construction, material, processing and assembly cost. Also with one less component to assemble the projected assembly time was reduced. We then did an analysis of the new part to compare with the old.

The second redesign suggestion was to build the cover so that it didn't have any internal undercuts. The cover is a plastic injection molded part and the die construction saw a large increase due to these internal undercuts. We offer a redesign suggestion which eliminate these undercuts and does not change the functionality of the part. The new design for the cover is then re-analyzed to show the advantages of this redesign.

By following this objective we were able to come up with some ideas on how to produce this part less costly. The redesigns are simple and yet they yield some significant savings in relative cost which can be seen in the analysis presented.

In addition to its excessive length and inclusion of detail inappropriate for an abstract, the writing can be criticized for at least the following additional faults:

1. Research Reports should not be written in the first person. (Business Technical Reports can and usually should be written in the first person.)
2. It is incredibly wordy. For example, the first sentence could be reduced to: "Simple objectives guided the analysis." It would be even better, however, to just state what the objectives were.
3. It is full of grammatical errors like "two redesign suggestion" instead of "two redesign suggestions."
4. It contains colloquial phrases like "came up with" that are inappropriate in a Research Report. And how about "the die construction saw . . ."!
5. There are awkward phrases such as "By following this objective" and "how to produce this part less costly." Less awkward alternatives might be: "To achieve this objective," and "to produce this part more economically."
6. In the third paragraph, there are many switches from first person to passive voice: "We offer . . ." is followed by "The new design is then re-analyzed . . ."
7. There is even a kind of technical error: apparently the writer doesn't know about (or doesn't care about) the difference between a part and an assembly. Note in the first sentence that the word part is used whereas the analysis was done on an assembly or subassembly that consisted of several parts. Readers can no doubt find other examples of subprofessional writing in the above piece.

The kind of writing illustrated by the above quote is totally unacceptable in the professional work environment.

14.3.4 Introductions

The Introduction of a Research Report repeats and expands on the Abstract's descriptive material, explains the motivation for the work, and provides an overview of the organization of the report. It should also provide a general context for the study and its results. An Introduction does not usually contain information about the results.

An effort should be made to make the Introduction short but interesting; if you bore your readers at the beginning, think how they will be dreading the task of reading on!

Suppose the purpose of a research project is to study a company product from an assembly viewpoint. Then the Introduction might be as follows:

Introduction

The current design of the ⟨product name⟩ was developed in 1981, and has been stable for the past eight years. There are four models consisting of a total of eighty-four parts. The product has not been studied or the design revised from an assembly standpoint since its inception. In the interim, new design for assembly (DFA) evaluation methods have been developed and new assembly technology has also become available.

Competition in the ⟨product name⟩ product line has recently focussed on price reduction, and there is now extreme pressure for cost reduction. Assembly costs currently amount to 37% of the total cost and thus present a prime opportunity for savings.

The current design is assembled manually using a progressive line and a two-shift operation. The proposed addition of two new models and the continued competitive pressure to reduce costs makes this an ideal time to investigate the possibility of improving the current line or of converting to an automatic line.

Therefore, the goal of this project is to analyze the assembly of the ⟨product name⟩. The specific objectives are:

1. *Evaluate the current design for both manual and automatic assembly;*
2. *Redesign the product for ease of assembly;*
3. *Compare the current design with the proposed new design on the basis of:*
   Assembly cost
   Line balancing efficiency for manual assembly
4. *Identify the tooling required for automatic handling of the parts in the new design via vibratory bowl feeders.*

14.3.5 Literature Review

The purpose of a literature review is to connect the research done in the report with previous work done on the same subject, especially by other researchers. Therefore, the related papers and publications are cited and summarized in an organized way, and it is shown how the research being reported is similar to, and how it differs from, the work done by others. Note that the idea of the literature review is not just to report on and review other related research, but also to place the new research in context with that research.

When there are no, or just a very few, references to work done by others in a Research Report, then it is either a profound work indeed, or else the authors have not done their homework in searching the literature. One also sometimes sees literature reviews that refer primarily or almost exclusively to the work of the same people who are writing the report. Though this may be quite proper occasionally (i.e., when these are the only people working on a subject in a field—a rare situation), probably this too indicates that the literature review is incomplete.

A research project should not be undertaken, and certainly a research paper should not be written or submitted for publication, unless the authors are reasonably confident that they are familiar with all the other relevant work previously done. Computer-based literature searches, and searches of the relevant research journals can help develop this confidence.

Here is an example of a good literature review. It is taken from a Research Report that describes an alternative approach to the rules-of-thumb-based design of metal stampings. The literature review is contained under "Related Work" and the references contained in the review are listed under "References."

RELATED WORK

To assist in the design of easier-to-produce stampings, several cost-estimating and/or group technology (GT) based systems have been developed. These systems were developed to replace the traditional qualitative rules of thumb found in handbooks and used by most part designers. Unfortunately, these cost-estimating and GT systems require knowledge not generally possessed by designers, and they fail to provide designers with quantitative information concerning difficult-to-produce features.

For example, sheet-metal cost estimators such as the Harig system (Harig, 1977) and the Bradley die estimating method (Bradley, 1980) require die design and process planning knowledge rarely found among product designers. These systems are important advances, but they require detailed input and so are not designed for use at the conceptual stage of design where parametric details have not yet been established. Moreover, the systems do not generate redesign suggestions based on part attributes.

The Opitz system (Hohmann et al., 1970), a GT based classification and coding system, uses a nine-digit code to classify sheet-metal parts based on similarities in processing. This technique rationalizes design procedures and enables the grouping of parts for production to reduce die costs and lead time. The GT based Salford system (Fogg et al., 1971) uses a six-digit code to classify sheet-metal parts based on the industrial sector that they serve, the type of die utilized, and similarities in processing. The objective of the Salford system is to simplify tool design by exploiting similarities in parts. The Hitachi sheet-metal CAD/CAM system (Shibata and Kunitomo, 1981) uses a four-digit structure code to access a database of sheet-metal configurations, each of which is associated with a graph representation. This improves the retrieval and modification of CAD models that represent sheet-metal plates with holes, notches, and flanges. None of these systems, however, provides cost data as feedback or assistance in the evaluation of design alternatives from the perspective of manufacturability.

In an effort to provide both quantitative measures of producibility and redesign suggestions, a system that analyzes planar stampings with holes was developed by Schmitz and Desa (1989). This system requires that the design be input to a non-manifold geometric modeler (NOODLES). Relevant part attributes and their parametric values are then extracted from the geometric model and conveyed to a domain mapper that generates a strip layout for a progressive tool. A series of manufacturing cost factors is then computed to provide a measure of producibility, as well as the basis for redesign suggestions. The system is limited to small planar stampings, with holes and notches, produced on progressive dies. It does not account for bent or formed parts, or parts with such features as extruded holes and embossings. The manufacturing cost factors are both quantitative (such as the number of punching tools) and qualitative (such as the need for pressure pads). The unification of these factors into a measure of manufacturability is subjective; it yields a value that is not an objective measure of producibility. Consequently, the importance of redesign suggestions is also dependent on the subjectivity of this measure.

Ulrich and Graham (1990) present a method for synthesizing simple blanked and bent parts used as supports. A solution map is constructed in three-dimensions representing a network of faces that connect two interface surfaces and avoid any intervening obstacles. Candidate configurations are automatically generated and pruned by a limited set of producibility constraints.

The approach adopted in this paper is an attempt to use a part coding and classification approach at the early concept stage of design, which communicates process knowledge in terms that are easily understood by product designers. This will permit quick evaluation of a broad range of stamped part designs from the perspective of relative part cost, and offer redesign suggestions that can achieve significant percentage savings in cost.

REFERENCES

Bradley, J. "Die Estimating." Pressworking: Stampings and Dies. *Dearborn, MI: Society of Manufacturing Engineers, 1980.*

Fogg, B., Jamieson, G. A, and Chisholm, A. W. J. "Component Classification as an Aid to the Design of Press Tools." Annals of the C.I.R.P, *XVIV, 1971.*

Harig, H. Estimating Stamping Dies. *Scottsdale, AZ: Harig Educational Systems Inc., 1977.*

Homann, H. W, Guhring, H., and Brankamp, K. "Ein Klassifiizerungsystem fur Stahlbaueinzelteik: Enwicklung und Beschreibung des Systems." Industrie Anziger, *vol. 93, 1970.*

Poli, C., and Divgi, J. "Product Design for Economical Injection Molding: Part Shape Analysis for Mold Manufacturability." Proceeding of ANTEC, 1987, *pp. 1148–1152.*

Poli, C., Dastidar, P., and Mahajan, P. "Design for Stamping: Coding System for Relative Tooling Cost." Mechanical Engineering Dept., Univ. of Massachusetts, Amherst, August 1990.

Poli, C., Dastidar, P., and Mahajan, P. "Design for Stamping: Part II—Quantification of Part Attributes on the Tooling Cost for Small Parts." Proceedings of the ASME Design Theory and Methodology Conference, Phoenix, 1992.

Sackett, P. J., and Holbrook, A. E. K. "DFA as a Primary Process Decreases Design Deficiencies." Assembly Automation, *December 1988.*

Schmitz, J. M, and Desa, S. "The Application of a Design for Producibility Methodology to Complex Stamped Parts." Technical Report, General Motors Systems Engineering Center, Dec. 21, 1989.

Shibata, Y., and Kunitomo, Y. "Sheet-Metal CAD/CAM System." Seiki Gakkai, Bulletin of the Japan Society of Precision Engineering, *vol. 15, no. 4, Dec. 1981.*

Ulrich, K. T., and Graham, P. V. "Using Producibility Constraints to Control the Automatic Generation of Sheet-Metal Structures." Proceedings of the ASME Design Theory and Methodology Conference, Chicago, Sept. 16–19, 1990.

Wozny, Michael J., et al. "A Unified Representation to Support Evaluation of Designs for Manufacturability." Proceedings of the 18th Annual NSF Grantees Conference on Design and Manufacturing Systems, Atlanta, GA, January 1992.

14.3.6 Main Body

In the main body of a Research Report, there are at least three subareas: (1) a section describing the goals, research issues, or hypotheses underlying the research together with the method used to explore or study them; (2) a description of the experiments, analyses, or simulations performed; and (3) a presentation of the results obtained.

Goals, Research Issues, Hypotheses, and Methods

Though not all research is done to prove or disprove an hypothesis in the formal scientific sense, there generally are goals and expectations that something will be learned about one or more research issues. This first section of the main body is where those goals and issues—and hypotheses if there are any—are stated and discussed.

The methods to be used to prove any hypotheses, achieve the goals, or to study or explore the research issues are also described here. It should be shown how the methods will in fact work to prove or disprove the hypotheses, to achieve the goals, or to elucidate the issues. That is, the methods should be shown to be connected to the hypotheses, goals, and issues to be studied.

Experiments, Analyses, or Simulations

If there are experiments (or the equivalent, such as surveys), then the experimental apparatus and procedures (including survey procedures) are described and discussed. Though appendices may also be needed, there should ideally be enough detail here for another researcher not only to evaluate what was done, but also to reproduce the experiments if desired.

Analyses and simulations are analogous in this discussion to physical experiments. Assumptions and simplification employed should be stated. If computer algorithms are constructed, their logic should be explained. Programming details, however, belong only in an appendix.

Results

The results of the experiments and/or analyses and simulations should be presented as completely and as clearly as possible. Actual data can be placed in an appendix.

As an example of unclear results, consider the following. The writing below is a real example except that the product names have been made fictitious.

> *For standard assumptions, the new WhizWidget assembly analysis indicated that the automatic assembly cost was 30% less expensive than the previous design. The competitor, CompWidget, was 12.7% more expensive than the new WhizWidget. Originally the CompWidget was 19.8% less expensive. The new WhizWidget assembly analysis indicated that the manual assembly cost was 42.6% less expensive than the previous WhizWidget design. The CompWidget was 31.9% more expensive to manually assemble than the new WhizWidget. Originally the CompWidget was 21.0% less expensive.*

Apparently this writer believes in the following Peter's Principle

> If you can't convince 'em, confuse 'em.

A reader of such confusing writing can only think, first, "*WHAT on earth does that say?*" and then, *"How could the study be of any use with such a confused person doing it?"* Indeed, not only is this writing unclear on a first reading, it is even more unclear after careful study! Here is a better presentation of results from a different study:

> *The results of the assembly design and cost study on the WollyWidget are as follows:*
>
> 1. *Automatic assembly is not practical for the present design.*
> 2. *For manual assembly of the current design (60 parts):*
>   *Total manual assembly time: 631 seconds/assembly*
>   *Total manual assembly cost: 2.84 $/assembly*
> 3. *The product can be redesigned to eliminate the following 31 parts:*
>   *4 main body screws, 1 charge end #1, 2 charge ends #2, 4 charge ends #3, 4 springs, 1 label sticker, 1 light cover, and 1 specification label.*
> 4. *For manual assembly of the redesigned product (29 parts):*
>   *Total manual assembly time: 276 seconds/assembly*
>   *Total manual assembly cost: 1.25 $/assembly*

Discussion of Results

In this section, the results are related back to the goals, issues, and hypotheses. What light do the results shed with respect to them? Also, the significance and implications of the results are described and discussed. Uncertainties and possible errors in the methodology or in its implementation are also described.

Sometimes a Research Report contains ideas or suggestions for continuing or future research work. This is all right, but in a business setting, one should not expect that such recommendations included at the end of a research paper are a substitute for preparing a Business Technical Report to make the recommendations effectively.

As an example of a good discussion consider the following, which is a discussion of the results comparing two alternative designs:

> DISCUSSION
>
> Tables XX, YY, and ZZ show a comparison of the alternative assembly system costs for the original design of the ABC-787 drive and for the alternate version of this drive, the ABC-787B. The analyses assumed that the ratio of faulty parts to acceptable parts is 1% and that both one-shift and two-shift operations are possible.
>
> *Table XX, which contains a parts list for both the original and alternate designs, shows that for the original version, 21 of the 36 total parts are screws, and the redesigned drive has only 7 screws of 20 total parts. In addition, the rolled point screws utilized in the 787 drive have been replaced by much easier to align "dog point" screws in the 787B version. These design changes, along with the additional changes to facilitate alignment of the various parts, result in a better than 50% reduction in manual assembly time.*
>
> *Table YY shows that for both designs the use of manual assembly is less costly than dedicated automatic assembly. For dedicated automatic assembly, however, the capital investment required for the 787B is about half that required for the 787.*
>
> *Table ZZ shows that although the anticipated production volume for the ABC-787 is insufficient to justify the use of dedicated automatic assembly, use of a product mix may justify the use of programmable automatic assembly. The results shown in Table ZZ also indicate that for the alternate 787B version, even the use of programmable assembly does not appear justified solely on the basis of cost.*

Appendices

Details of all kinds (e.g., of experimental apparatus, of analyses, of data, worksheets, etc.) belong in Appendices.

14.4 BUSINESS TECHNICAL REPORTS

14.4.1 Overview

The majority of reports written by design engineers in industry are Business Technical Reports. The main purpose of these reports is to recommend a course of action, or to provide results and conclusions relevant to short-term decisions and actions. Thus the recommendations, and/or the relevant conclusions and results, are placed first in the report. That is, when there is an action or decision recommendation, then it is stated concisely in the most prominent position in the report—at the beginning.

After actual recommendations, next in importance in a Business Technical Report are the conclusions and logic on which the recommendations are based. Next are the results on which the conclusions are based. Data and analyses are included only in Appendices in these reports. Note that this is more or less the reverse of the order in a Research Report.

Here is a recommended format for a Business Technical Report:

Title
Summary Abstract (1/4 to 1 page, absolute maximum)
- Short summary of recommendations
- Short summary of critical results, conclusions, and logic on which recommendations are based
- Short summary of important risks, benefits, and costs associated with the recommendations

Body of Report (2 to 4 or 5 pages)
- Complete recommendations
- Complete supporting arguments, including important conclusions and results
- Basis for and elaboration on the relevant risks, costs, and benefits associated with the recommendations

Appendices (as many as needed, as long as needed)
- Descriptions and documentation of what was done
- Data
- Analyses
- Methods for generalizing the results from the data and analyses
- Any Research Reports whose results are not familiar and that are critical to support conclusions or recommendations.

We will now discuss each of these sections in more detail beginning with the body of the report.

14.4.2 The Body

In the body of a Business Technical Report the recommendations come first, conclusions next, and then results. Data and analyses, unless there is something remarkable about them, come last, usually in one or more appendices.

Recommendations

A recommendation is some action you believe the company should take, or some decision that you think should be made. After a study of customer preferences you might recommend, for example, that a marketing study be initiated that would determine the expected added sales if your firm's coffeemakers had a customer-controlled variable temperature feature. You might also recommend that a parallel engineering design feasibility study be initiated that would determine the technical feasibility and manufacturing costs of producing such a suitable variable temperature coffeemaker. Or, you might recommend that no changes be made in the design.

Note that recommendations require consideration of business and perhaps other technical issues that may not be a direct part of the project being reported. Thus summaries of the benefits, costs, risks, and uncertainties associated with the recommendations are also included. It is very important that risks and costs be included; the last thing you want to do is recommend some action without informing the readers of the associated risks and costs as well as the benefits.

Note also that if all you are doing is reporting the results of the study, with no recommendations or conclusions relevant to pending decisions, then the

Research Report format is the better choice. If, however, you are making recommendations for action (especially actions that cost time or money), then the Business Technical Report format is better because you want those recommendations and the business reasons for them to be up front. You don't want readers to have to read through all of the detail of a Research Report in order to get to the business recommendation or critical conclusions related to pending business decisions.

Conclusions

Conclusions are generalizations of results from the reported project. For example, in thc coffee temperature case, a conclusion might be that the best temperature to provide is 165°F. Or we might conclude that an adjustable, variable temperature between 160° and 180° would be the most desirable design goal.

Results

Results are generalizations made from the data and analyses. For example, in a coffee temperature preference study, the results might say, for example, that 20% of 150 people surveyed prefer the temperature to be less than 160°F, that 50% prefer a temperature between 160°F and 170°F, and that 30% prefer greater than 170°F. The exact form of these summary results will of course depend on the nature of the experiments or analyses done. Note that such results are generalizations of the data, not the actual data. Data belong only in an Appendix, unless there is some very important conclusion to be made about the data itself that requires it to be included.

Results are included primarily to explain how and why the conclusions were determined. That is, results included are selected to show the basis for and to support the conclusions. It may also be necessary to refer to the results, conclusions, or recommendations of other studies to support your recommendations.

Appendices

In a Business Technical Report, there may be any number of Appendices that include the data taken, details of analyses or experiments performed, and whatever other details might be needed by someone at some future date. The Appendices are, in effect, a detailed record of thc activities of the project or study. Anyone questioning, for example, certain of the results could find the actual data, computations, and analyses in the Appendices on which the more general results were based.

14.4.3 Titles

Everything said in Section 14.3.2 about titles for Research Reports applies to titles for Business Technical Reports as well. Of course, in addition, the title of a Business Technical Report should clearly state that there is a recommendation or conclusions relevant to short-term actions or decisions. For example:

> *Recommendations for redesign of the ABC-786 Widget in order to reduce manufacturing costs*

On the other hand, here is an example of a poor title:

> *Analysis of the ABC-786 Widget*

The first title tells us exactly what to expect in the report. The second title fails to give us even a clue as to what the report is about. It tells us that it deals with the 786 Widget but we don't know if the analysis is one concerning manufacturing, functionality, marketing, or something else.

14.4.4 Abstracts

Every Business Technical Report must begin with a short Summary Abstract, preferably one-quarter to one-half page, and never more than one page. The purpose of this Summary Abstract in a Business Technical Report is to provide a short summary of the recommendations, conclusions, and results for readers who do not need or want to know more of the details. The same order (i.e., recommendations first) that is used in the body of a Business Technical Report is also used in its Summary Abstract.

The Summary Abstract of a Business Technical Report contains no descriptive material; that is, it does not describe what was done, or even why. Whatever is necessary for descriptive material to introduce the subject of the report can be included in a brief sentence or two in the covering letter or memo. (But remember KISS.) In other words, the Summary Abstract of a Business Technical Report gives only summaries of only the most important recommendations, and if needed and appropriate, summaries of the most important conclusions, results, logic, benefits, and risks. Whatever descriptions there are of what was done must be somewhere else, mostly in the Appendices.

Here is an example of a good summary Abstract:

The following redesign recommendations are made with regard to the current ABC-786 Widget:

1) *Replace the current motor, toggle switch, and loose wiring used to connect the switch and motor with the new CW-34 motor that incorporates the motor and toggle switch into a single unit;*
2) *Incorporate the energy pack into the handle subassembly; and*
3) *Incorporate the latch button into the energy pack portion of the handle subassembly, rather than assembling the button as a separate part.*

These recommendations are made based on the design for manufacturing and assembly analysis recently carried out by the Advanced Design and Manufacturing Group. Their analysis shows that a total savings in assembly time of over 50% can be achieved by incorporating these redesign suggestions. The suggestions can be implemented without the need to redesign any of the special purpose parts that are contained in the 786 Widget.

Here is an example of a poor abstract:

An analysis is done by decomposing the ABC-786 Widget into its component parts and subassemblies. The motor and its associated wires and switches are not considered here, as the motor is a standard part. The energy pack subassembly includes a snap-fit coupling to the nozzle subassembly and is attached to the handle subassembly with screws.

Each subassembly and part is analyzed for ease of manual assembly and manufacturability. Manual assembly requires 2 min, 13 sec. Relative manufacturing costs are calculated as 20.9 where the reference part is a 1-mm thick flat washer with inner and outer diameters of 60 mm and 72 mm, respectively.

A savings in assembly time of up to 52.3% can be achieved by:

1. *Selecting a standard part motor with a toggle ignition switch and tangle-free wiring;*
2. *Incorporating the energy pack into the handle subassembly; and*
3. *Incorporating the latch button into the energy pack portion of the handle subassembly, rather than assembling the button as a separate part.*

Notice the difference in the two example abstracts above. The first starts out, as it should, with concrete recommendations. The second does not even contain any recommendations. The first abstract does not contain any descriptive material concerning the product. The second abstract contains descriptive material concerning both the product and the procedure used to analyze the product. Finally, the second paragraph in the second abstract is so unclear that it probably makes the reader suspicious of everything contained in the abstract or the report itself.

14.5 HYBRID TECHNICAL REPORTS

Business recommendations are often made on the basis of research results. When this is the case, one can always prepare a Business Technical Report; that format permits the research results to be included as described above. But if the research requires a lengthy explanation, then such a report can become too long. There are also other ways to present both the business recommendations and the Research Report together.

One way is to prepare a Research Report and send or deliver it with a covering memo or letter that contains the recommendations. In this case, the reader gets the recommendations first in the short covering document, and then the abstract of the Research Report will be the next thing read. If that abstract is well done, then those readers with no need for more details can stop after reading it. This is thus a reasonably efficient format to follow.

Another way is to prepare a Research Report, but use an abstract that combines the information usually found in the abstracts of Research Reports and Business Technical Reports. This way the recommendations and short-term conclusions or results are up front as in a Business Technical Report.

Both of these hybrid forms are best used only when the Research Report is fairly short. If the Research Report is long and involved, then it might be necessary to prepare it separately, and then also to prepare a Business Technical Report or covering letter or memo for the recommendations and/or conclusions that are needed immediately.

Selecting the best format for a report should be done pragmatically: What is the best format for the report given its content, purpose, and the intended audience? Any format will do if it properly meets the purposes of the message and the audience(s). Thus the formats described as above should be modified and adapted as needed to each particular communication situation.

14.6 ORAL REPORTS

14.6.1 Preparation

Oral presentation of both Research Reports and Business Technical Reports is commonplace in industry, and Research Reports are almost always presented in person at conferences. Therefore, the ability to do a presentation well is important. In this section, we provide some guidelines that will help in the preparation and delivery of oral reports.

Begin Preparing Well in Advance

By "well in advance," we do not mean hours, or even days. We mean weeks, or even months in advance! As soon as you know you have an oral presentation to make, that very moment is the time to begin to prepare.

Preparation includes many of the same things that are required in preparing for a written communication. These are:

1. Explicitly state for yourself what is the purpose of the presentation. What is it that you want to accomplish? Is there something you want the listeners(s) to do? Is there information you want them to have, or that they need to know? In other words, understand for yourself why you are making the presentation.
2. Explicitly describe the audience for yourself. Who are they? Why are they interested in or wanting to hear your presentation? What words and symbols do they understand and not understand?

   Often, there are several classes of receivers in an audience, each with different needs. Each class must then be described, and the presentation constructed and organized so that each group has its needs met.
3. Decide what the desired take-aways are. In an oral presentation, there can be only a very few, perhaps only one, and no more than two or three.
4. Assuming your presentation will include a set of slides or transparencies (it normally will), make a tentative title for each.

   There can be no more than one slide or transparency for each minute you have been allotted for your presentation. And if you want to allow time for questions and discussion, then you must have fewer than one slide or transparency per minute. Remember, any message can be made any length. There is no better way to make a lousy impression on an audience than by having a presentation that is too long for the time allowed. Conversely, there hardly is any better way to make a good impression than to have a talk that is short, emphasizes what is important (the take-aways), is well organized, and allows time for questions and discussion.

   Suppose, for example, that we had an opportunity to make an oral presentation to a group of students concerning the complete content of this chapter in ten minutes. We would want to prepare about eight transparencies, a plan that should allow about two minutes for questions. The two main take-aways we could choose are: (1) the importance of keeping communications short, and (2) the importance of rewriting and rehearsing extensively. The title of the eight slides might then be:

   #1 Title Slide
   #2 The Importance of Communications in Engineering
   #3 Preparing to Write or Speak
   #4 The Writing Process: Rewriting
   #5 Format of Business Technical Reports
   #6 Format of Research Reports
   #7 Guidelines for Oral Presentations: Rehearsing to keep it short.
   #8 Summary (Review the take-aways)

   We would also certainly supplement such a short talk by handing out an example or two of actual reports that were exemplary.
5. Plan for any handouts or pieces of hardware that you can use to supplement your talk. Circulating examples of something real is always a good way to involve an audience in what you are saying.

14.6.2 Guided Iteration

An oral presentation, like a design and like a written communication, is prepared by a process of guided iteration. The steps just described have given you an initial design for your talk. Next prepare a first draft, try it, evaluate it, try it, and improve it repeatedly until you are satisfied with the content and comfortable presenting it.

14.6.3 Planning Slides or Transparencies

The first part of this process of iterative improvement is completing the slides (or transparencies) that initially were given only a title. Slides or transparencies must not be cluttered with words or complex drawings. The best slides will have three to five bullets (i.e., main subtopics) with no more than one very short phrase or sentence attached. If a slide or transparency has more than about forty or fifty words, it is probably too detailed. Shorten it to just the basic ideas and outline. Plan on elaborating while you speak.

The size of the letters on your slides and transparencies must be large enough for people in the rear of the meeting room to read easily. The minimum letter size is a 1/4 inch, and 3/8 inches or even 1/2 inches may be necessary for larger rooms. Also make sure that the layout of material on your slides is logical, balanced, and interesting. Though not everyone agrees, color is generally unnecessary, and may even detract from your message.

Never, ever, read what is on your slides to your audience. Not only is this dull and boring, but in addition they can read faster than you can speak. The information on the slides or transparencies is not there for you to read. Its purpose is to provide an outline (only) for both you and your audience to follow. Your speaking will explain and elaborate on the brief points made on the slides.

For example, here is a possible way to prepare a presentation for Slide #7 listed earlier:

Guidelines for Oral Presentations
\> *Prepare Well in Advance*
\+ *Purpose: Analyze Audience and Take-Aways*
\+ *One Slide per Minute: Prepare Titles First*
\> *Use Guided Iteration*
\+ *Complete the Slides—Fifty Words Each Maximum*
\+ *Rehearse Out Loud, Again and Again*

An alternative slide for Slide #7, and one that uses too many words, is the following:

\> *Prepare Well in Advance*
    Not hours or days but weeks or months
\> *State for yourself the purpose of the presentation*
    Why am I making the presentation?
    \+ *What you want to accomplish*
    \+ *What you want listeners to do*
    \+ *What information you want listeners to have*
\> *Analyze the audience*
    \+ *Who are they?*
    \+ *Why should they be interested?*

> Prepare Transparency Titles First
+ Keep them Brief
> No more than one slide per minute
More slides/minute = Talk will be too long
— no time for questions
— angry audience
> Use Guided Iteration
Draft, evaluate, redo, evaluate, redo
+ Complete the slides—do not use more than 50 words
+ Rehearse out loud

Here the slide becomes crowded and the audience—instead of listening to you—will simply read the slide. The more detailed ideas on the crowded slide are things you can say to the audience. The slide is an outline for the audience and you; they get the details and interesting sidelights from your talk.

14.6.4 Rehearsing

In addition to getting the slides or transparencies right by iteration, the other thing you must do is rehearse out loud—again and again and again. You begin rehearsing as soon as you have a rough draft of all the slides. You may need to rehearse out loud dozens of times (not an exaggeration!) over a period of weeks or even months, especially when this whole process is new to you, or if you are nervous about making a particular presentation. Then when the time comes to deliver your presentation, you are so familiar with it, and so confident, that you will do it in a relaxed and polished manner. You will also be able to think clearly about questions when they are asked.

When you feel you are about ready, but long before the actual presentation, practice your presentation before some friends and colleagues. They can give you valuable feedback on how your presentation can be improved.

When you rehearse, you must keep track of the elapsed time. Your actual presentation will usually take longer than when you rehearse. Cut out slides if necessary to keep within time limits. Better to have fewer ideas presented than to exceed your time. Remember, audiences will love you if you use less than your allotted time, and hate you for using more.

14.6.5 The Presentation

Whenever possible, visit the room where your presentation is to take place before the time of your talk. Make sure the projector works, that you know how to control it (e.g., on/off and focus), and that there is a table for your transparencies both before and after you show them. Find the pointer (better yet, bring your own) and think about where you will stand so the audience can see both you and your slides or transparencies.

An oral presentation is a very personal communication. You are face to face with your listeners, sometimes quite close. For small groups in a small room, there is a lot of intimacy. It is always a good idea to begin by saying something that will help you develop a positive and friendly personal relationship with the audience. In planning this, you have to consider the size of the audience, and the nature of the meeting, but it can always be done. At the least, for example, you can thank people for coming, especially considering their busy schedules. If you have been specially invited, you can express your appreciation for the honor of being asked.

A little humor is fine here, but jokes (if used) should be short and clean and funny. Otherwise, forget them. If you can tie some good, short joke or light remark to the local situation, that is good. The purpose of any humor interjected is to make the audience relate more personally and positively to you. If your joke won't do that, omit it. It is enough that you are genuinely pleased to be there, and to have them there; so just say so personally and sincerely and you will be off to a good start.

Never, ever read an oral presentation. Don't read the slides, and don't even read your own notes. By rehearsing sufficiently, and with the slides for your cues, you will know what to say without reading anything. If you absolutely must, it is acceptable (though not especially desirable) to have a few cue cards (e.g., 3 × 5 inches) with sentence-starting phrases on them to help you get started with what you want to say about each of your slides. For example, suppose we were to find in rehearsing the above slide that, for some reason, we have trouble getting started with what is to be said about "Rehearsing Out Loud." Then we might have on a cue card the phrase, "It is not enough to rehearse silently to one's self . . ." That cue on the card would be enough to get us started on this subject during the actual talk, and we could then go on elaborating from memory.

14.7 GRAPHICAL COMMUNICATIONS

14.7.1 Introduction

A picture, they say, is worth a thousand words. Maybe. It certainly is true that graphical communications are an important and efficient way for design engineers to communicate with each other, with customers, with manufacturing engineers, and with managers. Thus we introduce some of the basic concepts of graphical communications here.

Our scope is limited by space and by the expectation that students and other readers interested in this book will have previously been introduced to the fundamentals of graphical communications. Therefore we will discuss only the briefest introduction to orthographic projection. We will say only a little more about hand sketching, limiting the brief discussion to oblique and isometric forms. Because professional working standards are much higher than most students seem to think they are, we include some examples illustrating the acceptable standards for drawings in a working environment.

The goal of graphical communications, like all communications, is the effective and efficient transfer of accurate information. Thus the language and grammar used for communication must be correctly done by the sender so it will be correctly interpreted by the receiver. Students and practicing design engineers must understand graphical language and grammar, and use both properly. Proper use includes also preparing drawings that are well laid out and neat.

As noted above, we assume that readers of this book are knowledgeable about the rules and conventions of orthographic projection. Nevertheless, the next section provides a brief reminder of the most basic type; readers not thoroughly familiar with orthographic projection should consult one or more of the many books on the subject.

14.7.2 Orthographic Projection

In orthographic projection, an object is imagined to be viewed from an infinite distance along mutually perpendicular (i.e., orthogonal) axes. The resulting views

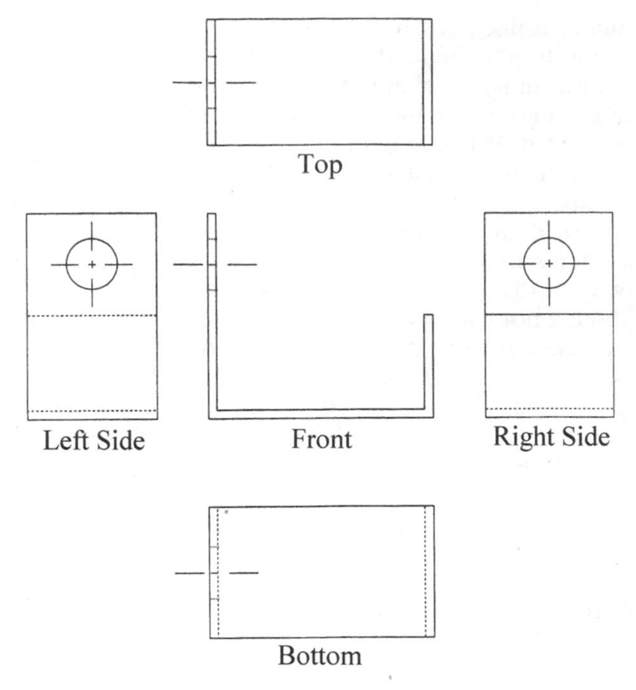

FIGURE 14.1 *An example part in five orthographic views.*

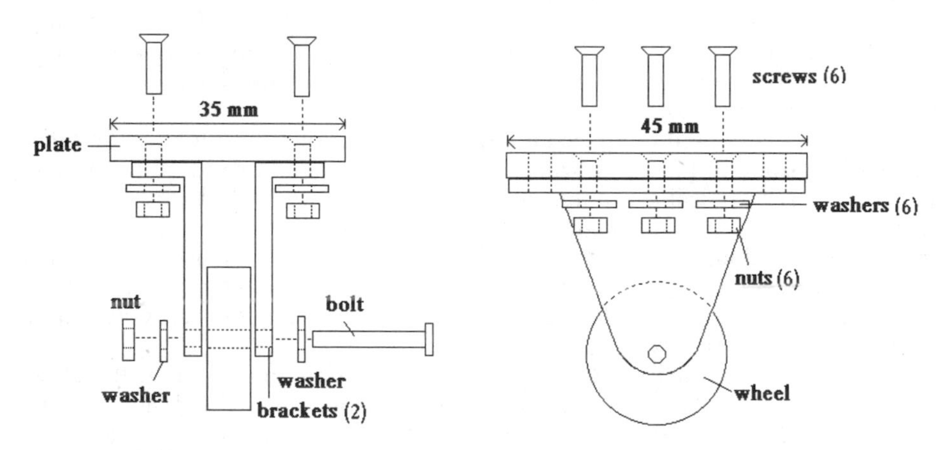

FIGURE 14.2 *An example assembly drawing in two orthographic views.*

are then drawn to scale and laid out on paper as shown in Figure 14.1. In the figure, five views are shown; a view from the rear of the part, if added, is drawn to the left of the left-side view. Often it is only necessary for a complete description of a part to show just two or three views.

Figure 14.2 shows an assembly drawing in two orthographic exploded views, and Figure 14.3 shows orthographic views of two parts of a proposed configuration for the food-carrying portions of a food scale.

Dimensioning

There are very definite rules and conventions for proper dimensioning that, again, we assume have been or will be learned elsewhere by our readers. Dimen-

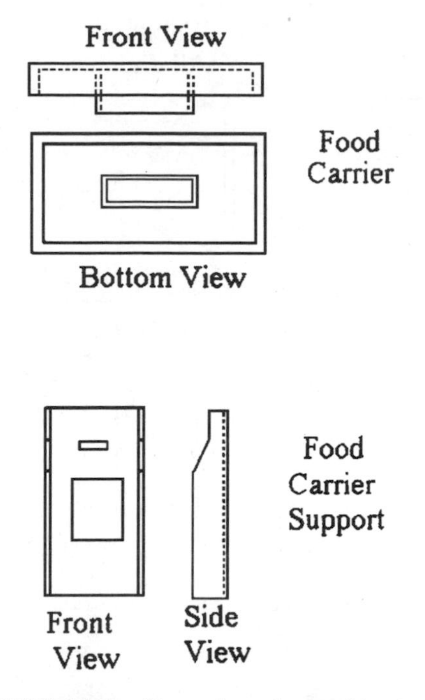

FIGURE 14.3 *Example orthographic views.*

sioning is an important subject. Not only does improper dimensioning (like improper spelling and grammar) indicate a serious degree of carelessness, but improper dimensioning may also lead to production of parts not in accordance with designers' expectations. Figure 14.4 shows a set of dimensions added for the part shown in Figure 14.1. Tolerances are not shown.

14.7.3 Pictorial Views

Orthographic projection is the most rigorous way to represent an object graphically. The representation can be as complete and detailed as desired. Often, however, it is helpful to the communication process to show a pictorial view that is more easily and quickly visualized. Designing and redesigning at early stages are also often done using pictorial sketches.

There are several types of pictorial views, but we will look at only the two most commonly used here. For more information on pictorial views, see Earle, James H., 1999, *Communication in Engineering Design.* The two we will discuss are (1) oblique and (2) isometric. Figure 14.5 shows an oblique view of the object shown in Figure 14.1. Figure 14.6 shows an isometric view of the same object.

Oblique Views

In an oblique view the orthogonal axes are as shown in Figure 14.7. The front face of the object is drawn in true view to full scale. Thus a one-inch solid cube is as shown in Figure 14.8a. Note that lengths on the front face are shown at full scale, whereas lengths along the oblique direction (at 30 degrees to the horizontal) are drawn to half scale. (If the oblique lengths are drawn full scale, the cube would look distorted; try it.)

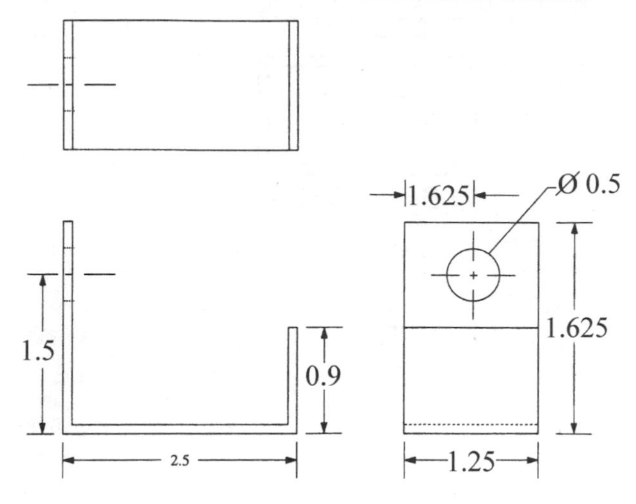

FIGURE 14.4 *Example orthographic view with dimensions added. Tolerances not shown.*

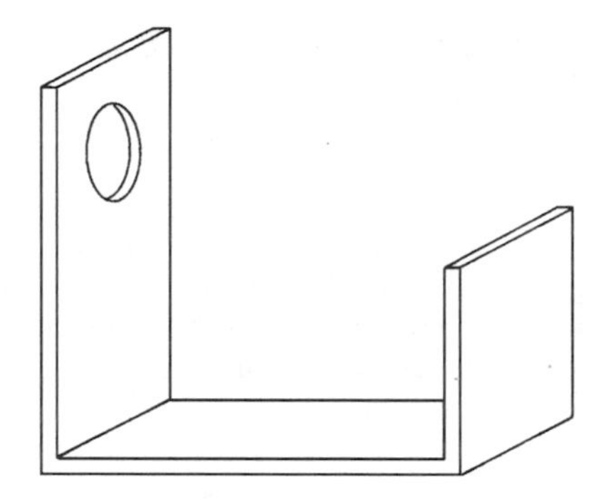

FIGURE 14.5 *An oblique view.*

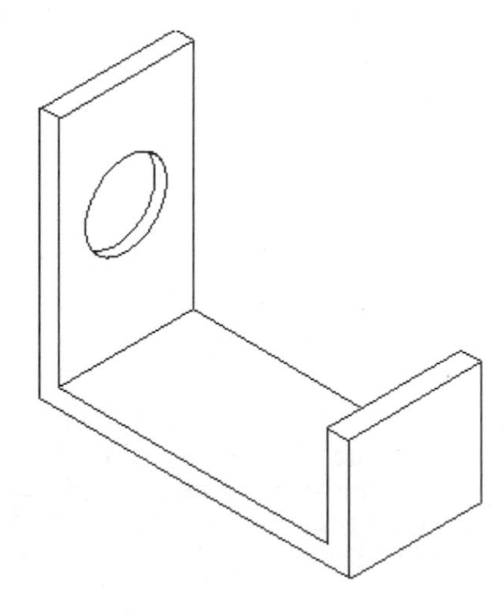

FIGURE 14.6 *An isometric view.*

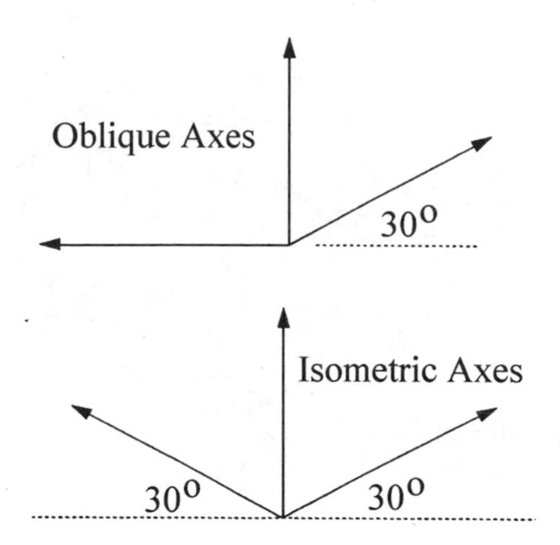

FIGURE 14.7 *Orthogonal axes for oblique and isometric views.*

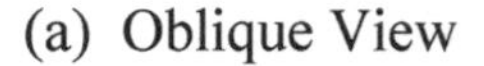

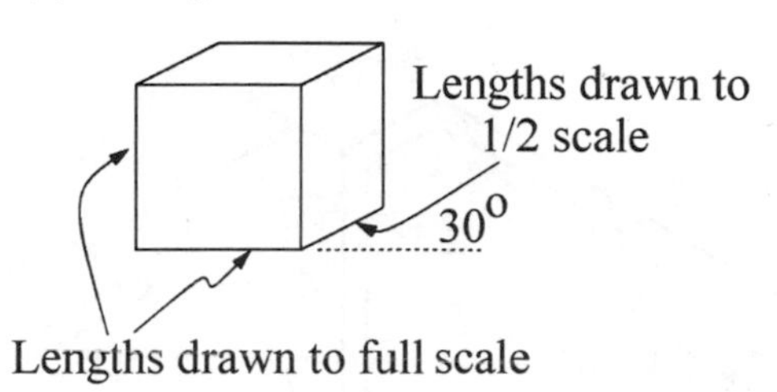

(b) Isometric View

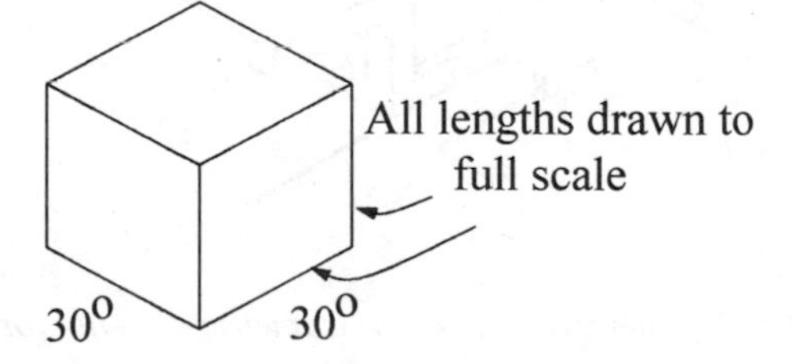

FIGURE 14.8 *Oblique and isometric views of a cube.*

One advantage to the oblique view is that circles, arcs, and any contours appear in their true shape in the front face of the sketch. In other directions, however, circles appear as ellipses, which are a bit more difficult to draw. However, often the object can be oriented in the pictorial so that the circular parts are parallel to the front face.

Isometric Views

In an isometric view the orthogonal axes are as shown in Figure 14.7. Dimensions along all the axes are drawn to full scale. Thus a one-inch solid cube is as shown in Figure 14.8b.

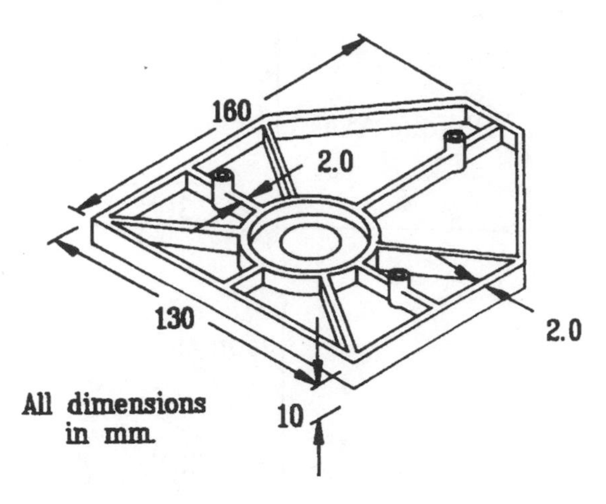

FIGURE 14.9 *An example of an isometric drawing of a production part.*

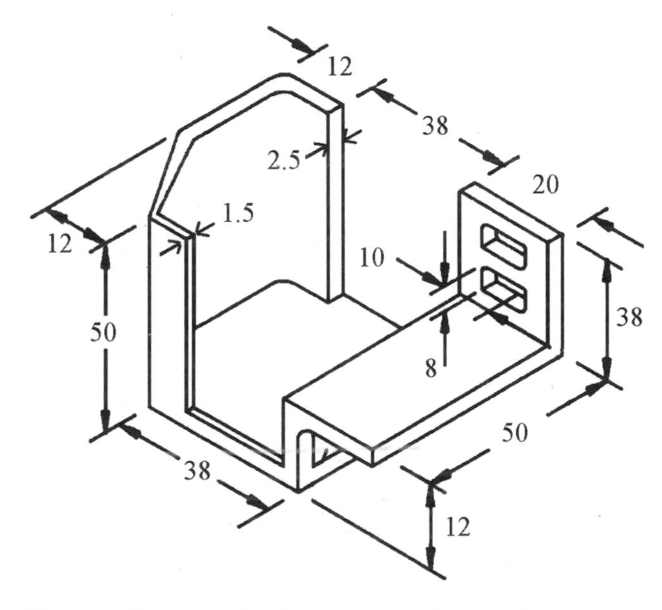

FIGURE 14.10 *Another example of an isometric drawing for a production part.*

Examples of parts drawn in an isometric view are shown in Figures 14.9, 14.10, and 14.11. Note in Figure 14.11 how the cut-away enables visualization of internal details.

14.7.4 Exploded Assembly Views

It is often helpful to develop drawings that illustrate graphically how the parts of an assembly relate spatially to one another. The best way to do this is with a pictorial called an exploded view of the assembly. In an exploded view, parts are shown individually but in a location and orientation along their axes of insertion

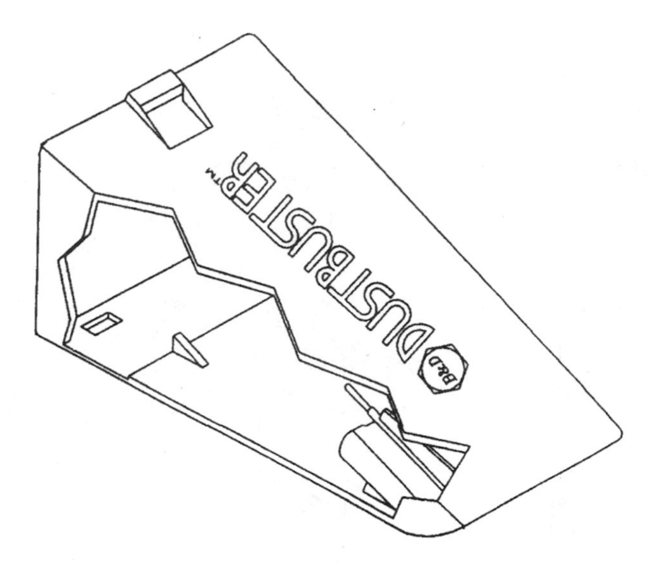

FIGURE 14.11 *Part of a hand vacuum cleaner (by G. Moodie).*

during assembly. As an example, see the isometric exploded view of a three-hole paper punch shown in Figure 14.12 and submitted as part of a class project. Note how the use of dashed center lines helps in the visualization of how the parts are assembled.

Other examples of isometric exploded assembly drawings submitted as part of class projects are shown in Figures 14.13 and 14.14.

14.7.5 Standards of Acceptability for Drawings and Sketches

The purpose of drawings and sketches, including those in CAD systems, is effective communication. For effective communication to take place, drawings and sketches must not only express their technical content in proper graphical language, they must also meet an appropriate level of quality in terms of their presentation. A drawing, and certainly a sketch, in an ongoing iterative design environment does not have to be finished so as to meet formal drafting standards. But it does have to be neat, of suitable size, and clear, and it has to communicate without extra effort on the part of others to overcome sloppiness, carelessness, and important omissions.

For example, lines that are supposed to be in line should look like they are in line. Lines that are supposed to be perpendicular should look like they are perpendicular. Circles should look like circles. Bigger parts should look bigger than the smaller ones. Straight lines should look essentially straight. Lines that are supposed to intersect should in fact intersect. And so on.

There is a tendency for students not familiar with professional standards in these regards to submit drawings and sketches that are far too sloppy or that fail to communicate properly. The drawings shown thus far in this chapter are all taken from student term project reports and meet at least a minimum standard for such communications. (Your instructor or boss may disagree; that is fine. Use his or her standards if they are higher than those exhibited here, but lower is not acceptable.)

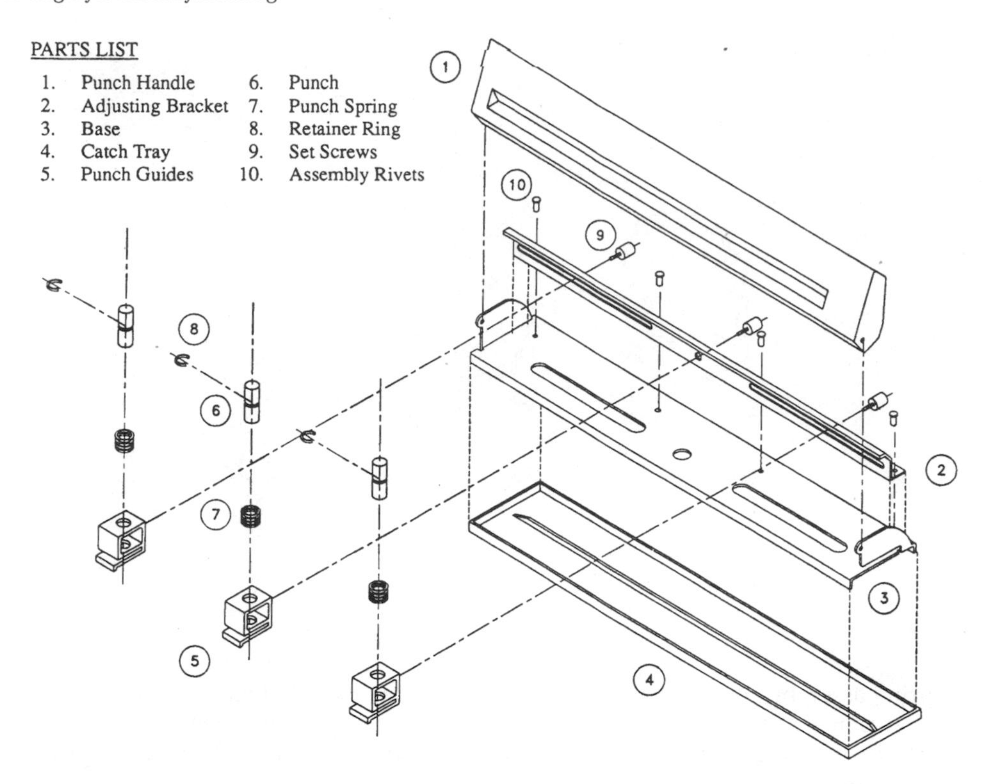

FIGURE 14.12 *Adjustable three-hole punch (by J. Harve and A. Tacke).*

Figures 14.12 to 14.14 are examples of student assembly drawings that are up to acceptable standards for a student project report. For examples of drawings and sketches that are not acceptable, see Figures 14.15 through 14.18. In none of these sketches is it possible to visualize the relationship of the parts. The dashed lines that show the connections are missing. Layouts are off balance. In some, the parts seem almost randomly placed. Notice that great effort went into preparing these very poor communications. However, the effort was not sufficiently supported by thought, critical evaluation, and redrawing to make the communications effective.

14.8 SUMMARY

- > Know the purpose of your communication, and the take-aways
- > Know the receivers—what words and symbols they understand, what their needs are
- > If a written message:
  - + Make an outline and a complete rough first draft as a first step
  - + Rewrite, rewrite, rewrite, rewrite, rewrite . . .
  - + Spelling and grammar (sentences, punctuation, paragraph structure, etc.) must be perfect
  - + Keep a copy
  - + Short is critical (KISS: Keep it Short, then Shorten it some more)
- > If a Research Report:

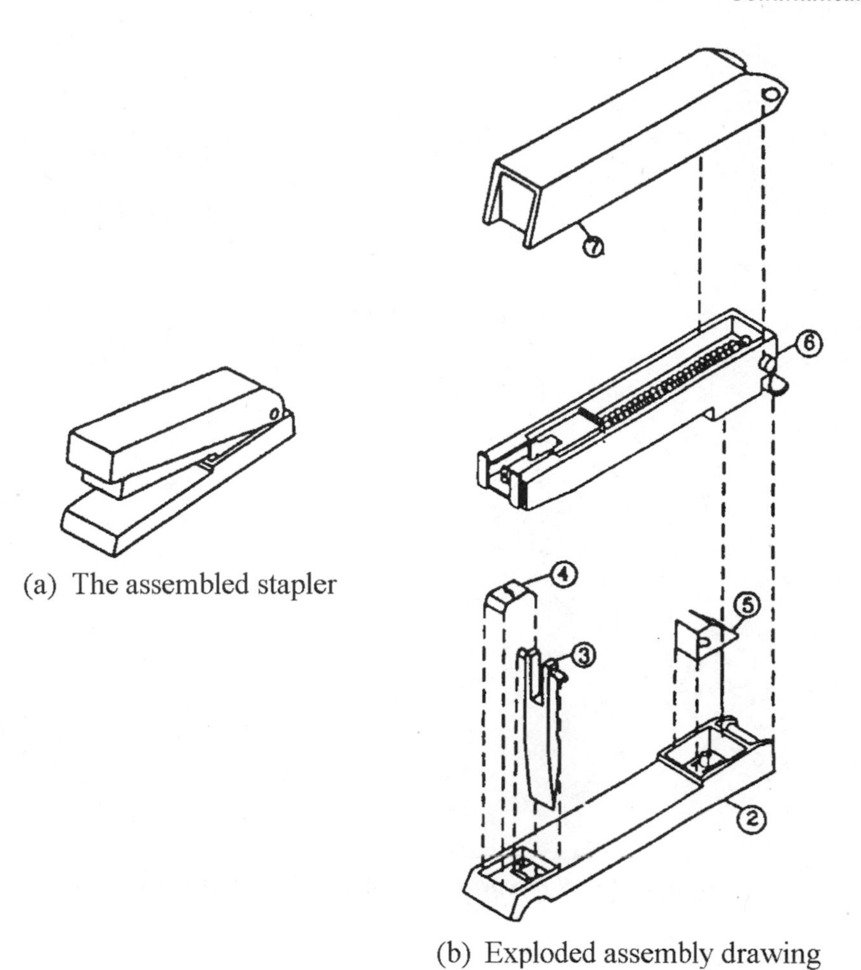

FIGURE 14.13 *A small stapler (by N. Renganath and R. Rajkumar).*

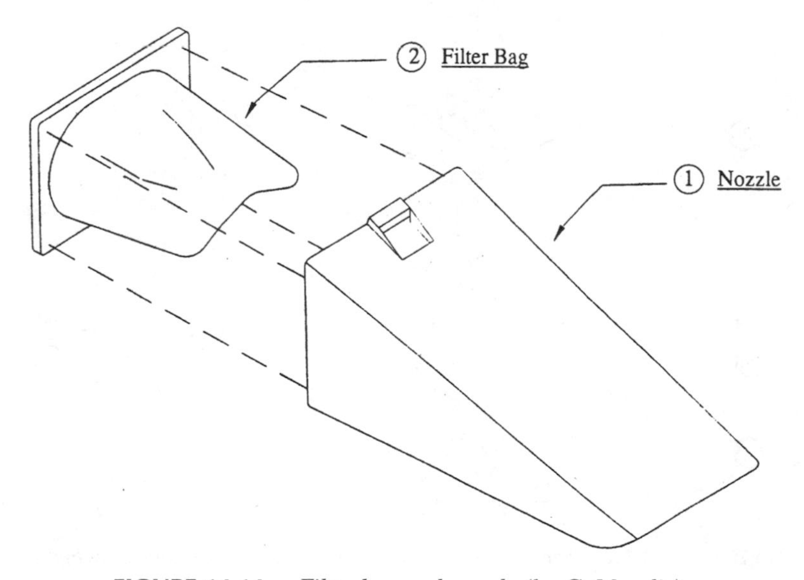

FIGURE 14.14 *Filter bag and nozzle (by G. Moodie).*

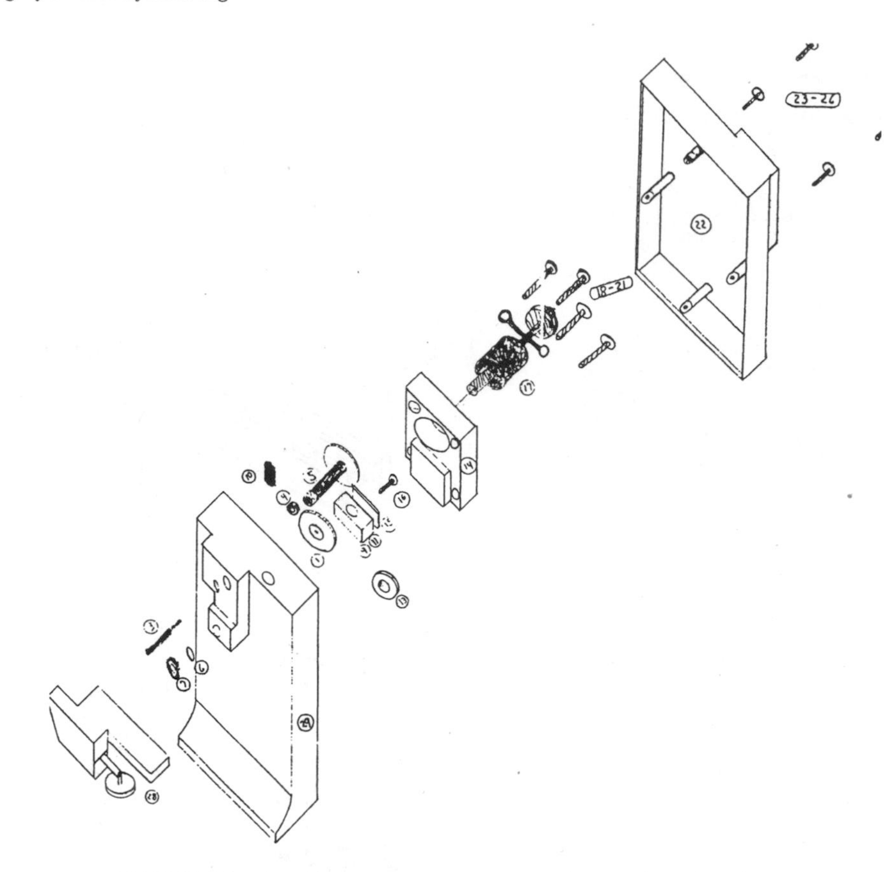

FIGURE 14.15 *Can opener: an unacceptable student drawing.*

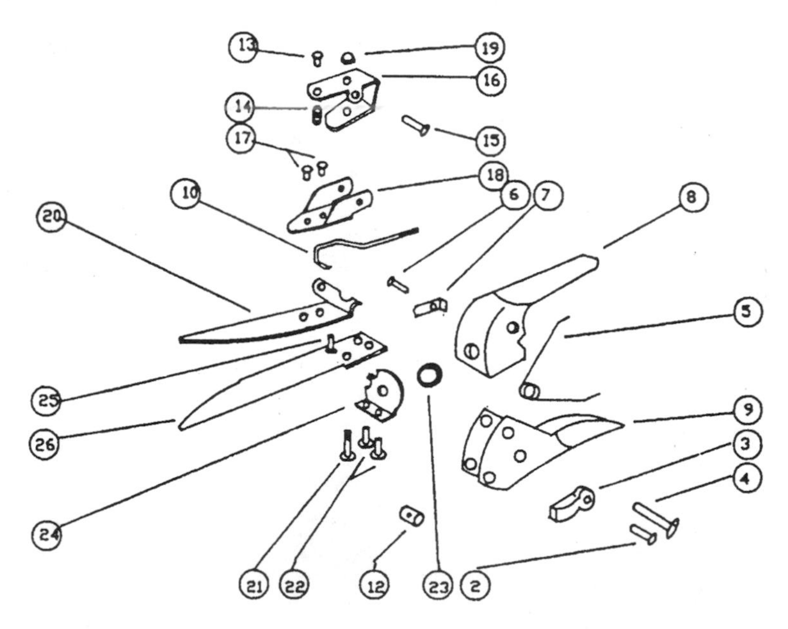

FIGURE 14.16 *Secondary shears: an unacceptable student drawing.*

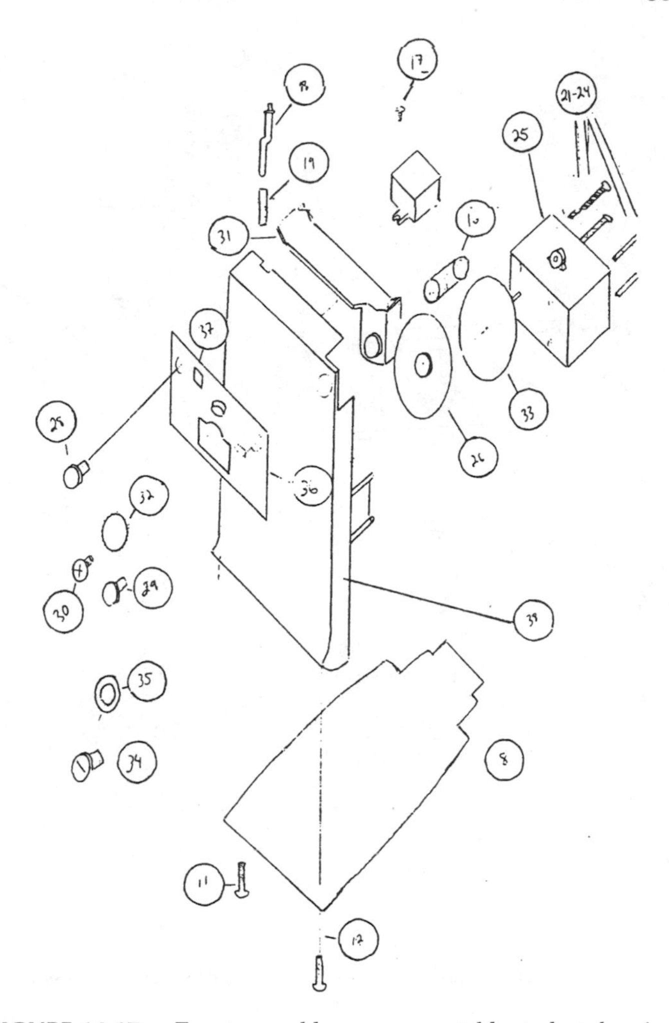

FIGURE 14.17 *Front assembly: an unacceptable student drawing.*

- + Use a descriptive plus summary-type abstract
- + Organization: Introduction, Literature Review, Body, Discussion, and Summary and Conclusions
- + Appendices

> If a Business Technical Report:
- + Use a summary-type abstract
- + Focus on the recommendations, conclusions, results, and data (in that order)
- + Details in appendices
- + Short (except appendices) is especially critical

> Make report titles complete and incisively descriptive of a report's contents

> If an oral report:
- + One slide or transparency per minute—maximum
- + Three or four bullets, and fifty words per slide—maximum
- + Never read from a slide or from notes
- + Rehearse, rehearse, rehearse, rehearse . . .

> Do not submit sloppy drawings or sketches. Maintain professional standards.

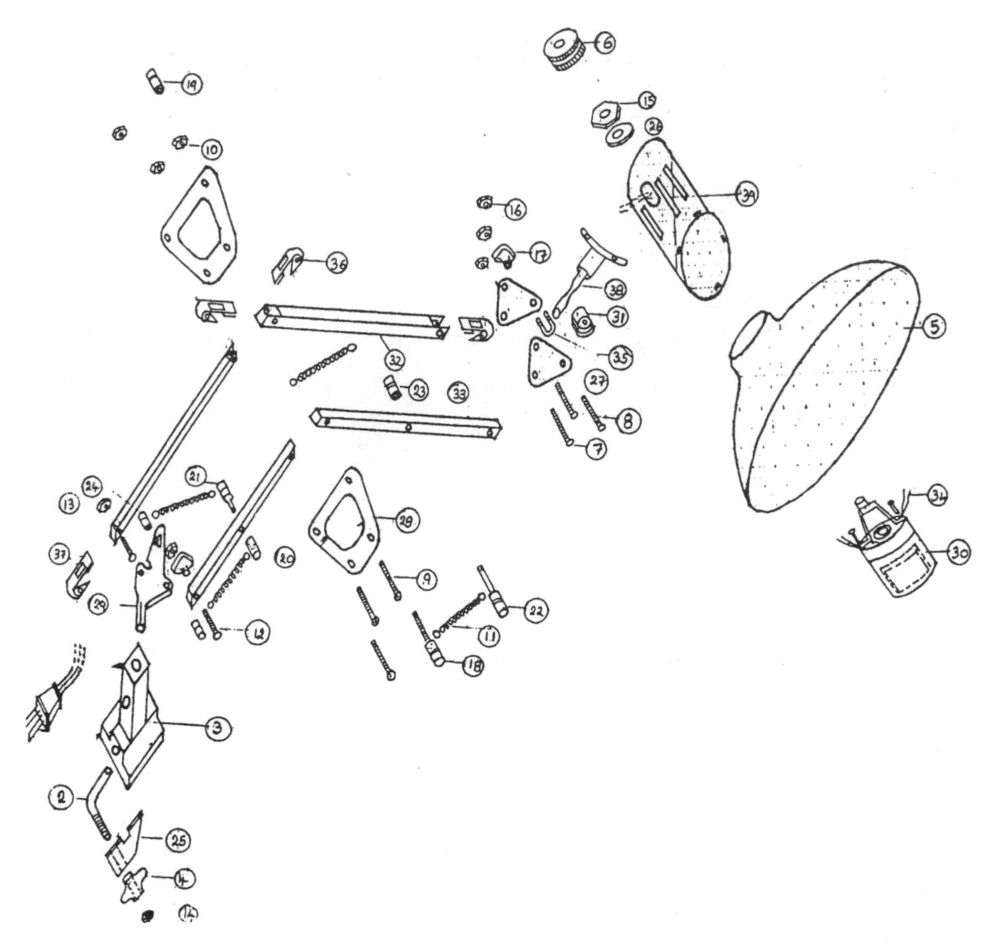

FIGURE 14.18 *Table lamp: an unacceptable student drawing.*

REFERENCES

Earle, James H. *Engineering Design Graphics*. Reading, MA: Addison-Wesley Longman, 1999.

Zinsser, Hans. *On Writing Well*, 5th edition. New York: HarperCollins, 1994.

QUESTIONS AND PROBLEMS

14.1 In Chaper 12, Problem 12.1, you were asked to analyze the assembly shown in Figure P12.1, make redesign suggestions in order to reduce assembly costs, and to estimate the approximate savings in assembly costs. Assume that you are a design engineer working for the ABC Manufacturing Company, which produces the assembly in question, and that you are writing a Business Technical Report that is to go to your immediate supervisor.

a) What would you use for a title for this report?
b) Write an Abstract of no more than 200 words for this report.
c) Prepare a Results section for the Main Body of this report.

d) What would you use for a title if this were a Research Report?
e) What would you use for an Abstract if this were a Research Report.

14.2 In Chaper 12, Problem 12.2, you were asked to analyze the assembly shown in Figure P12.2, make redesign suggestions in order to reduce assembly costs, and to estimate the approximate savings in assembly costs. Assume that you are a design engineer working for the ABC Manufacturing Company, which produces the assembly in question, and that you are writing a Business Technical Report that is to go to your immediate supervisor.
a) What would you use for a title for this report?
b) Write an Abstract of no more than 200 words for this report.
c) Prepare a Results section for the Main Body of this report.

14.3 Assume that you work for the ABC Manufacturing Company and that you are involved in a two-week training program for new engineering hires. Pretend that as part of this training program you have written Section 14.7 "Graphical Communications" as a stand-alone report that is given to the trainees.
a) Write a title for this report that differs from the one used for Section 14.7.
b) Write an Abstract of no more than 200 words for this report.
c) Write an Abstract of no more than 100 words for this report.

14.4 Pretend that you are the Manager of the Design Group within the ABC Manufacturing Company, which is responsible for the design of the floppy disk drive analyzed in Chapter 12, Exercise 12.2. Prepare a memo to be sent to A. B. Curtis, Vice-President of Engineering indicating
a) What the preliminary results of Exercise 12.2 indicate, and
b) What you expect to do as a result of these preliminary results. How would your memo differ if instead of the product used in Exercise 12.2, your product was the one considered in Exercise 12.2?

14.5 In Chapter 12, Figure 12.14, an Assembly Advisor was presented. Pretend you are writing a Research Report discussing the origins and use of this DFA advisor. Prepare the following:
a) A title for this report.
b) An Abstract for this report.
c) A Results section for the Main Body of this report.

14.6 Pretend that you are the author of this book. Write a Research Report Abstract for the following chapters:
a) Chapter 4, "Injection Molding: Relative Tooling Cost"
b) Chapter 5, "Injection Molding: Total Relative Part Cost"
c) Chapter 9, "Stamping: Relative Tooling Cost"

14.7 Pretend that you are the author of this book. Write a Business Technical Report Abstract for the following chapters:
a) Chapter 4, "Injection Molding: Relative Tooling Cost"
b) Chapter 5, "Injection Molding: Total Relative Part Cost"
c) Chapter 9, "Stamping: Relative Tooling Cost"

14.8 In Chapter 4, Figure 4.1, a classification system for the basic tool complexity of injection-molded parts was presented. Assume that you are preparing a Research Report titled, "Design for Injection Molding: An Analysis of Part Attributes that Impact Die Construction Costs for Injection-Molded Parts." Write the Results section for this report, indicating what the data presented in Figure 4.1 demonstrates.

14.9 In Section 4.11 of Chapter 4, it was shown that if the injection-molded part shown in Figure 4.26 was redesigned as shown in Figure 4.27, a 25% reduction in mold construction costs would result. Prepare a Business

Technical Report describing these two alternative designs. Be certain to include a recommendation in your report.

14.10 In Section 9.14 of Chapter 9, it was shown that a savings in tooling cost could be achieved for the part shown in Figure 9.43, if it were redesigned as shown in Figure 9.45. Prepare a Business Technical Report describing the three alternative designs and recommend what action should be taken.

14.11 In Section 14.2.2 there is a paragraph used as an example of poor writing. Rewrite it to eliminate the wordiness, and fix the spelling and grammar.

14.12 In Chapter 12, "Assembly," you were introduced to the idea of design for ease of assembly (DFA) and to the Assembly Advisor as an aid designing for ease of assembly. Select a product, preferably one you are involved in designing as part of a semester-long project, and prepare an oral presentation describing your use of DFA and the Assembly Advisor.

14.13 In Chapter 9, Tables 9.1 and 9.2, algorithms for determining the total number of active stations required for shearing, wipe forming and side-action features of stamped parts were presented. Prepare an oral presentation in which you describe to the audience the effect of part features on the tooling cost of stamped parts.

14.14 Prepare an oral presentation to accompany the Business Technical Report that you prepared for Exercise 14.9 above.

14.15 Prepare an oral presentation to accompany the Business Technical Report that you prepared for Exercise 14.10 above.

14.16 In place of the memo you wrote as part of Exercise 14.4, prepare a brief 5-minute oral presentation to be made to Vice-President Curtis.

14.17 Draw the top, front, and right side view for the parts shown in Figure P14.17.

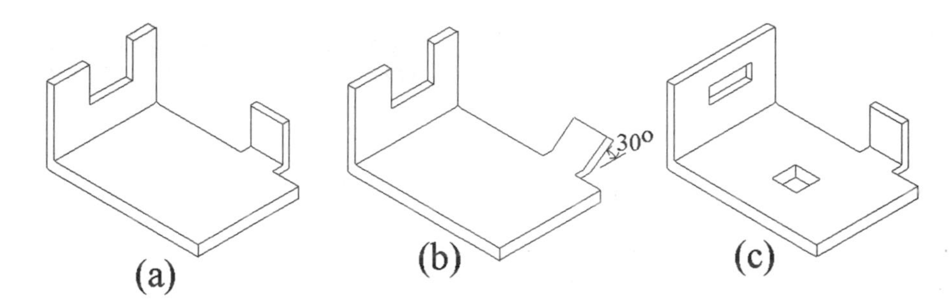

FIGURE P14.17 *For Problem 14.17.*

14.18 Draw the top, front, and right side view for the part shown in Figure P14.18.

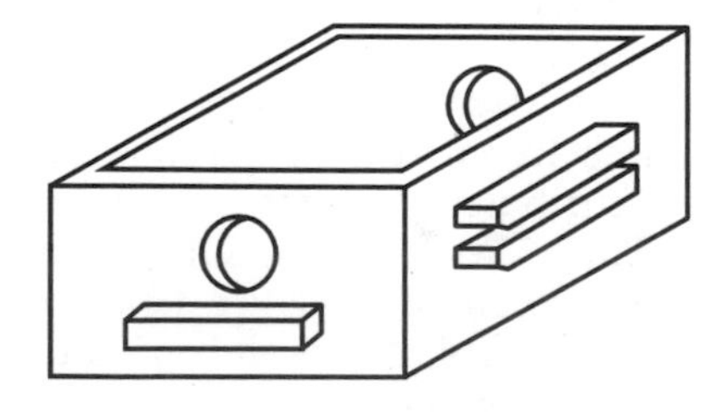

FIGURE P14.18 *For Problem 14.18.*

14.19 Draw the top, front, and right side view for the part shown in Figure P14.19.

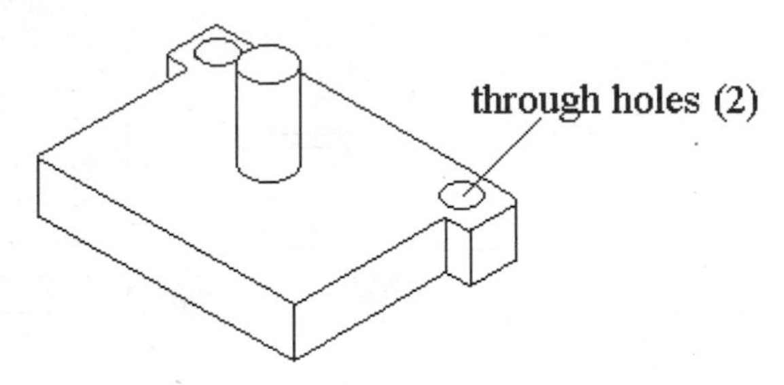

FIGURE P14.19 *For Problem 14.19.*

14.20 Draw the top, front, and right side view for the part shown in Figure P14.20.

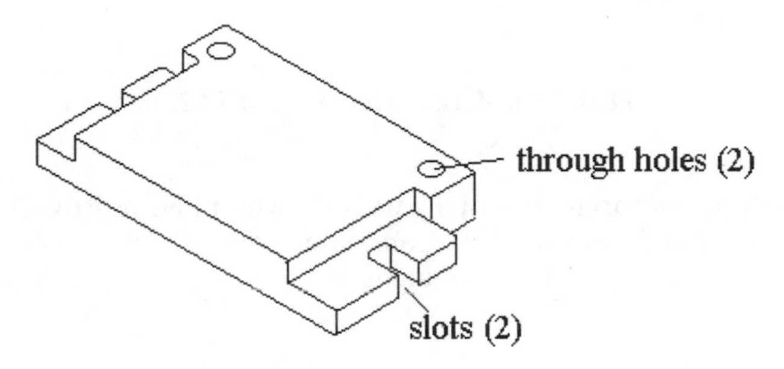

FIGURE P14.20 *For Problem 14.20.*

14.21 Draw the top, front, and right side view for the part shown in Figure P14.21.

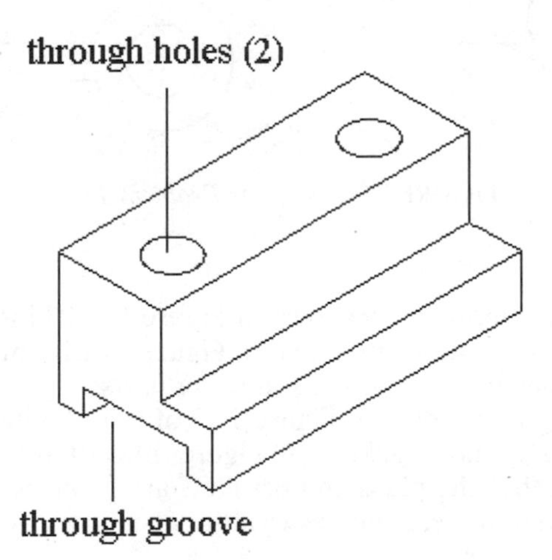

FIGURE P14.21 *For Problem 14.21.*

14.22 Shown in Figure P14.22 are orthographic projections for various parts. Are the top, front, and right side views shown correctly? What changes are required to make them correct?

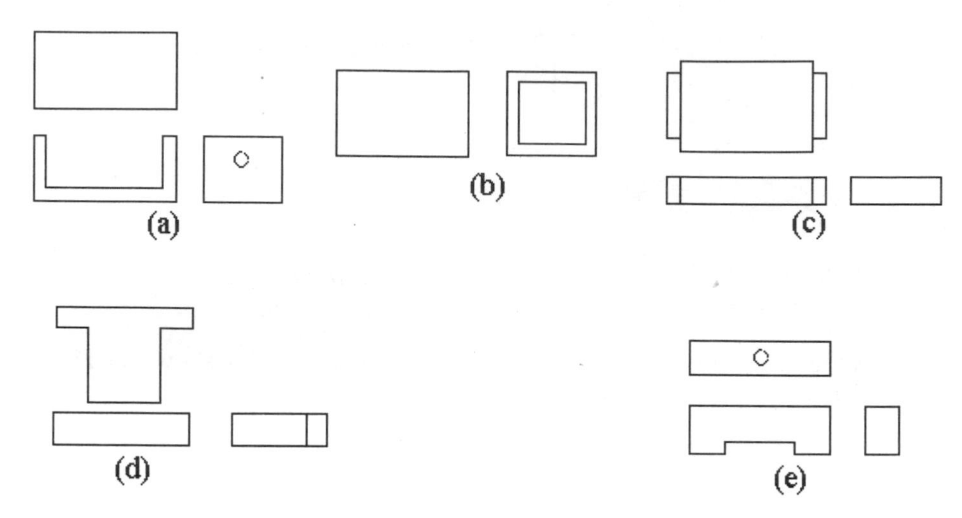

FIGURE P14.22 *For Problem 14.22.*

14.23 An oblique pictorial view of a part is shown in Figure P14.23. Also shown in Figure P14.23 are the top, front, and right side views of this part. Are these views correct? What changes are required to make them correct?

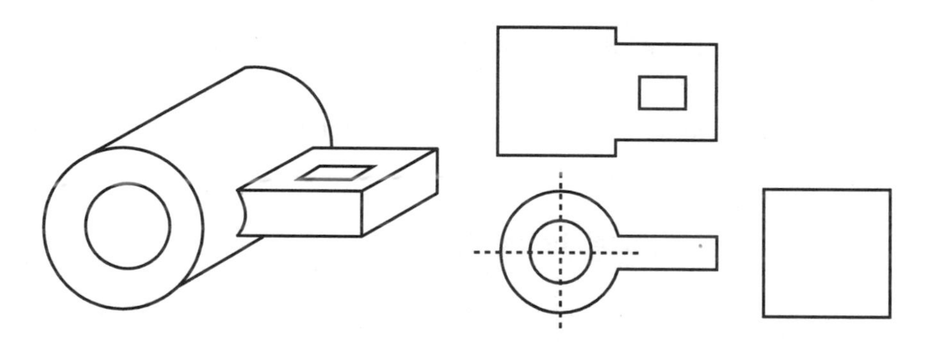

FIGURE P14.23 *For Problem 14.23.*

14.24 Convert the oblique view shown in Figure P14.23 into an isometric view.
14.25 Convert the oblique view shown in Figure P14.18 into an isometric view.
14.26 Create isometric views of the following parts:

Food carrier shown in Figure 14.3 of this chapter.

The plate and brackets of Figure 14.2 of this chapter, under the assumption that the plate and brackets are produced as a single component via some process such as casting.

14.27 Figure P14.27 shows a part cut by a plane. Create a full sectional view of the part at the location where the plane cuts the part.

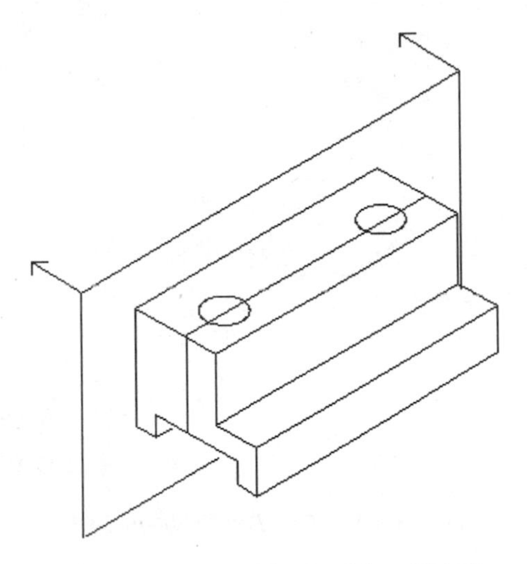

FIGURE P14.27 *For Problem P14.27.*

14.28 Figure P14.28 shows a part cut by a plane. Create a full sectional view of the part at the location where the plane cuts the part.

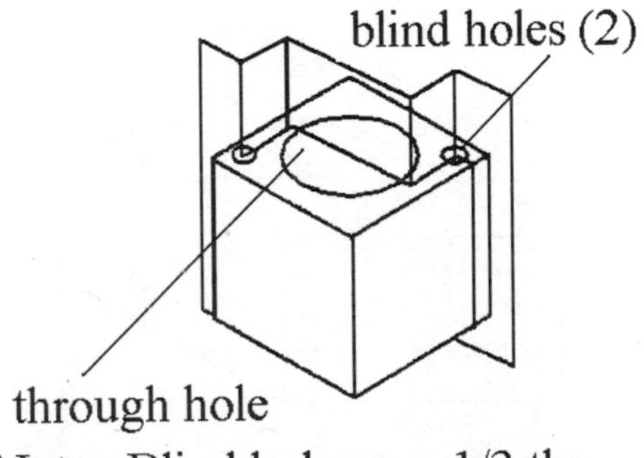

FIGURE P14.28 *For Problem P14.28.*

14.29 Figure P14.29 shows a part cut by a plane. Create a full sectional view of the part at the location where the plane cuts the part.

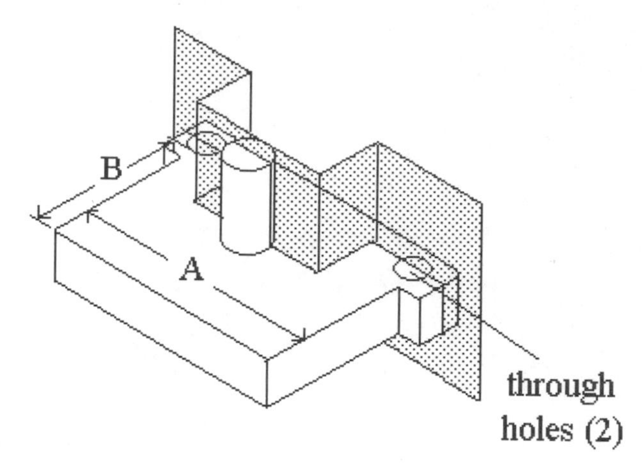

FIGURE P14.29 *For Problem 14.29.*

14.30 Figure P14.30 shows a part cut by a plane. Create a full sectional view of the part at the location where the plane cuts the part.

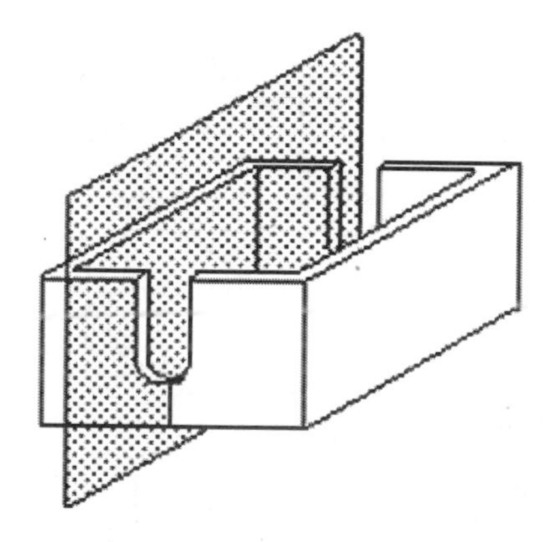

FIGURE P14.30 *For Problem 14.30.*

14.31 As part of a student project, the drawing shown in Figure P14.31 was submitted. The drawing was as an exploded assembly drawing of a small handheld hair dryer. What suggestions, if any, would you make to improve this drawing?

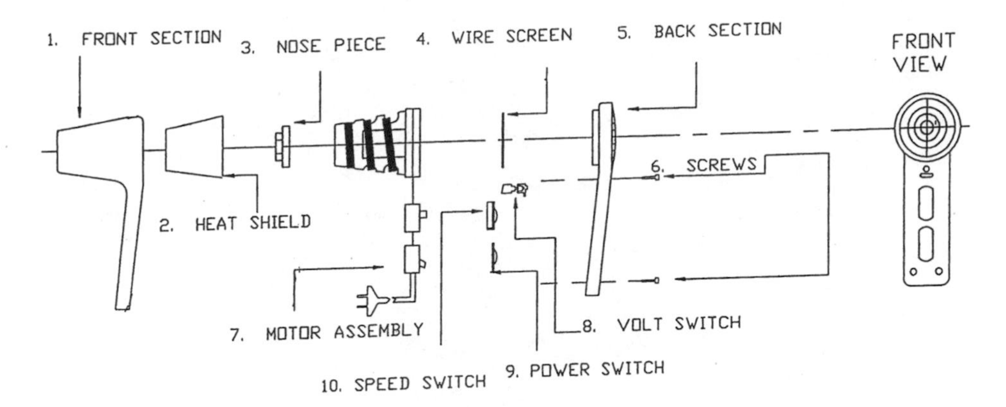

FIGURE P14.31 *Hair dryer assembly drawing for Problem 14.31.*

14.32 As part of a student project, the drawing shown in Figure P14.32 was submitted. The drawing was as an exploded assembly drawing of a battery-powered pencil sharpener. What suggestions, if any, would you make to improve this drawing?

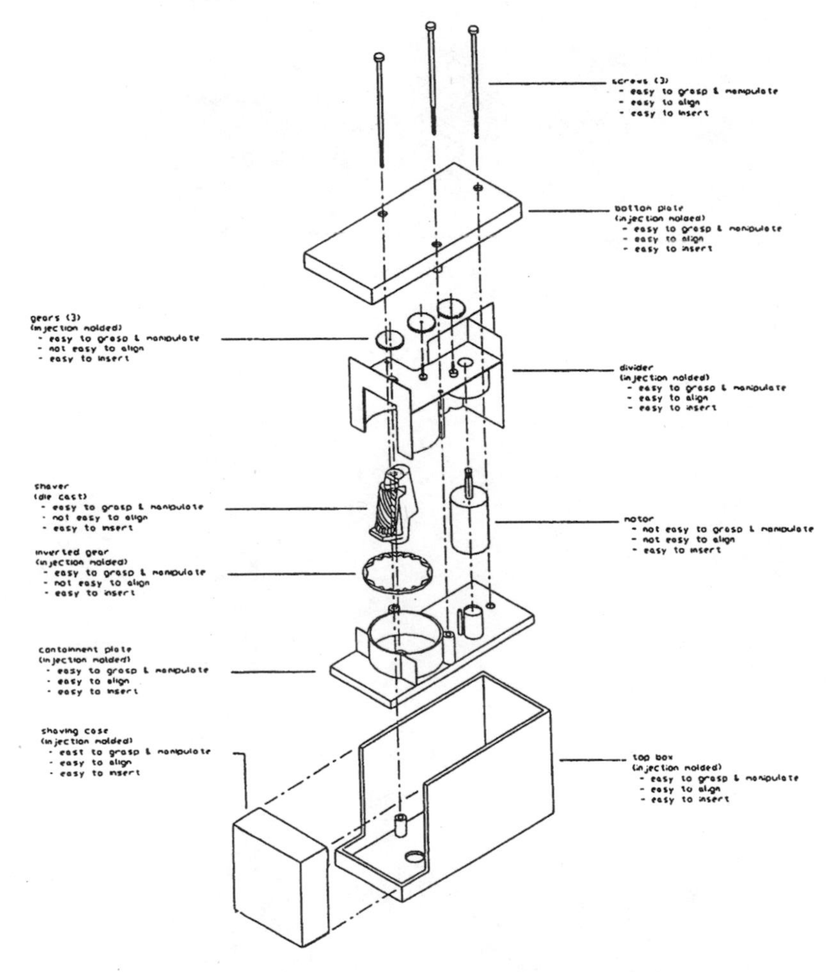

FIGURE P14.32 *Assembly drawing of a battery-powered pencil sharpener for Problem 14.32.*

Nomenclature

A = Die set length (for a progressive die); $K_{dmo}/(K_{dmo} + K_{dco})$
A_p = Projected area of a part normal to the direction of mold closure
a = Dimension, length; constant
B = Die set width (for a progressive die); $K_{dco}/(K_{dmo} + K_{dco})$
B_m = Width of the part in a direction normal to the direction of mold closure
B_u = Width of an injection-molded or die-cast part with the axis straight = H
b = Dimension, length, thickness, width; rib width; boss width
C = Value obtained from Figures 4.23 or 7.7
C_{an} = Cost to assemble n parts
C_b = Approximate relative tooling cost for an injection-molded or die-cast part due to size and basic complexity
$C_d = K_d/K_{do}$ = Tooling cost for a part relative to the tooling cost for the reference part
$C_{dc} = K_{dc}/K_{dco}$ = Tool construction cost for a part relative to the tool construction cost for the reference part (= C_b C_s C_t for an injection-molded or die-cast part; = t_{dc}/t_{dco} for a stamped part)
C_{dci} = Relative tool construction cost for the ith part.
C_{dcx} = Relative die construction cost for a single functionally equivalent part to replace n parts
$C_{dm} = K_{dm}/K_{dmo}$ = Tool material cost for a part relative to the tool material cost of the reference part. Value obtained from Figures 4.24 or 7.8 for injection-molded or die-cast parts. Value obtained from Equation 9.15 for stamped parts.
C_{dmi} = Relative die material cost for the ith part
C_{dmx} = Relative die material cost for a single functionally equivalent part to replace n parts
$C_{dp} = K_{dp}/K_{dop}$, Tooling cost of a thermoplastic injection-molded part relative to the tooling cost of the injection-molded reference part
$C_{ds} = K_{ds}/K_{dos}$, Tooling cost of a stamped part relative to the tooling cost of the stamped reference part
C_{dx} = Total relative die cost for a single functionally equivalent part
$C_e = K_e/K_{eo}$ = Processing cost for a part relative to the processing cost for the reference part
$C_{ea} = K_{ea}/K_{eao}$, Processing cost for an aluminum die-cast part relative to the processing cost for the aluminum die-cast reference part
$C_{ep} = K_{ep}/K_{eop}$, Processing cost for a thermoplastic injection-molded part relative to the processing cost for the injection-molded reference part
$C_{es} = K_{es}/K_{eos}$, Processing cost for a stamped part relative to the processing cost for the stamped reference part
C_h = Machine hourly rate
C_{ho} = Machine hourly rate ($/hr) for the reference part
C_{hr} = Relative machine hourly rate = machine hourly rate for a part relative to the machine hourly rate of the reference part = C_h/C_{ho}

C_{hra} = Relative machine hourly rate for aluminum die-casting machines
C_{hrp} = Relative machine hourly rate for injection-molding machines
$C_m = K_m/K_{mo}$, Material cost of a part relative to the material cost of the reference part
$C_{mp} = K_{mp}/K_{mop}$, Material cost of an injection-molded part relative to the material cost of the injection-molded reference part
$C_{mr} = K_p/K_{po}$ = material cost per unit volume of a part relative to the material cost per unit volume of the reference part
$C_{ms} = K_{ms}/K_{mos}$, Material cost of a stamped part relative to the material cost of the stamped reference part
C_r = Total cost of a part relative to the total cost for the reference part
C_{rp} = Total relative cost of a thermoplastic injection-molded part
C_{rs} = Total relative cost of a stamped part
C_s = For injection-molded and die-cast parts, a multiplier to account for the cavity detail (subsidiary complexity) for relative tooling cost.
C_t = For injection-molded and die-cast parts, a multiplier to account for the tolerance and surface finish requirements for relative tooling cost.
E = Modulus of elasticity
F = Estimated total force in tons required to form an injection-molded, die-cast, or stamped part
F_b = Bending force required (stamping)
F_{dm} = Factor to account for the effect of sheet thickness on die material cost for a stamped part
F_{eff} = Press efficiency
F_{out} = Force required to separate the stamped part from the trip
F_p = Machine tonnage required to mold or die cast a part; total press force required to stamp a part
F_{st} = Stripper force required to remove the strip from around the punches
F_s = Total shearing force required to stamp the part $(F_{out} + F_{st})(1 + X_{dd})$
F_t = Factor to account for the effect of sheet thickness on die construction costs for a stamped part
$f_d = K_{do}/K_o$, Tooling cost of the reference part relative to the total cost of the d reference part
$f_{dp} = K_{dop}/K_{op}$, Tooling cost of the injection-molded reference part relative to the total cost of the injection-molded reference part
$f_e = K_{eo}/K_o$, Processing cost of the reference part relative to the total cost of the reference part
$f_{ep} = K_{eop}/K_{op}$, Processing cost of the injection-molded reference part relative to the total cost of the injection-molded reference part
$f_m = K_{mo}/K_o$, Material cost of the reference part relative to the total cost of the reference part
$f_{mp} = K_{mop}/K_{op}$, Material cost of the injection-molded reference part relative to the total cost of the injection-molded reference part
H_m = The height of the part in the direction of mold closure
h = Rib height; boss height; height
K_d = Total tooling cost (mold or die cost) for the part
K_d/N = Tooling cost per part
K_{da} = Tooling cost for an aluminum die casting
K_{dc} = Die construction cost for a part
K_{dco} = Die construction cost for the reference part
K_{di} = Total tooling cost for the ith component
K_{dm} = Die material cost for a part
K_{dmo} = Die material cost for the reference part
K_{do} = Tooling cost for the reference part

K_{dop} = Tooling cost for the thermoplastic injection-molded reference part
K_{dp} = Tooling cost for a thermoplastic injection molding
K_{ds} = Tooling cost for a stamped part
K_{dx} = Total tooling cost for the single functionally equivalent part to replace the n parts
K_e = Processing cost per part (approximately the product of the machine hourly rate and the cycle time)
K_{ea} = Processing cost for an aluminum die casting
K_{eao} = Processing cost for the die-cast aluminum reference part
K_{ei} = Processing cost for the ith component
K_{eo} = Processing cost per part for the reference part
K_{ep} = Processing cost for a thermoplastic injection molding
K_{epo} = Processing cost for the thermoplastic injection-molded reference part
K_{es} = Processing cost for a stamped part
K_{ex} = Processing cost for the single functionally equivalent part to replace the n parts
K_m = Material cost per part
K_{ma} = Material cost for an aluminum die casting
K_{mi} = Material cost for the ith component
K_{mo} = Material cost per part for the reference part
K_{mop} = Material cost per part for the thermoplastic injection-molded reference part
K_{mp} = Material cost for a thermoplastic injection-molded part
K_{ms} = Material cost for a stamped part
K_{mx} = Material cost for the single functionally equivalent part to replace the n parts
K_o = Manufacturing cost of the reference part
K_{op} = Manufacturing cost of the thermoplastic injection-molded reference part
K_p = Material cost per unit volume
K_{po} = Material cost per unit volume for the reference part
K_t = Total production cost of a single part
K_{ta} = Total cost to produce a part as an aluminum die casting
K_{ti} = Total cost of producing the ith component
K_{tp} = Total cost to produce a part as a thermoplastic injection molding
K_{tx} = Total cost of producing the single functionally equivalent part to replace the n parts
L, B, H = Lengths of the sides of the basic envelope ($L > B > H$)
L_{bt} = Sum of all straight bend lengths
L_{db} = Die block width (progressive die)
L_{dl} = Die block length (progressive die)
L_m = The length of the part in a direction normal to the direction of mold closure
L_{out} = Peripheral length of an unfolded part
L_u = Maximum length of an injection-molded or die-cast part with the axis straight = L + B
L_{ub} = Envelope dimension of a stamped part parallel to the direction of strip feed
L_{ubn} = Envelope dimension of a stamped part perpendicular to the direction of strip feed
L_{ul} = Flat envelope length of a stamped part
L_{uw} = Flat envelope width of a stamped part
l = Length
M_a = Projected area of the mold base normal to the direction of mold closure
M_t = Thickness of the mold base
M_{wf} = Thickness of core plate
M_{ws} = Thickness of the mold's side walls

N = Production volume
N_a = Number of active stations for progressive die
N_{a1}, N_{a2} = Number of active stations required for shearing and local forming, and wipe forming and side-action features, respectively
N_{i1}, N_{i2} = Number of idle stations required for shearing and local forming, and wipe forming and side-action features, respectively
N_o = Production volume of reference part
N_s = Total number of stations required for a progressive die = $N_a + N_i$
n = Strain hardening coefficient; number of parts in an assembly
n_c = Number of cavities in a multiple cavity mold
R_a = Surface finish
r = Radius
S_{ds} = die set area = AB
T_a = Tolerance; difference between the maximum and minimum limit of size
t = Machine cycle time (processing time) for a part; sheet thickness for a stamped part; thickness; time
t_b = Basic relative cycle time for an injection-molded or die-cast part; tool build hours for a stamped part
t_d = Tool design hours for a stamped part
t_i = Build hours for a progressive die station
t_{dc} = Total die construction hours for a stamped part
t_{cy} = Effective cycle time of a press (stamping)
t_e = Additional relative cycle time due to inserts and internal threads
t_{eff} = Effective cycle time (injection molding) = machine cycle time, t, divided by the production yield, Y
t_o = Cycle time (processing time) for the reference part
t_p = A time penalty factor to account for surface requirements and tolerances on the relative cycle time
$t_r = t/t_o$ = Relative cycle time of a part
t_{rp} = Relative cycle time for a thermoplastic injection-molded part
t_{ra} = Relative cycle time for a die-cast aluminum part
t_{setup} = Approximate setup time in hours (stamping)
V = Part volume
V_o = Volume of the reference part
w = Wall thickness; width
X_{dd} = A force factor to account for die detail in stamping
Y = Production yield = usable parts/total parts produced
$\Delta C_{dm} = \Sigma C_{dmi} - C_{dmx}$
$\Delta C_{dc} = \Sigma C_{dci} - C_{dcx}$
ΔK = Cost to produce and assemble n parts minus the cost of producing one functionally equivalent part
ΔK_{ap} = Difference in cost to produce a part as an aluminum die casting and as a thermoplastic injection molding
ΔK_e = Difference in processing cost
ΔK_{ps} = Difference in cost to produce a part as a thermoplastic injection molding and as a stamping
$\varepsilon = \ln\left(\frac{l}{l_o}\right)$; l is the final length of the specimen, and l_o is the initial length of the specimen
$\varepsilon_e = \ln\left(\frac{l - l_o}{l_o}\right)$; l is the final length of the specimen, and l_o is the initial length of the specimen

$\sigma = \frac{P}{A}$; P is the tensile load applied to the specimen, and A is the cross-sectional area of the specimen

$\sigma_e = \frac{P}{A_o}$; P is the tensile load applied to the specimen, and A_o is the initial cross-sectional area of the specimen

σ_y = yield stres

Index